W9-CRL-686

J. G. Sanchez Marcano and
Th. T. Tsotsis

Catalytic Membranes and Membrane Reactors

 WILEY-VCH

Also of Interest

W. Ehrfeld, V. Hessel, H. Löwe
Microreactors – New Technology for Modern Chemistry (2000)
ISBN 3-527-29590-9

S. P. Nunes and K.-V. Peinemann (Eds.)
Membrane Technology – in the Chemical Industry (2001)
ISBN 3-527-28485-0

K. Sundmacher and A. Kienle (Eds.)
Reactive Distillation – Status and Future Directions
(planned 2002)
ISBN 3-527-30579-3

H. Schmidt-Traub (Ed.)
Chromatographic Separation – Fine Chemicals and
Pharmaceutical Agents (planned 2003)
ISBN 3-527-30643-9

José G. Sanchez Marcano and
Theodore T. Tsotsis

Catalytic Membranes and Membrane Reactors

Chemistry Library

T P
248
.25
M45
S36
2002
CHEM

Dr. José G. Sanchez Marcano
Institut Européen des Membranes
IEMM, UMII
Place Eugene Bataillon
34095 Montpellier cedex 5
France

Professor Theodore T. Tsotsis
University of Southern California
Department of Chemical Engineering
University Park, HED 210
Los Angeles, California 90089-1211
USA

This book was carefully produced. Nevertheless, authors and publisher do not warrant the information contained therein to be free of errors. Readers are advised to keep in mind that statements, data, illustrations, procedural details or other items may inadvertently be inaccurate.

Cover Illustration: Background: Catalytic Membrane; Foreground: Packed-Bed Membrane Reactor.
Source: IEM (M. Didier Cot and José G. Sanchez Marcano.)

Library of Congress Card No.: Applied for.

British Library Cataloguing-in-Publication Data:
A catalogue record for this book is available from the British Library

Die Deutsche Bibliothek – CIP Einheitsaufnahme
A catalogue record for this book is available from Die Deutsche Bibliothek

ISBN 3-527-30277-8

© Wiley-VCH Verlag GmbH, Weinheim, 2002

Printed on acid-free paper.

All rights reserved (including those of translation in other languages). No part of this book may be reproduced in any form – by photoprinting, microfilm, or any other means – nor transmitted or translated into machine language without written permission from the publishers. Registered names, trademarks, etc. used in this book, even when not specifically marked as such, are not to be considered unprotected by law.

Composition: M-O-P-S, Kirsten Pfaff, Hennef
Printing: Strauss Offsetdruck, Mörlenbach
Bookbinding: Großbuchbinderei J. Schäffer GmbH & Co. KG, Grünstadt

Printed in the Federal Republic of Germany.

Dedication

To Betty, Christine, Laura, Benjamin and Anaïs, for their love and ongoing support

JSM/TTT

Preface

The study of catalytic membranes and membrane reactor processes is a multidisciplinary activity, which in recent years has attracted the attention of scientists and engineers in a number of disciplines, including materials science, chemistry and biology, and chemical and biochemical engineering. For a process to qualify as a membrane reactor system, it must not simply combine a membrane separation unit with a chemical reactor; it must integrally couple them in such a fashion that a synergy is created between the two units, potentially resulting in enhanced performance in terms of separation, selectivity and yield. Often the membrane separation module and reactor are physically combined into the same unit. Such combination promises to result in a process, which is more compact and less capital intensive, and with potential substantial savings in processing costs. During the last twenty years this technical concept has attracted substantial worldwide research and process development efforts. This is the first book, we are aware of, that is completely dedicated to the topic of membrane reactors. It aims and hopes to introduce the topic and serve as reference for a wide audience of scientists and engineers.

The book is divided into six chapters, each covering a different aspect of the topic. Chapter 1 provides an introduction of key concepts. For the readers who are completely unfamiliar with membrane processes, some relevant basic concepts and definitions are introduced, followed by a concise review of some of the basic aspects and definitions of membrane reactors. Chapter 2 is dedicated to catalytic and electrochemical membrane reactors. Though notable exceptions exist, most of the catalytic membrane reactor processes are high temperature applications. Membrane reactors have been applied to many common classes of catalytic reactions including dehydrogenation, hydrogenation, and partial and total oxidation reactions; all are reviewed in Chapter 2. There are several configurations of catalytic membrane reactors which utilize catalytically active or inactive membranes of various shapes and types; they may operate, for example, under a transmembrane pressure gradient or through the use of a sweep gas under co-current or counter-current operation; the membrane participates in the removal/addition of various species. The discussion on electrochemical membrane reactors in Chapter 2 is limited to those systems which produce a valuable chemical with or without the co-generation of electricity. Fuel cells are deliberately excluded as there are already several books and detailed reports on the topic.

Chapter 3 is devoted to the topic of pervaporation membrane reactors. These are unique systems in that they use a liquid feed and a vacuum on the permeate side; they also mostly utilize polymeric membranes. Chapter 4 presents a survey of membrane bioreactor processes; these couple a biological reactor with a membrane process. Reactions studied in such systems include the broad class of fermentation-type or enzymatic processes, widely used in the biotechnology industry for the production of amino acids, antibiotics, and other fine chemicals. Similar membrane bioreactor systems are also finding application in the biological treatment of contaminated air and water streams. Chapter 5 presents modelling studies of the different membrane reactor configurations and processes described previously in chapters 2 to 4. Membrane reactors often show unique and different behavior from their conventional counterparts; efficient modelling is key in understanding the behavior of these systems and important in scale-up and optimization activities. Finally, Chapter 6 discusses a number of issues which determine the technical feasibility and economic viability of such reactors. Membrane bioreactors have already found commercial application. This is, still, not true for high temperature catalytic membrane reactors, though they have been studied equally as intensively and long. Some of the key barriers still hindering their commercial application are discussed in the same chapter.

A number of people have contributed to this book and are acknowledged here. They include Ms. Christine Roure-Sanchez and Ms. Karen Woo for their typing work, Mr. Didier Cot for his kind help with the cover figure, and Mr. Seong Lim for his assistance with various aspects of the book. Theo Tsotsis and Jose Sanchez Marcano gratefully acknowledge the ongoing support of the U.S. National Science Foundation, the Centre National de la Recherche Scientifique, the U.S. Department of Energy and the European Commission for their research efforts in the area.

Contents

1   Introduction

Membrane-based reactive separation (otherwise also known as membrane reactor) processes, which constitute the subject matter of this book, are a special class of the broader field of membrane-based separation processes. In this introduction we will first provide a general and recent overview on membranes and membrane-based separation processes. The goal is to familiarize those of our readers, who are novice in the membrane field, with some of the basic concepts and definitions. A more complete description on this topic, including various aspects of membrane synthesis can be obtained from a number of comprehensive books and reviews that have already been published in this area [1.1, 1.2, 1.3, 1.4]. We will then follow in this introduction with an outline of some of the generic aspects of the field of membrane-based reactive separations, which our reader will, hopefully, find useful while navigating through the rest of this book.

1.1   Principles of Membrane Separation Processes

Membrane-based separation processes are today finding widespread, and ever increasing use in the petrochemical, food and pharmaceutical industries, in biotechnology, and in a variety of environmental applications, including the treatment of contaminated air and water streams. The most direct advantages of membrane separation processes, over their more conventional counterparts (adsorption, absorption, distillation, etc.), are reported to be energy savings, and a reduction in the initial capital investment required.

A membrane is a permeable or semi-permeable phase, often in the form of a thin film, made from a variety of materials ranging from inorganic solids to different types of polymers. The main role of the membrane film, as shown schematically in Figure 1.1, is to control the exchange of materials between two adjacent fluid phases. For this role, the membrane must be able to act as a barrier, which separates different species either by sieving or by controlling their relative rate of transport through itself. The membrane action, as shown in Figure 1.1, results in a fluid stream (defined as the retentate), which is depleted from some of its original components, and another fluid stream (defined as the permeate), which is concentrated in these components. Transport processes across the membrane are the result of a driving force, which is typically associated with a gradient of concentration, pressure, temperature, electric potential, etc. The ability of a membrane to effect separation of mixtures is determined by two parameters, its permeability and selectivity. The permeability is defined as the flux (molar or volumetric flow per unit membrane area) through the membrane scaled with respect to the membrane thickness and driving force; for the case where transport is due, for example, to a partial pressure gradient the units of permeability are mol (or m^3)$\cdot$m$\cdot$m$^{-2}\cdot$Pa$^{-1}\cdot$s^{-1}. Often the true membrane thickness is not known and permeance, which is defined as the flux through the membrane scaled with respect to driving force (with units mol (or m^3)$\cdot$m$^{-2}\cdot$Pa$^{-1}\cdot$s^{-1}), is, instead, utilized. The second important parameter is the membrane selectivity, which characterizes the ability of the membrane to separate two given molecular species, and which is, typically, defined as the ratio of the individual permeabilities for the two species.

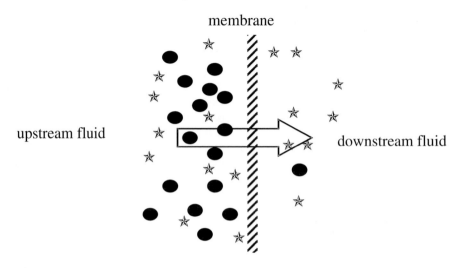

Figure 1.1. Schematic drawing of the basic membrane separation principle.

Membranes are classified by whether the thin permselective layer is porous or dense, and by the type of material (organic, polymeric, inorganic, metal, etc.) this membrane film is made from. The choice of a porous vs. a dense film, and of the type of material used for manufacturing depends on the desired separation process, operating temperature and driving force used for the separation; the choice of material depends on the desired permeance and selectivity, and on thermal and mechanical stability requirements. For membrane reactor applications, where the reaction is coupled with the separation process, the thin film has also to be stable under the reaction conditions.

Porous membranes are made of polymers (they include those used for dense membranes and, in addition, many others including polysulphones, polyacrylonitrile, polypropylene, etc.), ceramics (alumina, silica, titania, zirconia, zeolites, etc.), and microporous carbons. Dense organic and polymeric membranes are commonly used for molecular scale separations involving gas and vapor mixtures. There is a great variety of synthetic polymers, including silicones, perfluoropolymers, polyimides, polyamides, etc., which have, so far, found application in the membrane field. In the membrane reactor area, in addition to polymeric membranes, use is made of metal and solid oxide dense membranes. Metallic membranes are made of precious metals, like platinum, palladium, or silver, and of different alloys containing at least one of these metals. These membranes are specifically used for hydrogen (Pd, Pt, and their alloys) or oxygen (Ag) separations. Dense solid oxide membranes are used for oxygen and hydrogen separation, and are made by different types of ionic conducting materials, such as modified zirconias and perovskites. The use and application of these dense membranes in catalytic membrane reactors will be further described in the following chapters of this book.

Membranes are also classified by whether they have a symmetric (homogeneous) or an asymmetric structure. Homogeneous or symmetric membranes are prepared when the membrane material has the necessary mechanical stability to be self-supporting. In many instances, however, the thin, permselective layer does not have enough mechanical strength to be self-supporting. In such cases the membrane layer is deposited on a porous support, which could conceivably be made from a different material resulting in asymmetric membranes; the support gives the necessary mechanical stability to the membrane without being an obstacle to the mass transport. Figure 1.2 shows two different electron micrographs, one representing a symmetric, self-standing polymer membrane, and the other representing an asymmetric, composite membrane consisting of a thin permselective, polymeric layer on a macroporous ceramic support. Composite, asymmetric membranes, can be prepared by a variety of techniques, including the incorporation into the separative layer, itself, of nanoparticles made from a different material [1.5]. Other techniques include the deposition of a polymer or metallic thin layer [1.6, 1.7, 1.8] on a ceramic porous support (Figure 1.2a), and also the infiltration or synthesis of a different material into the porous structure of a support [1.9, 1.10].

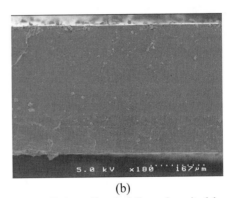

(a)   (b)

Figure 1.2. Electron micrograph (a) of a composite (polydiethoxydimethylsilane deposited by plasma on a porous alumina support [1.11]) membrane, and (b) a self-standing elastomeric membrane.

Membranes are manufactured in a diverse range of geometries; they include flat, tubular, and multi-tubular, hollow-fiber, and spiral-wound membranes. The type of geometry the membrane is manufactured into depends on the material the membrane is made from. Ceramic membranes, generally, come in tubular, multi-tubular and flat geometries, whereas spiral-wound and hollow-fiber membranes seem, for the most part (with a few notable examples), to be made from polymers.

For porous membranes the molecular size of the species to be separated plays also an important role in determining the pore size of the membrane to be utilized, and the related membrane process. According to the IUPAC classification, porous membranes with average pore diameters larger than 50 nm are classified as macroporous, and those with average pore diameters in the intermediate range between 2 and 50 nm as mesoporous; microporous membranes have average pore diameters which are smaller than 2 nm. Current membrane processes include microfiltration (MF), ultrafiltration (UF), nanofiltration (NF), gas and vapor separation (GS), and pervaporation (PV). Figure 1.3 indicates the type and molecular size of species typically separated by these different processes.

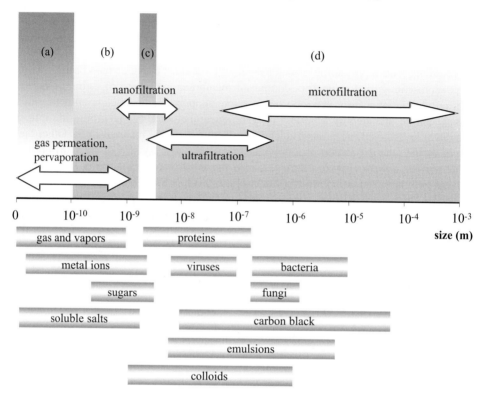

Figure 1.3. Various membrane processes and the different types of membranes and molecular species involved (a) dense and ultramicroporous, (b) microporous, (c) mesoporous, (d) macroporous.

The mass transfer mechanisms through membranes vary, depending on many factors like the membrane structure, the specific interactions between the membrane and the fluid, and the overall operating conditions. Mass transfer through dense polymeric membranes has been, typically, described in the engineering literature by a solution diffusion mechanism [1.1, 1.12, 1.13]. According to this mechanism, the molecular species must first adsorb on the surface and dissolve in the bulk of the polymer, where transport occurs by diffusion through the free volume in the polymer structure. Similar mechanisms taking into account a surface reaction (oxidation or hydride formation) and diffusion (hydride or ionic) in the membrane bulk have been utilized to describe diffusion through metallic or solid oxide membranes. Understanding transport through such membranes requires measuring both, the sorption and transport characteristics of these materials. The mathematical techniques for the analysis of the experimental data, taking into consideration the effect of the membrane's geometry, have been developed by Crank some years ago [1.14].

For porous membranes the mass transport mechanisms that prevail depend mainly on the membrane's mean pore size [1.1, 1.3], and the size and type of the diffusing molecules. For mesoporous and macroporous membranes molecular and Knudsen diffusion, and convective flow are the prevailing means of transport [1.15, 1.16]. The description of transport in such membranes has either utilized a Fickian description of diffusion [1.16] or more elaborate Dusty Gas Model (DGM) approaches [1.17]. For microporous membranes the interaction between the diffusing molecules and the membrane pore surface is of great importance to determine the transport characteristics. The description of transport through such membranes has either utilized the Stefan-Maxwell formulation [1.18, 1.19, 1.20] or more involved molecular dynamics simulation techniques [1.21].

For membrane processes involving liquids the mass transport mechanisms can be more involved. This is because the nature of liquid mixtures currently separated by membranes is also significantly more complex; they include emulsions, suspensions of solid particles, proteins, and microorganisms, and multi-component solutions of polymers, salts, acids or bases. The interactions between the species present in such liquid mixtures and the membrane materials could include not only adsorption phenomena but also electric, electrostatic, polarization, and Donnan effects. When an aqueous solution/suspension phase is treated by a MF or UF process it is generally accepted, for example, that convection and particle sieving phenomena are coupled with one or more of the phenomena noted previously. In nanofiltration processes, which typically utilize microporous membranes, the interactions with the membrane surfaces are more prevalent, and the importance of electrostatic and other effects is more significant. The conventional models utilized until now to describe liquid phase filtration are based on irreversible thermodynamics; good reviews about such models have been reported in the technical literature [1.1, 1.3, 1.4].

1.2 The Coupling of the Membrane Separation Process with a Catalytic Reaction

Membrane-based reactive separation processes, which seek to combine two distinct functions, i.e. reaction and separation, have been around as a concept since the early stages of the membrane field, itself, but have only attracted substantial technical interest during the last decade or so [1.22]. There is ongoing significant industrial interest in these processes, because they promise to be compact and less capital intensive, and because of their promise for potential substantial savings in the processing costs [1.23].

Membrane-based reactive separation processes (also known as membrane reactor processes) are attracting attention in catalytic reactor applications. In these reactor systems the membrane separation process is coupled with a catalytic reaction. When the separation and reaction processes are combined into a single unit the membrane, besides providing the separation function, also often results in enhanced selectivity and/or yield. Membrane-based reactive separations were first utilized with reactions, for which the continuous extraction of products would enhance the yield by shifting the equilibrium. Reactions of this type that have been investigated, so far, include dehydrogenation and esterification. Reactive separations also appear to be attractive for application in other types of reactions, including hydrogenation, and partial and total oxidation. In many reactor studies involving these reactions the use of membranes has been shown to increase the yield and selectivity. Published accounts on the application of membrane-based reactive separations in catalytic processes report the use of catalytically active and inactive membranes of various types, shapes, and configurations [1.24, 1.25, 1.26]. Reactor yield and reaction selectivity are found to be strongly dependent on the membrane characteristics, in addition to the more conventional process parameters.

Biotechnology is another area in which membrane-based reactive separations are attracting great interest. There, membrane processes are coupled with industrially important biological reactions. These include the broad class of fermentation-type processes, widely used in the biotechnology industry for the production of amino acids, antibiotics, and other fine chemicals. Membrane-based reactive separation processes are of interest here for the continuous elimination of metabolites, which is necessary to maintain high reactor productivity. Membranes are also increasingly utilized as hosts for the immobilization of bacteria, enzymes, or animal cells in the production of many high value-added chemicals. Similar reactive separation processes are also finding application in the biological treatment of contaminated air and water streams. Many of these emerging applications will be reviewed and evaluated in Chapter 4 of this book.

Reactor modelling has proven valuable for understanding the behavior of these systems. It will continue to serve in the future as an important tool for predicting and optimizing the behavior, and for improving the efficiency of these processes. Membrane reactor modelling will be discussed in Chapter 5. In this chapter, we will review key aspects of process modelling, design, and optimization that have been applied to the different membrane reactor types, that have been utilized in reactive separation studies.

In the early stages of the membrane-based reactive separations field, the coupling of the two functions happened by simply connecting in series two physically distinct units, the reactor and membrane separator (see Figure 1.4a). The membrane reactor concept, shown in Figure 1.4b, which combines two different processing units (i.e., a reactor and a membrane separator) into a single unit, was the result of natural process design evolution from the concept of Figure 1.4a. There are obvious advantages resulting from the design configuration of Figure 1.4b relating to its compact design, and the capital and operating savings realized by the elimination of intermediate processing steps. Other advantages relate to the synergy that is being realized between separation and reaction. This synergy is immediately obvious for reactions limited by thermodynamic equilibrium considerations, as is frequently the case with catalytic hydrocarbon dehydrogenation or esterification reactions. There, the continuous separation of one or more of the products (e.g., hydrogen or water) is reflected by an increase in yield and/or selectivity. Despite the obvious advantages of the reactor concept of Figure 1.4b over the design in Figure 1.4a, the more conventional membrane-based reactive separation concept of Figure 1.4a, because of its simplicity, is often the design of choice in biotechnological applications. In the membrane reactor concept in Figure 1.4b the presence of the membrane (tubular, hollow-fiber, or plate) helps to define two different chambers. These are the retentate chamber, where the reactants are fed and the reaction often takes place, and the permeate chamber. The latter is either swept by an inert gas or evacuated in order to maintain a differential pressure or concentration gradient for mass transfer between the two compartments.

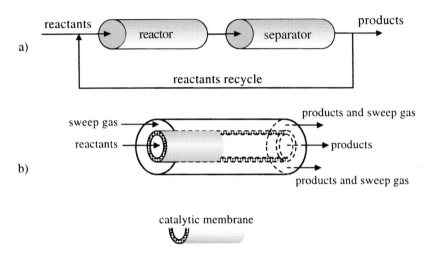

Figure 1.4. A conventional membrane reactor system (a) and an integrated membrane reactor system (b).

Publications discussing the membrane reactor concept first started appearing in the late 1960's [1.27, 1.28]. Most of the real progress in this area, however, has happened in the last twenty years [1.22]. This parallels progress in the overall field of membrane processes. There, as noted above, the development of new membrane materials during the last twenty years has opened the world of membrane technology to a broader range of applications beyond the classical ones, which, typically, involve low temperature microfiltration or ultrafiltration of liquids using polymeric membranes. The development of membranes made with a variety of inorganic materials has provided the opportunity to apply the catalytic membrane reactor concept for a much broader set of operating conditions. Inorganic membranes offer advantages in this regard over organic membranes, because they are stable at relatively high temperatures (> 373 K), and have good chemical and mechanical resistance [1.23]. The progress realized, over the last ten years, in the synthesis of stable microporous or dense inorganic materials for the preparation of membranes, has been the key factor motivating, for example, the application of membrane-based reactive separations in the catalysis field. Inorganic membranes are also more frequently being utilized in biotechnology for the production of fine chemicals via the use of both enzyme and whole-cell bioreactors [1.29], and for large-scale environmental clean-up type applications [1.30]. For these processes, which take place under milder conditions, organic membranes still remain the option of choice. It should be noted, however, that polymeric membranes still find extensive use in hybrid processes involving coupling of reaction with pervaporation (these are discussed in detail in Chapter 3), and in membrane bioreactors (Chapter 4). Recent developments in polymeric membranes (e.g., polyimides) are pushing the envelope for their application (T~300 °C). A number of studies, as a result, have also appeared discussing the use of such membranes in high temperature catalytic membrane reactor applications [1.31, 1.32].

As noted previously, many of the earlier applications involved equilibrium limited reactions. In a rather short time, the field has progressed far beyond this application. In some of the most recent studies (e.g., membrane bioreactors), for example, the membrane separates intermediates and products from the reacting zone, so that they do not deactivate the catalyst or undergo further undesirable reactions. In some of the catalytic applications the membrane is not even required to be permselective; it only acts as a controlled reactive interface between reactants flowing on opposite sides of the membrane [1.33, 1.34]. This significant progress in the field is reflected in the increasing number of publications on membrane reactors, which have grown exponentially over the last few years [1.35]. Detailed descriptions of the state of the art on the topic of catalytic membrane reactors, for example, have been published along the way by Catalytica [1.36], Armor [1.37, 1.38, 1.39, 1.40], Ilias and Govind [1.41], Hsieh [1.22, 1.42], Zaspalis and Burggraaf [1.43], Tsotsis *et al.* [1.44], Zaman and Chakma [1.45], Saracco and Specchia [1.46, 1.47], Saracco *et al.* [1.35, 1.48], Sanchez and Tsotsis [1.24], and, most recently, Dixon [1.25], Sirkar *et al.* [1.49], and Sanchez and Tsotsis [1.26].

Both, dense and porous membranes have found use in reactive separation applications. As was explained above, dense membranes are made of polymers, metals and their alloys, and solid oxides. Dense polymeric membranes are finding extensive use in pervaporation membrane reactors. For catalytic applications, except for some notable cases [1.31, 1.32], their use has been limited to low temperature reactions [1.50, 1.51, 1.52], because they are perceived to have poor thermal resistance. Dense metal membranes consist mostly of noble metals Pd, Pt, Ru, Rh, Ir, Ag and their alloys, but most recently also from other hydrogen storage alloys (e.g., LaNi$_5$, [1.53]). These membranes (particularly Pd) have found extensive use in the early stages of the reactive separations field (starting as early as 1966, for the study of the ethane dehydrogenation reaction [1.54]). They were used extensively in the former Soviet Union by Gryaznov and co-workers [1.55, 1.56, 1.57]. These membranes have high selectivity towards hydrogen (Pd and its alloys) or oxygen (Ag) but their high cost, relatively low permeability, and limited availability have hindered their extensive industrial application. There is also concern with their mechanical properties, particularly, embrittlement and fatigue as a result of repeated thermal cycling at high temperatures. There are ongoing significant efforts in this area for improving the characteristics of these membranes [1.58]. The approach here is to deposit thin metallic films on underlying porous substrates, aiming to improve permeance and cost, without unduly impacting selectivity [1.59, 1.60, 1.61, 1.62, 1.63].

Another type of dense membranes, which have attracted significant attention in recent years [1.64], consist of solid oxides (ZrO_2, Y_2O_3, Bi_2O_3) as well as of solutions of mixed solid oxides (perovskites, brownmillerites, etc.). These materials act as solid electrolytes allowing the transport of oxygen or hydrogen. They are finding use in catalytic membrane reactor applications involving partial and total oxidation reactions. They are discussed in Chapter 2.

Porous membranes, composed of glass, ceramic materials, as well as polymers, have also found use in membrane-based reactive separations. The earlier studies in the area of membrane catalysis made use of commercially available mesoporous membranes (Vycor®glass, alumina, titania or zirconia [1.43]). Recent efforts have focused on trying to increase the selectivity of these membranes towards small molecules. This has been accomplished either by decreasing the mean pore size and/or narrowing the pore size distribution of existing mesoporous membranes [1.35], or by synthesizing microporous membranes from new materials altogether (e.g., zeolite membranes [1.65]). Porous polymeric membranes have been the design choice in membrane bioreactors; they have also been recently used for low temperature catalytic applications [1.66].

There is a multitude of different configurations that have been proposed in the literature in order to combine the membrane separation module and the reactor into a single unit (Figure 1.4b). Sanchez and Tsotsis [1.24] have classified these configurations for catalytic membrane reactors into six basic types, as indicated in Table 1.1 and Figure 1.5. This classification and acronyms are also applicable to other types of membrane reactors, and will be used throughout this book.

Table 1.1. Classification of Membrane Reactors (Adapted from Sanchez and Tsotsis, [1.24]).

Acronym	Description
CMR	Catalytic membrane reactor
CNMR	Catalytic non-permselective membrane reactor
PBMR	Packed-bed membrane reactor
PBCMR	Packed-bed catalytic membrane reactor
FBMR	Fluidized-bed membrane reactor
FBCMR	Fluidized-bed catalytic membrane reactor

The most commonly utilized catalytic membrane reactor is the PBMR, in which the membrane provides only the separation function. The reaction function is provided (in catalytic applications) by a packed-bed of catalyst particles placed in the interior or exterior membrane volumes. In the CMR configuration the membrane provides simultaneously the separation and reaction functions. To accomplish this, one could use either an intrinsically catalytic membrane (e.g., zeolite or metallic membrane) or a membrane that has been made catalytic through activation, by introducing catalytic sites by either impregnation or ion exchange. This process concept is finding wider acceptance in the membrane bioreactor area, rather than with the high temperature catalytic reactors. In the latter case, the potential for the catalytic membrane to deactivate and, as a result, to require subsequent regeneration has made this technical concept less than attractive. On occasion, the membrane used in the PBMR configuration is also, itself, catalytically active, often unintentionally so (frequently the case for metallic membranes), but, on occasion, purposely so in order to provide an additional catalytic function. The membrane reactor in this case is named PBCMR. For better control of the process temperature some authors have suggested that the packed-bed should be replaced by a fluidized-bed (FBMR or FBCMR). In the CNMR configuration the membrane is not typically permselective. It is only used to provide a well-defined reactive interface.

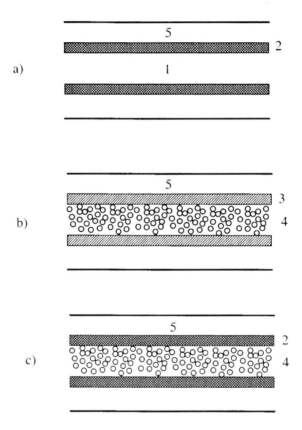

Figure 1.5. Different MR configurations. 1: tubeside, 2: catalytic membrane, 3: inert membrane, 4: catalyst bed, 5: shellside. a) CMR, CNMR, b) PBMR, FBMR, c) PBCMR, FBCMR.

As previously noted, a broader classification of membrane reactors can be made relevant to the role the membrane plays with respect to the removal/addition of various species [1.25, 1.49, 1.67]. Membrane reactors could be classified as reactive membrane extractors when the membrane's function is to remove one or more products. Such action could result in increasing the equilibrium yield, like in the catalytic dehydrogenation reaction applications described previously. Reactive membrane extractors are also finding broad application in biotechnology, where the membrane removes one of the products that are inhibitory to the bioreaction. Membranes could also serve as a distributor of one of the reactants. Such membrane reactors find application for consecutive and parallel reactions. Partial oxidation of hydrocarbons is the most relevant application here. Controlling the addition of the oxidant through the membrane lowers its partial pressure within the reaction vessel, and results, in general, in a better yield of the intermediate oxidation products. Another advantage obtained with this type of membrane reactor is the potential to avoid the thermal runaway phenomena, typically associated with highly exothermic reactions, like partial oxidations. The membrane's role in a membrane reactor could also be to improve the contact between different reactive phases; here, the membrane acts as a medium providing the intimate contact between different reactants, which are fed separately in either side of the membrane; reactions requiring a precise stoichiometry are an example of such an application. Intimate contact between the reactants can also be realized by forcing them to pass together through a membrane, which is endowed with catalytic sites for the reaction. Multiphase reactions involving a catalyst and liquid and gaseous reactants can also be studied in reactive membrane contactors. Here, the primary advantage of the use of the membrane reactor is in decreasing the mass transfer limitations, frequently encountered with such reactions in slurry- or trickle-bed reactors.

There are examples of unique membrane reactor systems reported in the technical literature, which may defy general classification. They include staged membrane reactors, membrane reactors with multiple feed ports, multi-membrane reactors [1.68, 1.69], etc. Such examples will be encountered throughout the book. Considerable attention in many studies has also been paid to the effect of various ways of operating these reactors, including the means to minimize reactant loss [1.70, 1.71], the use of sweep gas under co-current or counter-current operation [1.72], or the use of a vacuum on the permeate side (see Chapter 3 on pervaporation membrane reactors). Again these issues will be discussed in greater detail in the corresponding chapters. Membrane reactors have found limited commercial application in the biotechnology area, and in various environmental applications. Despite considerable technical attention, in the last 15 years, industrial application of membrane-based reactive separations to large-scale catalytic reaction processes is still lacking. Insight why this is so is provided in Chapter 6 of this book.

1.3 References

[1.1]   M. Mulder, *Basic principles of membrane technology*, Kluwer Acad. Publ., Dordrecht, The Netherlands, 1991.

[1.2]   R.R. Bhave, Ed., *Inorganic Membranes: Synthesis, Characteristics and Applications*, Reinhold, New York, USA, 1991.

[1.3]   A.J. Burgraaf and L. Cot, Ed., *Fundamentals of inorganic membrane science and technology*, Elsevier, Amsterdam, The Netherlands, 1996.

[1.4]   K. Scott and R. Hughes, Eds., *Industrial membrane separation technology*, Blackie Acad. and Prof. Glasgow, UK, 1996.

[1.5]   C. Joly, S. Goizet, J.C. Schrotter, J. Sanchez, and M. Escoubes, *J. Membr. Sci.* **1997**, 130, 63.

[1.6]   J. Shu, B.P.A. Granjean, A. van Neste, and S. Kaliaguine, *Can. J. Chem. Engng.* **1991**, 69, 1036.

[1.7]   S. Uemiya, T. Matsuda, and E. Kikuchi, *J. Membr. Sci.* **1991**, 56, 325.

[1.8]   S. Uemiya, Y. Kunde, K. Sugino, N. Sato, T. Matsuda, and E. Kikuchi, *Chem. Lett.* **1988**, 1687.

[1.9]   J. Ramsay, A. Giroir-Fendler, A. Julbe, and J.A. Dalmon, French Patent Appl. 94 055652, 1994.

[1.10] A. Julbe, D. Farrusseng, G. Volle, J. Sanchez, and C. Guizard, *Proc. Fifth International Conference on Inorganic Membranes*, Nagoya, Japan, June 22-26, 1998.

[1.11] S. Roualdes, *Elaboration par PECVD et caracterisation de membranes polyxiloxanes plasma pour la permeation gazeuse,* PhD dissertation, Universite de Montpellier II, France, 11/23, 2000.

[1.12] R.M. Barrer, *J. Phys. Chem.* **1957**, 61, 178.

[1.13] H.L. Frisch, *J. Phys. Chem.* **1957**, 61, 93.

[1.14] J. Crank, *The Mathematics of Diffusion*, 2nd ed., Oxford University Press, Oxford, 1975.

[1.15] P.C. Carman, *Flow of gases in porous media*, Butterworth, London, UK, 1956.

[1.16] M. Tayakout, B. Bernauer, Y. Toure, and J. Sanchez, *J. Simul. Pract. and Theory* **1995**, 2, 205.

[1.17] E.A. Mason and A.P. Malinauskas, *Gas transport in porous media: The dusty gas model*, Elsevier, Amsterdam, 1983.

[1.18] L.J.P. van den Broeke, W.J.W. Bakker, F. Kapteijn, and J.A. Moulijn, *Chem. Engng. Sci.* **1999**, 54.

[1.19] J. Sanchez, C.L. Gîjiu, V. Hynek, O. Muntean, and A. Julbe, *Sep. Purif. Technol.* **2002**, 25, 467.

[1.20] J. Romero, S. Lecam, J. Sanchez, A. Saavedra, and G.M. Rios , *Chem. Engng. Sci.* **2001**, 56, 3139.

[1.21] M.G. Sedigh, W.J. Onstot, L. Xu, W.L.. Peng, T.T. Tsotsis, and M. Sahimi, *J. Phys. Chem. A* **1998**, 102, 8580.

[1.22] H.P. Hsieh, *Inorganic Membranes for Separation and Reaction,* Elsevier Science B.V., Amsterdam, The Netherlands, 1996.

[1.23] R. Soria, *Catal. Today* **1995**, 25, 285.

[1.24] J. Sanchez and T.T. Tsotsis, "Current Developments and Future Research in Catalytic Membrane Reactors", in: *Fundamentals of Inorganic Membrane Science and Technology*, A.J. Burggraaf and L. Cot (Eds.), Elsevier Science B.V., Amsterdam, 1996.

[1.25] A. Dixon, "Innovations in Catalytic Inorganic Membrane Reactors", in: *Catalysis*, Vol. 14, The Royal Society of Chemistry, Cambridge, 1999.

[1.26] J. Sanchez and T.T. Tsotsis, "Reactive Membrane Separation", in: *Reactive Separation Processes*, S. Kulprathipanja, Ed., Taylor & Francis, USA, 2001.

[1.27] B.J. Wood and H. Wise, *J. Catal.* **1968**, 11, 30.

[1.28] V.M. Gryaznov, V.S. Smirnov, L.K. Ivanova, and A.P. Mishchenko, *Dokl. Akad. Nauk. SSR.* **1970**, 190, 144.

[1.29] H.N. Chang and S. Furusaki, *Adv. Biochem. Eng. Bioeng.* **1991**, 44, 27.

[1.30] P.R. Brookes and A.G. Livignston, *Water Res.* 28, 13, 1994.

[1.31] M.E. Rezac, W.J. Koros, and S.J. Miller, *J. Membr. Sci.* **1994**, 93, 193.

[1.32]   M.E. Rezac, W.J. Koros, and S.J. Miller, *Ind. Eng. Chem. Res.* **1995**, 34, 862.

[1.33]   H.J. Sloot, G.F. Versteeg, and W.P.M. van Swaaij, *Chem. Eng. Sci.* **1990**, 45, 2415.

[1.34]   M. Torres, J. Sanchez, J.A. Dalmon, B. Bernauer, and J. Lieto, *Ind. Eng. Chem. Res.* **1994**, 33, 2421.

[1.35]   G.H.W. Saracco, J.P. Neomagus, G.F. Versteeg, and W.P.M. van Swaaij, *Chem. Engng. Sci.* **1999**, 54, 1997.

[1.36]   Catalytica Study Division, *Catalytica Study* No. 4187, Mountain View, CA, 1989

[1.37]   J.N. Armor, *Appl. Catal.* **1989**, 49, 1.

[1.38]   J.N. Armor, *Chemtech.* **1992**, 22, 557.

[1.39]   J.N. Armor, *Catal. Today* **1995**, 25, 199.

[1.40]   J.N. Armor, *J. Membr. Sci.* **1998**, 147, 217.

[1.41]   S. Ilias and R. Govind, *A.I.Ch.E. Symp. Ser.* **1989**, 85, 268, 18.

[1.42]   H.P. Hsieh, *Catal. Rev. Sci. Eng.* **1991**, 33, 1.

[1.43]   V.T. Zaspalis and J. Burggraaf, "Inorganic Membrane Reactors to Enhance the Productivity of Processes," in: *Inorganic Membranes: Synthesis, Characteristics and Applications*, R.R. Bhave, Ed., Reinhold, New York, 1991.

[1.44]   T.T. Tsotsis, R.G. Minet, A.M. Champagnie, and P.K.T. Liu, "Catalytic Membrane Reactors", in: *Computer Aided Design of Catalysts*, R. Becker and C. Pereira, Eds., Dekker, New York, p. 471, 1993.

[1.45]   J. Zaman and A. Chakma, *J. Membr. Sci.* **1994**, 92, 1.

[1.46]   G. Saracco and V. Specchia, *Cat. Rev.-Sci. Eng.* **1994**, 36, 304.

[1.47]   G. Saracco and V. Specchia, "Inorganic Membrane Reactors", in *Structured Catalysts and Reactors*, A. Cybulski and J.A. Moulijn, Eds., Marcel Dekker Inc., New York, 1998.

[1.48]   G. Saracco, G.F. Versteeg, and W.P.M. van Swaaij, *J. Membr. Sci.* **1994**, 95, 105.

[1.49]   K.K. Sirkar, P.V. Shanbhag, and S. Kovvali, *Ind. Eng. Chem. Res.* **1999**, 38, 3715.

[1.50]   J. Feldman and M. Orchin, *J. Mol. Catal.* **1990**, 63, 213.

[1.51]   J.S. Kim and R. Datta, *AIChE J.* **1991**, 37, 1657.

[1.52]   L. Troger, H. Hunnefeld, S. Nunes, M. Oehring, and D. Fritch, *J. Phys. Chem. B* **1997**, 101, 1279.

[1.53]   Y. Uemura, Y. Ohzuno, and Y. Hatate, "A Membrane Reactor Using Hydrogen Storage Alloy for CO_2 Reduction", *Proc. 5th ICIM, Nagoya, Japan,* p. 620, 1998.

[1.54]   W.C. Pfefferie, U.S. Patent App. 3,290,406, 1966.

[1.55]   V.M. Gryaznov, *Platinum Met. Rev.* **1986**, 30, 68.

[1.56]   V.M. Gryaznov, V.S. Smirnov, and M.G. Slin'ko, "Binary Palladium Alloys as Selective Membrane Catalysts", G.C. Bond, P.B. Wells, and F.C. Tompkins, Eds., *Proc. 6th Intl. Cong. Catal.*, **1976**, 2, 894.

[1.57]   V.M. Gryaznov and A.N. Karavanov, *Khim.-Farm.-Zh.* **1979**, 13, 74.

[1.58]   V.M. Gryaznov, *Platinum Met. Rev.* **1992**, 36, 70.

[1.59]   R.E. Bauxbaum and A. B. Kinney, *Ind. Eng. Chem. Res.* **1996**, 35, 530.

[1.60]   V. Jayaraman, Y.S. Lin, M. Pakala, and R.Y. Lin, *J. Membr. Sci.* **1995**, 99, 89.

[1.61]   N. Jeema, J. Shu, S. Kaliaguine, and B.P.A. Grandjean, *Ind. Eng. Chem. Res.* **1996**, 35, 973.

[1.62]   K.L. Yeung and A. Varma, *AIChE J.* **1995**, 41, 4823.

[1.63]   K.L. Yeung, J.M. Sebastian, and A. Varma, *Catal. Today* **1995**, 25, 232.

[1.64]   H.J.M. Bouwmeester and A.J. Burggraaf, "Dense ceramic membranes for oxygen separation", in: *Fundamentals of Inorganic Membrane Science and Technology*, A.J. Burggraaf and L. Cot, Eds., Elsevier Science B.V., Amsterdam, 435, 1996.

[1.65]   J.N. van de Graaf, F. Kapteijn, and J.A. Moulijn, "Catalytic Membranes", in *Structured Catalysts and Reactors,* A. Cybulski and J.A. Moulijn, Eds., Marcel Dekker Inc., New York, 1998.

[1.66]   D. Fritsch and J. Theis, *Proc. Fourth Workshop: Optimisation of Catalytic Membrane Reactors Systems*, R. Bredesen, Ed., Oslo, Norway, May; 1997, pp. 109.

[1.67]   A. Julbe, D. Farrusseng, and C. Guizard, *J. Membr. Sci.* **2001**, 181, 3.

[1.68]   M.N. Tecik, R.N. Paunovic, and G.M.Ciric, *J. Membr. Sci.* **1994**, 96, 213.

[1.69]   R.P. Omorjan, R.N. Paunovic, and M.N. Tekic, *J. Membr. Sci.* **1998**, 138, 57.

[1.70]   J.S. Wu and P.K.T. Liu, *Ind. Eng. Chem. Res.* **1992**, 31, 322.

[1.71]   F. Tiscareno-Lechuga, C.G. Hill Jr., and M.A. Anderson, *J. Membr. Sci.* **1996**, 118, 65.

[1.72]   N. Itoh, *Catal. Today* **1995**, 25, 351.

2   Catalytic Membrane Separation Processes

2.1   Dehydrogenation Reactions

The catalytic dehydrogenation of light alkanes is, potentially, an important process for the production of alkenes, which are valuable starting chemical materials for a variety of applications. This reaction is endothermic and is, therefore, performed at relatively high temperatures, to improve the yield to alkenes, which is limited, at lower temperatures, by the thermodynamic equilibrium. Operation at high temperatures, however, results in catalyst deactivation (thus, requiring frequent reactivation), and in the production of undesired by-products. For these reasons, this reaction has been from the beginning of the membrane reactor field the most obvious choice for the application of the catalytic membrane reactor concept, and one of the most commonly studied reaction systems.

Light alkane (C_2–C_4) dehydrogenation was the reaction studied by Gryaznov and co-workers in their pioneering studies [2.1, 2.2]. In their dehydrogenation reaction studies, they used Pd or Pd-alloy dense membranes, which were 100 % selective towards hydrogen permeation. The choice of these membranes in many of the early studies is because they were commercially available at that time in a variety of compositions, and their metallic nature allows the construction of multitubular and other complex-shaped membrane reactor systems. Comprehensive review papers on Pd membrane reactors have been published by the same group [2.1, 2.2], and also by Shu *et al.* [2.3].

In recent years, various groups have focused their attention on optimizing membrane reactor design aiming to compensate for the relatively low membrane permeability.

Wolfrath *et al.* [2.4], for example, have studied the production of propene via the non-oxidative catalytic dehydrogenation of propane in a PBMR. This is an important reaction, because of the increasing demand for propene and propene derivatives. It has attracted the attention of several groups working with membrane reactors, because it faces technological constrains, as it is highly endothermic and the thermodynamic equilibrium limits the conversion; at the temperature required by the reaction thermodynamics, thermal cracking occurs lowering the selectivity, and resulting in catalyst deactivation due to coke deposition on the catalyst surface. The use of membrane reactors has the promise to lower the reaction temperature and, therefore, eliminate the aforementioned difficulties. Wolfrath *et al.* [2.4] have proposed the use of a PBMR, operating in a cyclic fashion, which combines a Pd-Ag membrane with a structured catalyst bed consisting of fibers packed inside the reactor, in order to optimize the reactor fluid dynamics, minimize pressure drop, and ensure laminar flow and narrow residence time distribution. During the PBMR operation, as Wolfrath *et al.* [2.4] envision it, the dehydrogenation takes place on one side of the membrane (zone I), with simultaneous coke formation on the catalyst surface, and diffusion of hydrogen through the membrane wall. On the other side of the membrane (zone II), flow of incoming oxygen (5 % in N_2) will convert the diffused hydrogen to water and oxidize the coke (created during the previous operating cycle), thus, regenerating the catalyst. The oxidation of carbon and hydrogen generates heat along the reaction zone and keeps a permanent hydrogen concentration gradient during continuous operation. After the reaction is under way, the flows of propane and oxygen are switched periodically from zone I to zone II.

Despite the success in the laboratory of the application of Pd membranes [2.5] to a variety of light (but also larger M.W. [2.6]) alkane dehydrogenation reactions, their large-scale industrial application still remains problematic because of a number of well-documented problems [2.7]. These problems include their high cost and limited commercial availability, questions concerning their mechanical and thermal stability, and in the case of alkane dehydrogenation, poisoning due to carbon deposition; these were discussed in some detail in Chapter 1.

As one may surmise, the emphasis in recent years has been on improving the properties of these membranes. As noted previously, the approach taken here, recently, is to deposit thin films of metals like Pd and its alloys over a porous support (ceramic, Vycor®glass, or metal) using a variety of methods, including electroless-plating, sol gel techniques, and deposition by magnetron sputtering. The aim of these efforts is to prepare a membrane with added mechanical strength due to the presence of the porous matrix, which maintains the high selectivity of the dense membrane, with improved hydrogen permeability, due to the presence of a very thin (1–20 μm) metal film. Such membranes have been applied to butane and isobutane dehydrogenation. Matsuda *et al.* [2.8], for example, obtained a considerable increase in isobutene yield (+600 %) compared to the calculated equilibrium value at 673 K using a PBMR. The membrane was prepared by covering a porous alumina tube with a film of platinum by the electroless-plating technique. The ethane dehydrogenation reaction was studied by Gobina and Hughes [2.9] using a porous Vycor® glass membrane modified with a thin film of Pd-Ag alloy deposited by a magnetron sputtering technique. They reported that the conversion increased from 2 % (equilibrium value) to 18 % at 660 K.

Reactive separations for light alkane dehydrogenation reactions to produce olefins have also been successfully performed using inorganic porous membranes. Champagnie *et al.* [2.10, 2.11], in one of the earliest investigations, studied ethane dehydrogenation using a commercial 40 Å alumina membrane (Membralox®, U.S. Filter) impregnated with platinum as catalyst (CMR), or using the membrane with a packed-bed of a commercial catalyst (PBMR or PBCMR). They reported increased yields in the conversion values from those attained in the absence of the membrane, and greater than reference equilibrium conversions calculated at the tubeside and shellside pressures and temperatures. Porous membranes have also been applied to the propane dehydrogenation reaction to propylene, which is another very important raw material in the petrochemical industry. First mention of the potential application of membrane reactors to this reaction took place in 1988 in a British patent [2.12], which proposed the use of γ-alumina membranes for the dehydrogenation of many organic compounds including propane. The first open literature application of these membranes for propane dehydrogenation is by Ziaka *et al.* [2.13], who studied the reaction in a PBMR using a γ-alumina membrane and a commercial Pt/alumina catalyst.

As discussed in the introduction, hydrogen transports through mesoporous γ-alumina membranes by Knudsen diffusion, and its selectivity is, therefore, not very high for light alkane dehydrogenation. γ-alumina, furthermore, is a metastable material, which at high temperatures and in the presence of steam evolves slowly towards the more stable α-alumina form (the latter is also a key concern for all microporous membranes, which use mesoporous γ-alumina membranes as a support). The emphasis in recent years has been on preparing porous ceramic membranes with higher hydrogen selectivity. A type of membranes that have attracted attention are microporous silica membranes deposited by sol gel (de Lange *et al.* [2.14], Raman and Brinker [2.15], Ayral *et al.* [2.16], de Voss and Verweij [2.17]) or chemical vapor deposition (CVD/CVI) techniques (Tsapatsis and Gavalas [2.18], Kim and Gavalas [2.19], Morooka *et al.* [2.20]) on underlying mesoporous alumina or silica substrates.

These membranes have been utilized in a number of membrane reactor studies [2.21, 2.5, 2.22, 2.23]. Weyten *et al.* [2.21, 2.5], studied the propane dehydrogenation reaction in a PBMR using a microporous, commercial silica/alumina membrane made by (CVD/CVI) with a H_2 permeance of $\sim 1.4 \times 10^{-9}$ mol/m^2.Pa.s, and a H_2/C_3H_8 permselectivity in the range of 70–90 at 500 °C. The conversion in the PBMR at 500 °C was twice as high as the equilibrium value. Weyten *et al.* [2.5, 2.21] report that the increase in conversion was, however, only significant for relatively small values of the propane feed; for high propane feeds, the membrane cannot remove the hydrogen fast enough to have a significant impact on conversion. Replacing the silica membrane with a Pd/Ag membrane with superior permselectivity and higher permeance provided better performance. Weyten *et al.* [2.5, 2.21] noted, as did previous investigators (Ziaka *et al.* [2.13]), that the selectivity in the membrane reactor is higher than in the plug-flow reactor, when they are run under similar conditions. This likely is because as H_2 is selectively removed from the reaction mixture, it is not available for competitive side reactions, like the production of methane, which limits the propene selectivity. Schaefer *et al.* [2.24] studied the same reaction in a PBMR using a Cr_2O_3/Al_2O_3 catalyst. They utilized a sol gel SiO_2 membrane, which they prepared using a mesoporous γ-Al_2O_3 membrane as a substrate. The membrane showed hydrogen permeance > 25 m^3/m^2.h.bar and a H_2/C_3H_8 permselectivity of 30–55 in the temperature region of 450–550 °C. Ioannides and Gavalas [2.23] utilized a dense silica membrane for isobutane dehydrogenation in a PBMR, with modest increases in isobutene yield over those attained in a conventional reactor in the absence of the membrane. Amorphous silica microporous membranes, unfortunately, have also been shown to be sensitive to steam and coke in the presence of light alkanes and olefins under temperature conditions akin to catalytic dehydrogenation.

More promising for reactive separations involving gas phase reactions appears to be the development and use in such applications of microporous zeolite and carbon molecular sieve (Itoh and Haraya [2.25]; Strano and Foley [2.26]) membranes. Zeolites are crystalline microporous aluminosilicate materials, with a regular three-dimensional pore structure, which are relatively stable to high temperatures, and are currently used as catalysts or catalyst supports for a number of high temperature reactions. One of the earliest mentions of the preparation of zeolite membranes is by Mobil workers (Haag and Tsikoyiannis [2.27]), who reported in a U.S. patent the synthesis of very thin (and likely to be fragile) self-supported membranes made of ZSM-5 zeolite. The field of zeolite membrane synthesis has since then become a very active area (van de Graaf *et al.* [2.28]). A variety of apparently defect-free zeolite membranes have been prepared by depositing thin films of zeolite (silicalite and ZSM-5, SAPOs, etc.).

Despite the significant recent interest in these types of membranes, there are relatively few reports of the application of such membranes in high temperature catalytic membrane reactor applications. Dalmon and coworkers (Casanave *et al.* [2.29, 2.30], Ciavarella [2.31]), for example, have reported the application of a MFI zeolite membrane supported on a mesoporous alumina tube to the isobutane dehydrogenation reaction in a PBMR, but with modest gains in yield. Yin Au *et al.* [2.32] have prepared MFI zeolite membranes with catalytic vanadium and titanium metals imbedded into their structural framework, potentially for catalytic membrane reactor applications. Incorporating the metals into the membrane significantly influenced its physicochemical characteristics as well as the permeation and adsorption properties. One explanation for the lack of many applications in high temperature membrane reactors is that the majority of the membranes synthetized, so far, are MFI-type zeolite membranes. These have pore diameters (~5Å), which are still too big to separate selectively small gaseous molecules. Another problem may be the apparent difficulty of controlling the membrane thickness and permeance, since for optimal PBMR design one must strive to balance membrane flux with reaction rate. A number of research efforts are currently under way aiming to synthesize other more permselective zeolite membranes with smaller diameters. Two International Workshops on the preparation of zeolite membranes and films were held, one in June 1998 in Gifu, Japan, and the other in July 2001 in Purmerend Netherlands, during which a multitude of new and exciting developments were reported. Several good papers were also presented at the 6th International Conference on Inorganic Membranes held in June 2000 in Montpellier, France.

In addition to light alkanes, membrane reactors have also been applied to the dehydrogenation of a variety of other hydrocarbons. The most notable among these applications is the dehydrogenation of ethylbenzene (EB) to styrene, which is among the most important monomers used in the polymer industry (Gallaher *et al.* [2.33], Tiscareno-Lechuga *et al.* [2.34, 2.35], Tagawa *et al.* [2.36]). A number of studies have reported the application of catalytic membrane reactors to this reaction using porous alumina or composite membranes, prepared by depositing thin films of Pd on porous ceramic or stainless steel substrates [2.37, 2.38]. In general, a PBMR has been used. The studies report an increase in the styrene yield, when compared to the yield attained in the absence of the membrane in a conventional packed-bed reactor. Tiscareno-Lechuga *et al.* [2.34] also reported that the continuous removal of hydrogen by the membrane had the additional beneficial effect of slowing down the undesirable hydrodealkylation side reaction. The presence of the membrane, thus, resulted in an increase in the selectivity as well. Jiang and Wang [2.39] in their studies used a PBMR with a microporous permselective membrane with a H_2 permeance in the range of ~10^{-6} mol/m^2.s.Pa and H_2/EB selectivity of over 75. In the temperature range of 560–600 °C, EB liquid hour space velocity (LHSV) in the range of 0.5–1.0/h, and water/EB molar ratio in the range of 9.86–16.42, the membrane reactor improves styrene yield to a maximum of 21.5 % and a top per-pass stryrene yield of 75 %, which is 10 % above that obtained in a fixed-bed reactor. The use of microporous

(Fe-MFI and Fe-Al-MFI) zeolite membranes during the catalytic dehydrogenation of ethylbenzene has been recently reported by Xiongfu *et al.* [2.40].

Cyclohexane dehydrogenation to benzene is another reaction that has been studied by a number of research groups (Mondal and Ilias [2.41]). This reaction takes place at relatively low temperatures and has well known kinetics, and, as a result, has been the reaction of choice in modelling studies of membrane reactors (see further discussions in Chapter 5). The reaction has also potential significance for hydrogen storage and renewable energy applications (Saracco *et al.* [2.7]). In a recent study Terry *et al.* [2.42] studied this reaction utilizing commercial (U.S. Filter) ceramic membranes of various pore sizes altered by addition of successive thin film layers of silica prepared with the aid of silica oxide particles in an iron(III) solution. Experiments with the coated membranes operating in the Knudsen regime showed a 300 % increase in yield from that, which was obtained in a conventional reactor under similar conditions. Interestingly, Terry *et al.* [2.42] report that the coated membranes resulted in a higher reaction yield than the uncoated membranes, whose transport mechanism was also Knudsen dominated. Itoh and Haraya [2.25] studied the cyclohexane dehydrogenation in a PBMR utilizing microporous, hollow-fiber carbon membranes prepared by the pyrolysis of polyimide hollow-fiber precursors. The carbon hollow fibers had a high hydrogen permselectivity vs. cyclohexane and benzene, but poor mechanical properties. In the experiments of Itoh and Haraya [2.25] they were protected from the packed-bed of catalysts by a porous tube of sintered metal. A 300 % enhancement in conversion with respect to the equilibrium was obtained. A schematic of the membrane reactor and experimental data of reactor conversion as a function of the feed rate are shown in Figure 2.1.

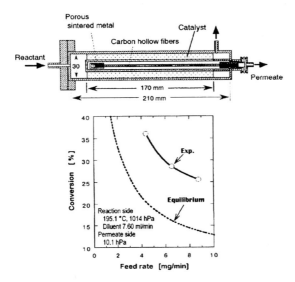

Figure 2.1. Carbon membrane reactor and cyclohexane conversion as a function of feed rate. From Itoh and Haraya [2.25], with permission from Elsevier Science.

The same reaction has been studied, in recent years, by Frisch and co-workers [2.43] using dense polymeric membranes. The dense polymeric membranes were prepared by blending polyethylacrylate with a 13X zeolite, which contained a dehydrogenation catalyst (Ti or Ni). They also prepared catalytic polymeric membranes by free radical polymerization of the monomer in the presence of the zeolite. These membranes were shown to be active for the cyclohexane dehydrogenation reaction at low temperatures. The recent development of thermally resistant polymeric membranes (Koros and Woods [2.44] and Rezac and Schoberl [2.45]) provides promise for the more widespread use of such membranes in CMR applications.

The dehydrogenation of various alcohols has been reported by a number of groups. Mouton *et al.* [2.46] studied the catalytic dehydrogenation of isopropanol using a Cu on SiO_2 catalyst in a membrane reactor utilizing a Pd-Cu alloy membrane. The membrane was prepared by electroless plating of alumina supports, followed by heat-treatment after electroplating. The same reaction has been studied by Trianto *et al.* [2.47] in a membrane reactor utilizing a modified Vycor® glass membrane. An improvement in selectivity was reported by Raich and Foley [2.48] during the dehydrogenation of ethanol to acetaldehyde in a PBMR utilizing a Pd membrane. For this reaction acetaldehyde formation is endothermic and overall unfavorable, while the secondary formation of ethyl acetate from the product and reactant is exothermic and favorable. Therefore, it is observed that the secondary product forms under catalytic reaction conditions, thereby, resulting in a diminished yield of the desired aldehyde product. In their study Raich and Foley have shown that the Pd membrane, by removing hydrogen, helps to shift further to the right the acetaldehyde-forming step before the product can react with ethanol, thereby, significantly increasing the yield of acetaldehyde. With the Pd membrane present, ethanol conversion increased from 60 % to nearly 90 % with a commensurate rise in selectivity to acetaldehyde from 35 % to 70 %, shifting the yield from 21 % to 63 %.

Ethanol dehydrogenation to acetaldehyde has also been studied by Liu *et al.* [2.49]. In their studies they utilized a membrane consisting of Ni-P amorphous thin metal alloy film on a ceramic membrane, prepared by an electroless Ni-plating technique, using a metal-activated paste. Two kinds of Ni-P alloy/ceramic membranes were utilized during the membrane reactor studies of ethanol dehydrogenation, those prepared by the conventional electroless Ni-plating technique, and those that were crystallized by a subsequent treatment. In the reactor studies the effect of the reaction temperature, argon sweep flow rate, and space-time on ethanol conversion and acetaldehyde yield were investigated. Ethanol conversion in the reactor with the conventionally prepared Ni-P amorphous alloy membranes was significantly higher than in the membrane reactor using the re-crystallized membranes. This group (Liu *et al.* [2.50]), in a study of the same reaction, also utilized Rh-modified γ-Al_2O_3 membranes; these were prepared by adding small amounts of rhodium to the boehmite sol, during the preparation of the γ-Al_2O_3 membranes. The Rh-modified γ-Al_2O_3 membranes exhibited a higher H_2/Ar separation factor than the pure γ-Al_2O_3 membranes in the range of 100–350 °C. The reactor studies of ethanol dehydrogenation showed that the ethanol conversion and acetaldehyde selectivity in the Rh-Al_2O_3 membrane reactor were higher than those in the γ-Al_2O_3 membrane reactor, owing to the former's higher hydrogen permeability and separation factor in the range of experimental temperatures. In their most recent study the same group (Xue *et al.* [2.51]) employed a new amorphous Ni-B alloy membrane, prepared by an electroless plating method, in a membrane reactor for ethanol dehydrogenation. This membrane exhibited a significant promotion effect for the reaction, due to its catalytic activity as well as its high permselectivity for hydrogen

The catalytic dehydrogenation of methanol to methyl formate has been studied by Lefu *et al.* [2.52]. The authors utilized a Pd/Ag dense membrane prepared by electroless plating on a mesoporous γ-Al_2O_3 membrane. A CuO-ZnO-ZrO_2/Al_2O_3 catalyst proved to be particularly active for this reaction. The PBMR was shown to have superior performance to that of a fixed-bed reactor. The same reaction was studied by Gorshkov *et al.* [2.53]. The diffusion of hydrogen through a Pd-Ru membrane was investigated both in the presence and the absence of catalyst when the reaction does not occur. In the presence of the catalyst, the membrane was not poisoned with carbon monoxide, even so it was present at a high concentration; interestingly, CO rapidly deactivated the membrane in the absence of the catalytic reaction. In a related study Amandusson *et al.* [2.54] studied methanol dehydrogenation in a catalytic membrane reactor containing a Pd membrane, which also acted as the catalyst for the reaction. Amandusson *et al.* [2.54] believe that at the temperature they studied the reaction (350 °C) methanol adsorbs on a clean palladium surface and decomposes leading to adsorbed hydrogen and a carbonaceous surface overlayer. The hydrogen can either desorb from the membrane surface or permeate through it. During a continuous supply of methanol, hydrogen permeation was observed to decrease and, eventually, totally stop. Amandusson *et al.* [2.54] interpret this to be due to the growing carbon monoxide/carbon coverage of the membrane's surface. Adding oxygen in the methanol supply could balance the increasing carbonaceous coverage through the production of carbon dioxide, allowing no CO bond scission to occur, and no build-up of a carbon surface overlayer. Choosing the appropriate methanol/oxygen ratio was crucial for maintaining the hydrogen permeation rate. Experiments utilizing isotope-labelled methanol (CH_3OH, CH_3OD, CD_3OH and CD_3OD) have shown that it is the hydrogen in the methyl (or methoxy) group that preferentially permeates through the membrane.

Experiments by Hara *et al.* [2.55] with methanol decomposition in a PBMR utilizing a Pd membrane have shown that carbon monoxide prevented hydrogen permeation through the membrane at temperatures less than 280 °C. Hydrogen permeation in the PBMR was also affected by concentration polarization. The drop in hydrogen permeation due to both factors had a significant influence on the performance of the PBMR. Similar observations were made by Amandusson *et al.* [2.56], who reported that the presence of CO below 150 °C completely inhibits hydrogen permeation for Pd membranes, while above 300 °C it has no effect. Hydrogen permeation through a Pd membrane in the presence of steam, methane, propane, and propylene was investigated by Jung *et al.* [2.57]. Methane and propane had a negligible effect on hydrogen permeation. When propylene was fed together with hydrogen, however, the permeance greatly decreased, and the decline increased with time. The decline in hydrogen permeation was again attributed to a carbonaceous matter on the membrane, which was formed by the decomposition of propylene. This carbonaceous matter could be slowly removed by treatment with hydrogen at 600 °C, the hydrogen permeation rate recovering to 80 % of its initial value after a week of treatment. Treatment with oxygen at 600 °C was more effective in removing the carbonaceous layer, and the permeation rate was fully recovered within 15 min.

A number of other published studies also report difficulties with the Pd membranes' mechanical stability during dehydrogenation reactions. In the work of She and Ma [2.37], for example, microcracks developed in the membrane (as observed in SEM micrographs) after the membrane had been used for about 500 h at a temperature between 500 and 625 °C during ethylbenzene dehydrogenation. These authors attribute the problems to a rearrangement of the Pd grains, possibly combined with carbon (resulting from hydrocarbon cracking) diffusion into the metal membranes. Both problems, as previously noted, are commonly reported with the use of Pd membranes in hydrocarbon dehydrogenations [2.7]. Gallaher *et al.* [2.33] also reported robustness problems with their γ-Al$_2$O$_3$ membrane resulting from the close proximity of the membrane with the alkali promoted iron oxide catalyst.

Non-oxidative coupling of CH$_4$ to produce valuable C$_2$-type products (ethane or ethylene) is also recently attracting attention (Liu *et al* [2.58]). The reaction to produce ethane, for example, is described as follows:

$$2CH_4 \leftrightarrow C_2H_6 + H_2 \tag{2.1}$$

This reaction is equilibrium limited and the use of membrane reactors has the potential for significantly improving its yield by shifting the conversion towards the product side through the removal of H$_2$. The concept was first tested by Anderson *et al.* [2.59] using a PBMR equipped with a Pd tubular membrane, and utilizing a conventional Pt-Sn/Al$_2$O$_3$ catalyst. At 300 °C they observed a concentration of C$_2$H$_4$ in the products in the range of 0.4–0.6 wt. %, very low to be of practical importance, but significantly higher (4 to 6 times) than the expected equilibrium composition. The problem with the use of Pd membranes for reaction (2.1) is that one is limited in their use to moderate temperatures (typically, less than 600 °C). Conventional dehydrogenation catalysts are, however, inactive towards reaction (2.1) at these moderate temperatures.

Proton conducting, solid-state membranes show more significant promise for application to this reaction [2.58]. Such membranes with reported chemical compositions BaCe$_{0.9}$Nd$_{0.1}$O$_{3-x}$, SrCe$_{0.95}$Yb$_{0.05}$O$_{3-x}$, and CaZr$_{0.9}$In$_{0.1}$O$_{3-x}$, having high proton conductivities at high temperatures ($\sim$ 1000 °C), were first developed by Iwahara and his coworkers (Iwahara *et al.* [2.60, 2.61, 2.62, 2.63]). The hydrogen permeance through such membranes can be enhanced by electrochemical pumping through the imposition of an external voltage (Iwahara *et al.* [2.62, 2.64]). Some of the initial studies in this area (Iwahara *et al.* [2.65]; Hamakawa *et al.* [2.66]) utilized exactly such a concept shown schematically in Figure 2.2. Hamakawa *et al.* [2.66], for example, used a SrCe$_{0.95}$Yb$_{0.05}$O$_{2.95}$ membrane as a proton, solid oxide electrolyte with two porous Ag electrodes attached to it. CH$_4$ was passed in the one compartment of the electrochemical membrane reactor (the anode) while Ar was flown into the other cell. C$_2$ products (C$_2$H$_4$ and C$_2$H$_6$) were detected in the anode compartment, while hydrogen was detected in the cathode compartment. Hamakawa *et al.* [2.66] report that the C_2 formation rate in the electrochemical membrane reactor is 7 times that in the absence of electrochemical hydrogen pumping. Stoukides and coworkers (Chiang *et al.* [2.67, 2.68, 2.69, 2.70, 2.71]) for the same system report that the presence of the electrochemical hydrogen pumping significantly enhances catalytic activity. This phenomenon is known as NEMCA [2.72], and is discussed further in Section 2.3 for oxygen conducting solid oxide membranes. If instead of Ar oxygen is fed in the cathode, one may be able to operate the membrane reactor in a "fuel cell" configuration with simultaneous production of C_2 products in the anode, water in the cathode, and electricity. This concept has been already tested by Iwahara *et al.* [2.61]. The primary hindrances appear to be the high ohmic resistance of the electrolyte, and the deposition of carbon on the electrolyte on the anode side [2.65]. A similar concept has also been tested using oxygen conducting solid oxide membranes (see further discussion to follow in Section 2.3). The use of proton conducting membranes appears advantageous, however, because no products of complete oxidation are formed in the anode compartment, and the C_2 products and water are formed in separate compartments.

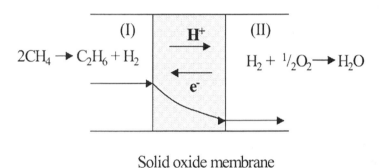

Solid oxide membrane

Figure 2.2. Schematic of an electrochemical MR using a proton conducting membrane.

The materials developed by Hamakawa *et al.* [2.66, 2.73] are, under the proper set of conditions, mixed conductors, i.e., they have both good proton and electron hole conductivities. Hamakawa *et al.* [2.66, 2.73] report that protons in these materials are generated as a result of the interaction of H_2 or H_2O from the gas phase with defects (i.e., electron holes) in their lattice. Protonic conductors are created through doping of ABO_3-type perovskites (A=Ca, Sr, or Ba; B=Ce, Tb, Zr, or Th), which, themselves, have no proton conducting properties, i.e., by substitution of B with an aliovalent cation C (C=Ti, V, Cr, Mn, Co, Ni, Cu, Al, Ga, Y, In, or Yb). For example, by substituting Yb^{3+} into the structure of the starting material $SrCeO_3$ for Ce^{4+} one creates a material (e.g., $SrCe_{1-x}Yb_xO_{3-x/2}$), which has oxygen vacancies, V_O. Interaction of this material with gaseous oxygen creates electron holes according to the following reaction

$$V_O + 0.5O_2 \leftrightarrow O_L + 2h \tag{2.2}$$

where O_L represents a lattice oxygen and h an electron hole. Hamakawa *et al.* [2.66, 2.73] report that interaction of the electron holes with either gaseous H_2O or H_2 creates the protons responsible for the material's ionic conductivity according to the following reactions:

$$H_2O + 2h \leftrightarrow 2H^+ + 0.5O_2 \tag{2.3}$$

$$H_2 + 2h \leftrightarrow 2H^+ \tag{2.4}$$

When exposing one side of these membrane materials to oxygen, while the other side is exposed to hydrogen, hydrogen spontaneously transports through the membrane without the need to impose an external electrical voltage. Simultaneously, because of the need to maintain electrical neutrality, an electron hole flow is established in the opposite direction to that of hydrogen (proton) transport.

The use of mixed proton-hole conductive membranes for CH_4 coupling has been reported by Hamakawa *et al.* [2.73]. With CH_4 flowing in one chamber of the membrane reactor and oxygen in the other, they observed C_2 forming almost exclusively, without any CO_x by-products formed. The reported rates were rather low (less than 2% conversion), however, and coke was also formed. Both were attributed to the low activity of the membrane surface towards methane coupling, and the low hydrogen transport rates due to the rather thick membranes that were utilized. Langguth *et al.* [2.74] have also studied the methane coupling reaction using a $SrCe_{0.95}Yb_{0.5}O_{3-x}$ membrane. As with the study of Hamakawa *et al.* [2.73], CH_4 passed on one side of the membrane and oxygen on the other. Selectivities as high as 77 % were reported. However, Langguth *et al.* [2.74] reported the presence of CO_X type products of total oxidation, which they attribute to the fact that their membrane is a mixed proton-oxygen-electron conductor, in contrast to the findings of Hamakawa *et al.* [2.66, 2.73]. Terai *et al.* [2.75] studied the oxidative coupling of methane using water as an oxidant in a membrane reactor using a $SrCe_{0.95}Yb_{0.5}O_{3-x}$ membrane to remove H_2 selectively.

$$2CH_4 + H_2O \leftrightarrow C_2H_6 + H_2 + H_2O \tag{2.5}$$

A $SrTi_{0.4}Mg_{0.6}O_{3-x}$ catalyst was used, which had been previously shown to be an effective catalyst for this reaction. The use of the membrane significantly increased the yield of C_2 hydrocarbons. This remains an area with significant unexplored potential. Progress can be made here by developing CMR systems with enhanced catalytic activities towards the CH_4 coupling reaction, and asymmetric-type proton-hole or proton-electron conducting membranes with significantly increased conductivities.

An interesting application of proton-electron conducting membranes has recently been reported by Li *et al.* [2.76]. These authors studied the conversion of CH_4 first to C_2H_4 and its subsequent direct catalytic aromatization to benzene and other valuable aromatic hydrocarbons. Their reactor configuration is shown schematically in Figure 2.3. The two distinct additional features of their work are the use of an active catalyst for the reaction itself, (Mo/H-ZSM5), and the use of asymmetric membranes with a thin (10–30 µm) dense solid oxide film on the top of a macroporous support made of the same material, and with a thermal expansion coefficient close to that of the dense film. Two types of membrane materials were tested, namely $SrZr_{0.95}Y_{0.05}O_{3-x}$, and $SrCe_{0.95}Y_{0.05}O_{3-x}$. The latter is reported to have a mixed protonic/electronic conductivity, while the former material has little electronic conductivity, and its use would require the application of an external voltage. A detailed kinetic/transport model utilizing both homogeneous and catalytic reactions was utilized (for further discussion see Chapter 5), which predicts yields to C_2–C_{10} hydrocarbons larger than 90 %. Some of preliminary experimental data reported appears promising. The key challenge here will be to limit the chain-growth reactions to undesired polyaromatic compounds that would tend to coat and deactivate the membranes.

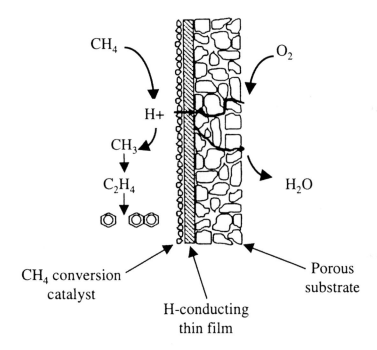

Figure 2.3. A solid oxide MR for CH_4 aromatization. Adapted from Li *et al.* [2.76].

2.2 Hydrogenation Reactions

Membrane reactors have also been utilized in catalytic hydrogenation reactions. For hydrogenation reactions involving liquid hydrocarbons the membrane's role is in separating the liquid from the gaseous reactant (e.g., hydrogen), and providing a means for delivering this reactant at a controlled rate. So doing, reportedly, helps to avoid hot spots in the reactor or undesirable side reactions. For CMR applications the porous membranes create a triple-point interface between the three different phases (gas, liquid, and the solid catalyst on the membrane); this decreases the mass transfer limitations typically encountered with classical slurry or trickle-bed reactors (a further discussion on modeling aspects of such reactors is given in Chapter 5).

Gryaznov and his coworkers [2.77] in the former Soviet Union were again among the first to report the application of catalytic membrane reactors to a hydrogenation reaction. They studied a number of hydrogenation reactions of value in the production of fine chemicals using Pd membranes. In such reactors hydrogen, which flows on one side of the membrane, diffuses selectively through the Pd membrane, and emerges on the other side in a highly reactive atomic form to react with the liquid substrate. To provide a source for the hydrogen, Gryaznov and coworkers proposed the coupling, through the membrane, of a dehydrogenation with a hydrogenation reaction. In this interesting concept the dehydrogenation reaction, occurring on one side of the membrane, produces the hydrogen, which, after diffusing through the membrane, participates in the liquid phase hydrogenation reaction [2.1, 2.77, 2.78]. This technical approach also provides a means for coupling the reactions energetically, in order to attain an autothermal operation. Several other novel concepts for providing the hydrogen for the hydrogenation MR or making use of the hydrogen from the dehydrogenation MR have also been proposed. Itoh *et al.* [2.79], for example, recently suggested directly integrating the hydrogenation of benzene to cyclohexane with water electrolysis in an electrochemical cell utilizing a polymer electrolyte (Nafion®, Du Pont). A Rh-Pt electrode, which was found to be active for the electrochemical hydrogenation of benzene, was formed on the polymer electrolyte by means of a soaking-reduction process. Water electrolysis took place at the anode of the electrochemical cell, and the hydrogen (proton), which was produced, was then pumped to the cathode to participate in the electrochemical hydrogenation of benzene at temperatures of 25–70 °C, and at atmospheric pressures. The same group also proposed coupling in a Pd membrane reactor the cyclohexane dehydrogenation reaction with *n*-hexane reforming [2.80]. They have shown, that removing the hydrogen through the membrane favorably impacted the conversion of the dehydrogenation reaction. The hydrogen that permeated through the membrane, furthermore, appeared to have a positive effect on the reforming reaction, itself, in that it allowed it to proceed without any significant deactivation.

A number of industrial applications of the technology have been reported. They include the synthesis of vitamin K from quinone and acetic anhydride [2.81], and the *cis/trans*-2-butene-1,4 diol hydrogenation to *cis/trans*-butanediol [2.82]. More recently membrane reactors have been applied to cyclopentadiene hydrogenation utilizing Pd-Ru ceramic-metallic composite membranes [2.83]. Often the hydrogen permeating through the membrane is reported to be more active towards the hydrogenation than hydrogen coming directly from the gas phase. Shirai *et al.* [2.84], for example, have investigated the reactivity of hydrogen atoms permeating through a Pd membrane towards thiophene hydrodesulfurization. They compared the reactivity of such hydrogen atoms to that of atoms dissociatively adsorbing on the palladium membrane from the gas phase. A reactor in which the thiophene reacted with hydrogen atoms permeating through the Pd gave higher desulfurization activities, when compared to a system with hydrogen atoms co-adsorbing with thiophene molecules on the membrane surface. The characteristics of the ammonia synthesis reaction on a Ru-Al$_2$O$_3$ catalyst in the presence of a Ag-Pd hydrogen permeable membrane in the temperature range from room temperature to 523 K under atmospheric pressure, were studied by Itoh *et al.* [2.85]. Ammonia was reported to form in the membrane reactor even at room temperature, which was attributed to the high reactivity of atomic hydrogen supplied from the membrane.

In the area of dense membrane applications for hydrogenation reactions a number of recent studies also report the use of proton conducting solid oxide membranes (Otsuka and Yagi [2.86], Panagos *et al.* [2.87], Marnellos and Stoukides [2.88]). As noted previously, this is an exciting class of new materials with significant potential applications.

Hydrogenation reactions have also been studied with catalytic membrane reactors using porous membranes. In this case the membrane, in addition to being used as a contactor between the liquid and gaseous reactants, could, potentially, also act as a host for the catalyst, which is placed in the porous framework of the membrane. As previously noted a triple-point interface between the three different phases (gas, liquid, and the solid catalyst) is then created in the membrane. The first application was reported by Cini and Harold [2.89] for the hydrogenation of α-methylstyrene to cumene. The authors report that this type of reactor presents the advantage of enhanced mass transfer rates between the reactant phases, and of a more efficient contact with the catalyst. A CMR was used by Torres *et al.* [2.90] for nitrobenzene hydrogenation using a commercial Membralox® membrane, which was catalytically activated with Pt by ion-exchange and impregnation. In this study the effects of the various operating reactor parameters were investigated in detail both experimentally and through the use of a theoretical model. Torres *et al.* [2.90] showed that diffusional and kinetic resistances could be well controlled by adjusting the experimental parameters. A detailed description of the theoretical model developed by these authors is given in Chapter 5. Monticelli *et al.* [2.91] reported the hydrogenation of cinnamaldehyde in a similar three-phase CMR. They concluded that with the CMR the selectivity towards hydrocinnamyl alcohol was better than the one attained with a classical slurry-reactor. This result was again explained by the absence of diffusional or other mass transport limitations when using the CMR. This is the result of the direct delivery of the gaseous reactant to the three-phase interface, located in a thin catalytic layer (~2 μm) at the external membrane surface. More recently, the same reaction was studied by Pan *et al.* [2.92] using a γ-Al$_2$O$_3$ membrane impregnated with Co. Again the liquid reactant was fed in one side of the membrane, and the gaseous reactant was introduced in the other. By capillary force the liquid contacted with the catalyst and the gas in the thin, catalytically active γ-Al$_2$O$_3$ layer. Activities as high as 1.53 (mol cinnamaldehyde/mol Co.h) were reported. Veldsink [2.93] studied the hydrogenation of sunflower seed oil in a similar type membrane reactor. The reactor consisted of a mesoporous zirconia membrane impregnated with Pd as the active catalyst, which provided a catalytic interface between the H$_2$ and the oil. Hydrogenation was carried out at different pressures, and temperatures, and the formation of trans isomers was monitored during the run. For the three-phase catalytic MR, interfacial transport resistances and intraparticle diffusion limitations did not influence the reaction. Under such kinetically controlled conditions, oleic and elaidic acids were not hydrogenated in the presence of linoleic acid. The catalytic membrane, unfortunately, showed severe catalyst deactivation. Only partial recovery of the catalyst activity was possible.

Polymeric and organic membranes have also found application in hydrogenation reactions as a contactor, and as hosts of valuable catalysts. Liu and coworkers [2.94] have prepared, for example, catalytic hollow-fiber membranes by supporting polymer anchored Pd catalyst on the inside wall of cellulose acetate (CA) hollow fibers. Using these membranes they studied the selective hydrogenation of propadiene and propyne in propene in catalytic membrane reactors at atmospheric pressure and 40 °C. A membrane reactor utilizing a polyvinylpyrrolidone (PVP) anchored Pd catalyst on a CA membrane was shown particularly effective in reducing the content of propadiene and propyne to <10 and 5 ppm, respectively; these hydrocarbons were very selectivily (97.8 %) converted to propene. The same group [2.95] has also studied the selective hydrogenation of butadiene to 1-butene by using hollow-fiber membranes containing Pd and bi-metallic Pd-Co catalysts. The objective of the study was purification of 1-butene stream by hydrogenating the diene molecule. The best results were obtained with the bi-metallic catalytic membranes, for which the content of butadiene was reduced to less than 10 ppm, with a 1-butene loss of about 2 %. The selective hydrogenation of cyclopentadiene to cyclopentene has also been reported by Liu *et al.* [2.96] by using the same type of bimetallic polymeric hollow-fiber membranes under mild conditions of 40 °C and 0.1 MPa. They found a synergistic effect between Pd and Co in the bimetallic PVP-Pd-0.5Co/CA hollow-fibers that resulted in a 97.5 % conversion of cyclopentadiene, and a 98.4 % selectivity for cyclopentene. Ziegler *et al.* [2.97] modified porous polymeric ultrafiltration membranes by impregnation with TiO_2 and treatment with palladium acetate to yield catalytically active membranes. The ultrafiltration membranes were made of polyamideimide, and contained an inorganic filler (up to 40 wt. %) incorporated during the casting stage. They proved stable to temperatures up to 200 °C. Ziegler *et al.* [2.97] used these membranes to study the hydrogenation of propene and propyne. For the most effective membranes 100 % conversion of propene at a maximum yield of 98 % of propane was obtained at the permeate side. In the selective hydrogenation of 5 % propyne in propene, a selectivity of 99 % to propene at 100 % conversion of propyne was achieved at a permeate flux of 0.08 m^3/m^2. h.bar.

Polymeric membranes also show potential for application in the area of chiral catalysis. Here metallocomplexes find use as homogeneous catalysts, since they show high activity and enantioselectivity. They are expensive, however, and their presence in the final product is undesirable; they must be, therefore, separated after the reaction ends. Attempts have been made to immobilize these catalysts on various supports. Immobilization is a laborious process, however, and often the catalyst activity decreases upon immobilization. An alternative would be a hybrid process, which combines the homogeneous catalytic reactor with a nanofiltration membrane system. Smet *et al.* [2.98] have presented an example of such an application. They studied the hydrogenation of dimethyl itaconate with Ru-BINAP as a homogeneous chiral catalyst. The nanofiltration membrane helps separate the reaction products from the catalyst. Two different configurations can be utilized, one in which the membrane is inserted in the reactor itself, and another in which the membrane is extraneous to the reactor. Ru-BINAP is known to be an excellent hydrogenation catalyst at ~60 bar of hydrogen. This, however, exceeds the upper operational pressure limit of most nanofiltration membranes (~30 bar). At these conditions the catalyst showed a lower activity, but still maintained its enantioselectivity. The polymeric membranes tested were shown to completely exclude the Ru-BINAP catalyst. The hybrid reactor system was tested for about a month, and showed very stable activity and enantioselectivity. Dwars *et al.* [2.99] used a micellar enlarged Rh-(*2S,4S*)-*N-tert*-butoxy-carbonyl-4-diphenyl-phosphino-2-diphenyl-phosphino-methyl-pyrrolidine (BPPM) catalyst in a membrane reactor, equipped with an ultrafiltration membrane for the enantioselective hydrogenation of α-amino acid precursors. The chiral *a*-amino acid derivatives were obtained with good enantioselectivity and yield. The catalyst, embedded in micelles, obtained from triblock copolymers as surfactants, was retained and reused several times without loss of activity and enantioselectivity. Of all membranes a regenerated cellulose acetate ultrafiltration membrane from AMICON was found most appropriate. With this membrane only a minimal leaching of the catalyst components was observed. Laue *et al.* [2.100] attached a homogeneous hydrogenation catalyst onto a polymer. They used a membrane reactor with a UF membrane, which was capable of retaining the soluble, polymer-bound catalyst. Their membrane reactor achieved high space-time yields of up to 578 g/l.d, and enantioselectivities of up to 94 %.

Electrochemical membrane reactors using solid polymer electrolytes (SPE) have also been used in liquid phase hydrogenations by An *et al.* [2.101, 2.102, 2.103] with water as the anode feed and source of hydrogen. The key component of the reactor was a membrane electrode assembly (MEA), composed of a RuO_2 anode, a Nafion®117 cation-exchange membrane, and precious metal-black cathodes that were hot-pressed as thin films onto the opposing surfaces of a Nafion® cation-exchange membrane. The SPE reactor was operated in a batch recycle mode at 60 °C and 1 atm pressure, using commercial-grade soybean oil as the cathode feed. During reactor operation at constant applied current, water was back-fed to the RuO_2 anode, where it was oxidized electrochemically. Protons migrated through the Nafion® membrane under the influence of the applied electric field and contacted the cathode, where they were reduced to atomic and molecular hydrogen. Oil was circulated past the back-side of the cathode and the unsaturated triglycerides reacted with the electrogenerated hydrogen species. The experiments focused on identifying cathode designs and reactor operating conditions that improved fatty acid hydrogenation selectivities. Increasing oil mass transfer into and out of the cathode catalyst layer (by increasing the porosity of the cathode carbon paper/cloth backing material, increasing the oil feed flow rate, and inserting a turbulence promoter into the oil feed flow channel) decreased the concentrations of stearic acid and linolenic acid in the oil products. When a second metal (Ni, Cd, Zn, Pb, Cr, Fe, Ag, Cu, or Co) was electrodeposited on the precious metal-black powder cathode, substantial increases in the linolenate, linoleate, and oleate selectivities were observed. In addition to soybean oil, An *et al.* also studied canola and cottonseed oils, and mixtures of fatty acids, and fatty acid methyl esters.

The presence of nitrates in drinking water is a serious problem in many agricultural areas in the USA and Europe, with a number of areas exceeding the acceptable safe standards. Conventional techniques involve the use of physicochemical techniques like reverse osmosis and ion exchange, and biological treatment. Both techniques have a number of disadvantages. The physicochemical techniques create a waste that must be disposed-off and the biological techniques require the use of a co-metabolite, and for drinking water applications also raise other safety concerns. Recently an alternative procedure involving the catalytic hydrogenation of nitrates to N_2 has been proposed (Vorlop and Prusse [2.104], Prusse and Vorlop [2.105], Pintar *et al.* [2.106], Berndt *et al.* [2.107]) providing a promising new alternative for denitrification of drinking water with several advantages over established technologies, such as reverse osmosis or biological denitrification. Unfortunately, N_2 is only the intermediate product of this reaction, with further hydrogenation leading to ammonium, which together with nitrite are undesired side products, both subject to a limiting value far below the admissible nitrate content. Studies have shown that the catalyst particle size, among other factors, is a key parameter determining the selectivity of the catalyst, pointing to a detrimental effect of pore diffusional limitations. Moreover, activity and selectivity depend on the concentration of dissolved hydrogen; high concentration means high activity, but at the same time this favors ammonium formation due to further hydrogenation. An alternative that has been suggested is the use of CMR as three-phase catalytic conductor devices between the catalytic surface, dissolved nitrate, and hydrogen gas; the local reaction environment is shown schematically in Figure 2.4. Both microfiltration and nanofiltration membranes (alumina or zirconia) and noble (Pd) and bimetallic (Pd/Cu and Pd/Sn) catalysts have been utilized by Dittmeyer and coworkers (Daub *et al.* [2.110, 2.108], Dittmeyer *et al* [2.109]). Such a design offers the potential to control the activity and selectivity of the process through controlled dosage of hydrogen.

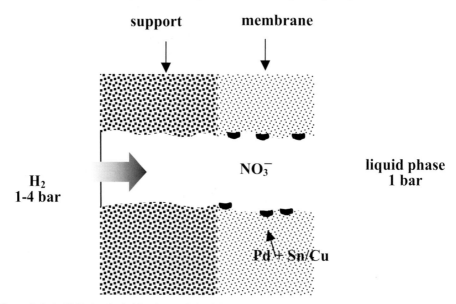

Figure 2.4. A CMR for denitrification. Adapted from Daub *et al.* [2.110].

Ilinitch and Cuperus [2.111] and Ilinitch *et al.* [2.112, 2.113] have also studied the same reaction using macroporous membranes impregnated with a Pd-Cu catalyst, and operating in a "flow through" configuration. They also studied the same reaction over a series of Pd-Cu/γ-Al$_2$O$_3$ supported catalysts. During the reduction of nitrate ions by hydrogen in water at ambient temperatures, pronounced internal diffusional limitations of the reaction rate were observed for the Pd-Cu/γ-Al$_2$O$_3$ catalysts. The use of the catalytic membrane with the Pd-Cu active component resulted in a multifold increase in the observed catalytic activity when operated in the forced flow configuration. These improvements are attributed to the intensification of the intraporous mass transfer, as a result of the reactants being forced to flow through the membrane pores. The same reactor configuration has been used by the authors to study the catalytic oxidation of sulfides (S^{2-} + O$_2$ $\Leftrightarrow$ S + SO$_3$$^{2-}$ + S$_2$O$_3$$^{2-}$ + SO$_4$$^{2-}$). For this reaction macroporous polymeric membranes were utilized, impregnated with a sodium salt of tetra(sulfophthalocyanine)cobalt(II) (Co-TSPC). Co-TSPC has been shown to be a very effective homogeneous catalyst for the reaction, but previous efforts to immobilize it on conventional supports had proven fruitless. The MR, utilizing this catalyst in the "flow through" mode, was reported to be very effective for this application. The same group (Ilinitch *et al.* [2.114]) recently also reported the preparation, and use in the catalytic denitrification of water, of a low-cost inorganic membrane via extrusion and thermal processing of a natural silicate-based material (mixed silicates of Ca, Al, Mg). The membrane has an average pore size of 1 μm, a pore volume of 0.2 cc/g, and very good water permeability (900 l mm/m^2·h·bar). Strukul *et al.* [2.115] studied the hydrogenation of nitrates in drinking water using Cu or Sn promoted Pd catalysts supported on zirconia and titania, either as powders or as membranes deposited on commercial alumina tubes. The catalytic performance of the powders and their selectivity towards the formation of nitrogen depended on the preparation procedure, the type of precursor, the Pd/Cu ratio, and the type of promoter. Their use under diffusion-controlled conditions allows significant reduction of the amount of ammonia formed, while retaining a high catalytic activity. The reactivity of the catalytic membranes was studied in a recirculation, and in a continuous flow reactor configuration; performance depended on the initial pH, the residence time, and the internal hydrogen pressure.

2.3   Oxidation Reactions

Selective catalytic hydrocarbon oxidation reactions are difficult to implement because, in general, the intermediate products are more reactive towards oxygen than the original hydrocarbons. The net result is, often, total oxidation of the original substrate. One of the ways to increase the selectivity towards the intermediate products is to control the oxygen concentration along the reactor length. This can be conveniently implemented by means of the membrane reactor concept. The use of a membrane allows for the oxygen and the hydrocarbon reactants to be fed in different compartments. The most preferable configuration is the CMR, where the membrane, itself, is rendered catalytic providing a reactive interface for the reaction to take place, while avoiding a long contact time between the desired products and oxygen. For the reasons previously outlined, however, the PBMR is the design of choice for most of the studies reported. With the use of the PBMR the oxygen concentration along the reactor length can be, within limits, carefully controlled to favorably influence the reactor selectivity. This, of course, is not possible in the case of the conventional fixed-bed reactors, where the oxygen concentration is maximum at the entrance, and decreases monotonically along the length of the reactor. In these reactors, as a result, the selectivity is typically low at the reactor inlet, where the reaction rate is the highest, and this negatively impacts on the overall yield. One potential additional benefit of the application of the membrane reactor concept to catalytic partial oxidation is that the separation of the oxidant and organic substrate creates reactor conditions less prone to explosions and other undesirable safety effects, that are, typically, associated with the oxidation of gaseous hydrocarbons (Coronas *et al.* [2.116]), thus, potentially broadening the range of feasible operation (Dixon, [2.117]). Of concern are diminished reaction rates, due to the decreased oxygen partial pressures, and reactant hydrocarbon back-diffusion.

The use of membranes to implement, through controlled addition of oxygen, selective catalytic hydrocarbon oxidation has attracted considerable attention in recent years. The studies reported, so far, have made use of both dense and porous membranes. Dense membranes are made, typically, of metallic silver and its alloys, various solid oxides and solid oxide solutions (like stabilized zirconias), as well as perovskites and brownmillerites. These materials are useful in preparing membranes, because they are capable of transporting oxygen selectively. Porous membranes that have been utilized include zeolite and alumina either intact or impregnated by a variety of catalytic materials, including LaOCl, various perovskites, etc. Depending on their pore size and pore size distribution, they have been used, with a varying degree of success, in order to maintain a controlled concentration of oxygen in the reaction side.

Silver and its alloys with other metals like vanadium are unique among metals in that they are very selective to oxygen transport. This phenomenon was first exploited by Gryaznov *et al.* [2.1], who applied non-porous Ag membranes to the oxidation of ammonia, and to the oxidative dehydrogenation of ethanol to acetaldehyde. Most of the recent research efforts in the area of catalytic partial oxidation reactions have made use of solid oxide dense membranes. The earlier dense solid oxide membranes were conventional solid oxide electrolytes (PbO, Bi_2O_3, etc.), or solutions of solid oxides, most commonly Ti, Y, Ca, or Mg stabilized zirconias (TiSZ, YSZ, CSZ, MSZ) (Steele [2.118], Stoukides, [2.119]). These materials have good mobility of the oxygen anion in their lattice, but lack sufficient electronic conductivity, thus, necessitating the use of an external circuit. Significant advances in this area have been made by the introduction of perovskites (Teraoka *et al.* [2.120]) and brownmillerites (Schwartz *et al.* [2.121], Sammels *et al.* [2.122]), which have both good ionic and electronic conductivity, and are also more thermally stable solids, better adapted to high temperature applications. More novel membranes have also been prepared. For example, a U.S. patent (Mieville, [2.123]) reports the synthesis of membranes composed of mixed oxides and metals as selective oxygen transporters and catalysts for the methane oxidative coupling or partial hydrocarbon oxidation reactions.

The direct conversion of methane into ethylene and ethane by its oxidative coupling with oxygen (commonly known as OCM) is the prime example of a reaction to which dense solid oxide membrane reactors have been applied. This reaction has received significant attention in the field of catalysis. Since the first detailed study for the screening of catalysts by Keller and Bhasin [2.124], the number of studies on this reaction has increased exponentially. Though much progress has been achieved, the obtainable yields using conventional, fixed-bed reactors still remain less than the level of 30 %, considered the minimum required for commercial application (Wolf [2.125]). Other challenges to be overcome include the high temperatures required (700–900 °C) for reaction, which shorten the life of catalysts, the high reaction exothermicity, which makes the design of heat transfer equipment technically challenging, and the low concentration of C_2 products in the reactor effluent, making recovery of such products by conventional techniques uneconomical. Since additional ongoing research on catalyst development has not born substantial results, emphasis in recent years has shifted on the application of novel reactors, among them membrane reactors.

The application of membrane reactors to this reaction has been extensively studied. The earlier efforts were reviewed by Eng and Stoukides [2.126], and the most recent efforts have been detailed in the many good review papers on the subject of membrane reactors (Stoukides [2.127]). Different dense membranes, with or without the application of an external electrical potential (Otsuka *et al.* [2.128], Hazburn [2.129], Nagamoto *et al.* [2.130], Nozaki *et al.* [2.131], Harold *et al.* [2.132], Hibino *et al.* [2.133], Langguth *et al.* [2.74], Zeng and Lin [2.134, 2.135, 2.136]) have been utilized. In addition, porous membranes have also been applied to this reaction (Lafarga *et al.* [2.137], Coronas *et al.* [2.138, 2.139], Herguido *et al.* [2.140], Kanno *et al.* [2.141]). The motivation to apply membrane reactors results from the observation that the reaction order in oxygen for the CH_4 coupling reaction is lower than that for the total oxidation, implying that lowering the partial oxygen pressure will result in higher C_2 selectivities (Harold *et al.* [2.142]), commonly defined as the fraction of moles of CH_4 that react which are converted to (C_2H_6 + C_2H_4). Distributing the oxygen feed along the length of the reactor through multiple feed-ports (or more conveniently through the use of membrane reactors) is expected to improve reactor selectivity and yield, as has been shown in many theoretical (Santamaria *et al.* [2.143, 2.144], Reyes *et al.* [2.145, 2.146], Cheng and Shuai [2.147], Wang and Lin [2.148]) and experimental studies (Finol *et al.* [2.149]). The earlier efforts involved the use of a single oxide (PbO) membrane on a porous MgO carrier (Omata *et al.* [2.150], Nozaki *et al.* [2.131, 2.151], Nozaki and Fujimoto [2.152]). Their reactor configuration is shown in Figure 2.5. The methane feed is in contact with the MgO phase, while the O_2 containing stream is in contact with the PbO layer. The methane that diffuses through the porous MgO phase comes in contact with the PbO, and converts into C_2 products with the aid of the PbO lattice oxygen. The PbO phase remains active with the aid of the oxygen stream that is in contact with it. The lattice oxygen in PbO appears very active towards oxidizing CH_4 into useful C_2 products rather than towards unwanted CO or CO_2 with a selectivity of better than 90 %. Due to the low oxygen transport through the PbO membrane the observed rates were rather low. The MgO/PbO membranes are, furthermore, rather difficult to prepare, and the thin PbO films are not very stable. Anshits and coworkers [2.153] used thin, dense Ag membranes, instead, in order to investigate the OCM reaction. As with other noble metal membranes, for the dense Ag membranes cost and mechanical and thermal stability remain the key concerns.

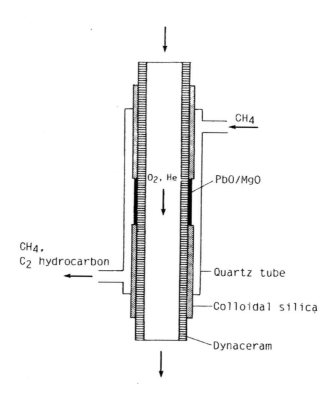

Figure 2.5. MR for the OCM reaction using a PbO membrane. From Omata *et al.* [2.150] with permission from Elsevier Science.

Stabilized zirconias (principally YSZ) have also been utilized in OCM studies (Otsuka *et al.* [2.128]; Nagamoto *et al.* [2.130]; Guo *et al.* [2.154]). Since YSZ has limited electronic conductivity, its application to OCM involves, typically, the use of an external voltage, which imposes an electrical current through the membrane (Eng and Stoukides [2.126]). A schematic of a generic electrochemical membrane reactor utilized is shown in Figure 2.6 (for the OCM reaction A is CH_4 and mB represents C_2 as well as total oxidation products). In addition to the YSZ electrolyte, one, commonly, utilizes an OCM catalyst in the anode, and a different catalyst at the cathode to improve the surface exchange kinetics for oxygen. By changing the polarity and the magnitude of the voltage drop across the membrane (case c) one may vary the flow of oxygen ions (O^{2-}) towards or away from the OCM catalyst. What has been observed often (Vayenas *et al.* [2.155, 2.156,

2.72]) is that the application of the voltage seems to be also favorably affecting the catalytic activity, a phenomenon known as NEMCA (non-faradaic electrochemical modification of catalytic activity), though the effect is much more moderate for the OCM application than what has been observed for the partial/total oxidation of various compounds (Eng and Stoukides [2.126], Stoukides [2.127]). The NEMCA effect is attributed to the electrochemically imposed spillover of ions from the solid electrolyte to the electrode-catalyst; these ions act as promoters for the catalytic reaction.

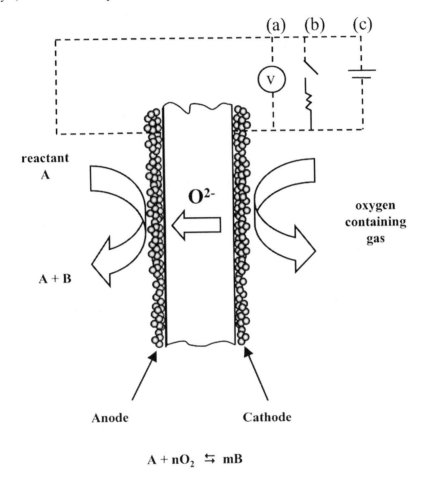

Figure 2.6. A generic electrochemical MR. Adapted from Stoukides [2.127].

Equally interesting effects were reported in a study by Hibino and coworkers [2.157, 2.158] who studied the OCM reaction using a Li^+ ion conducting electrolyte $[Li_2O]_{0.17}[BaO]_{0.07}[TiO_2]_{0.76}$ using Au as the electrodes. They utilized two different electrochemical reactor configurations. In the first configuration the membrane and the attached electrodes were immersed in a single reactor chamber, in which CH_4 and O_2 were allowed to flow through. The other system utilized was a conventional electrochemical membrane reactor (Figure 2.6). Hibino and coworkers used an alternating (AC) current instead of a direct current. For an AC voltage of 3 V the methane conversion was twice as high as that obtained under open circuit conditions, and the C_2 selectivity was 50 % higher. The authors studied the effect of varying the current frequency on the C_2 formation rate, and they observed, interestingly enough, that there is an optimal frequency of about 1 Hz, for which the C_2 formation rate is maximized. This rate enhancement was attributed to the Li^+ that may accumulate on the electrode surface, and which may, in turn, enhance the rate of CH_3^- radical formation.

An interesting concept is to use the OCM reaction in a "fuel cell" (case b), chemical co-generation mode to convert the reaction heat into useful electrical energy, while simultaneously producing a useful C_2 product. Pujare and Sammels [2.159], Eng and Stoukides [2.126], Tagawa *et al.* [2.160], and Guo *et al.* [2.161], are among those that have carried out studies of the OCM reaction with the goal of producing both valuable products and electricity. Other reactions that have been studied in the chemical co-generation context include the oxidation of NH_3 to NO (Vayenas and Farr [2.162], Farr and Vayenas [2.163], Sigal and Vayenas [2.164], Manton *et al.* [2.165], Sammes and Steele [2.166]), the oxidative dehydrogenation of CH_3OH to formaldehyde (Neophytides and Vayenas [2.167]), ethylbenzene to styrene (Michaels and Vayenas [2.168, 2.169]), ethane to ethylene (Iwahara *et al.* [2.170]), and the Andrussov reaction to produce HCN (Kiratzis and Stoukides [2.171], McKenna and Stoukides [2.172]).

The use of stabilized zirconias (like YSZ), which at the typical OCM temperatures are practically only pure oxide conductors (Stoukides [2.127] defines as a "pure" ionic conductor a material, whose ion transference number is two or more orders of magnitude higher than that of electrons; otherwise the material is a mixed conductor), has the downside that one requires the use of electrodes and external electrical connections, which (the fuel cell co-generation type applications, discussed previously, notwithstanding) adds significant complexity to the reactor design of a large scale operation, the OCM MR application is envisioned to be. The development of mixed ionic conductors like perovskites and brownmillerites, which, in addition to having good ionic conductivity also have a sufficiently high electronic conductivity at typical OCM conditions, significantly simplifies matters. The oxygen fluxes reported, for example, by Teraoka *et al.* [2.173, 2.174, 2.120] for the $LaCoO_3$-type perovskite membranes they developed are 2–4 orders of magnitude higher than those of stabilized zirconias under the same conditions. Several theoretical and experimental studies have been published, in which such membranes have been applied to the OCM reaction (Wang and Lin [2.148], Kao *et al.* [2.175], Ten Elshof *et al.* [2.176, 2.177], Lin and Zeng [2.178] Zeng and Lin [2.134, 2.135], Zeng *et al.* [2.136]; Xu and Thompson [2.179]). A general analysis of the oxygen transport through such membranes, particularly in the context of reactive applications, has been presented by Lin *et al.* [2.180]. Two additional modelling studies by the same group have analyzed the behavior of CMR (Wang and Lin [2.148]) and PBMR systems for the OCM reaction (Kao *et al.* [2.175]) using such membranes. Their studies predict C_2 yields as high as 84 % under optimal conditions, for which the CH_4 flow into the reactor, the permeation of oxygen through the membrane, and the intrinsic reaction rate all match each other. These yields are significantly higher than what has been accomplished experimentally with these membranes. Matching the oxygen permeation rate with the catalytic activity is of particular importance for the application of membrane reactor to the OCM reaction. This reaction involves both gas phase and surface reaction steps. The activation of CH_4 with the aid of lattice oxygen to produce methyl radicals ($CH_3 \cdot$) is a surface reaction requiring a highly active catalyst (most dense membranes used to date in OCM CMR are, unfortunately, very poor OCM catalysts). The production of C_2 products or undesirable CO_x total oxidation products occurs in the gas phase. In the CMR configuration an imbalance in the oxygen permeation rate will result either in reduced catalytic activity or excess gas phase oxygen, which may accelerate the undesirable total oxidation reactions.

An experimental investigation using a CMR configuration has been reported by Ten Elshof *et al.* [2.176, 2.177]. They utilized $La_{0.6}Sr_{0.4}Co_{0.8}Fe_{0.2}O_3$ or $La_{0.8}Ba_{0.2}Co_{0.8}Fe_{0.2}O_3$ dense disk membranes, which are, themselves, catalytically active towards the OCM reaction. They report high C_2 selectivities (70 %), though the overall yields were rather small. Zeng *et al.* [2.181] in their studies used a $La_{0.8}Sr_{0.2}Co_{0.6}Fe_{0.4}O_3$ membrane. They compared several different synthesis methods for the preparation of (LSCF) powders. The co-precipitation method was found most suitable for preparation of these powders with respect to their processibility into dense ceramic membranes. The oxygen permeation flux through a 1.85 mm thick LSCF membrane, exposed to an O_2/N_2 mixture on one side and helium on the other, was reported to be about 1×10^{-7} mol/cm$^2 \cdot$s at 950 °C, increasing sharply around 825 °C due to an order-disorder transition of the oxygen vacancies in the membrane. Oxidative coupling of methane (OCM) was performed in the LSCF membrane reactor with one membrane surface exposed to an O_2/N_2 stream, and the other to a CH_4/He stream. At temperatures higher than 850 °C, high C_2 selectivity (70–90 %) and yield (10–18 %) were achieved with a He/CH_4 feed ratio in the range of 40–90. The C_2 selectivity dropped to less than 40 % as the He/CH_4 ratio decreased to 20, of interest since it is opposite to the behavior reported with co-feed type conventional OCM reactors. The same group (Zeng and Lin [2.182]) used disk-shaped dense membranes made of 25 mol % yttria doped bismuth oxide (BY25) fabricated by using a press/sinter method from the citrate-derived BY25 powders. Oxygen permeation fluxes through these membranes were measured at different temperatures and oxygen partial pressures. C_2 selectivity and yield were, respectively, in the range of 20–90 % and 16–4 %. At the same C_2 yield, the membrane reactor selectivity is about 30 % higher than the selectivity observed with the co-feed fixed-bed reactor packed with BY25 pellets under similar conditions. After OCM the BY25 membrane remained in good integrity, but the reaction caused formation of some impurity phases on the membrane surface that was exposed to the methane stream. To improve on the stability Zeng and Lin [2.183] doped their membranes with samarium. The resulting samarium-doped yttria-bismuth oxide (BYS) membranes have high oxygen permeability and good OCM catalytic properties. OCM studiess were conducted in a membrane reactor made of the disk-shaped BYS membrane with CH_4 and O_2 containing streams fed into the opposite sides of the BYS membrane. The MR gave C_2 yields of

4–11 %, with C_2 selectivity up to 74 %. The MR effectiveness improved by increasing the membrane surface area to reactor volume ratio, with the highest C_2 yield and selectivity achieved being 17 % and 80 % correspondingly.

Xu and Thompson [2.179] used perovskite membranes with the overall formula $La_{1-x}A_xCo_{0.2}Fe_{0.8}O_3$ (A=Sr or Ba) for the study of OCM. When compared to the performance of more conventional reactors under identical experimental conditions, the membrane reactor selectivity and C_2 production rate were shown to be 4 and 7 times higher correspondingly. Lu *et al.* [2.184, 2.185] studied OCM using a La/MgO catalyst and a distributed oxygen feed through mixed-conducting dense membrane tubes in a shell-and-tube membrane reactor configuration. A $SrFeCo_{0.5}O_3$ membrane was tested first, but tests without the catalyst being present showed that it acted as a total oxidation catalyst producing no C_2 products. Attempts to render the membrane unreactive by coating it with a non-combustion material ($BaCe_{0.6}Sm_{0.4}O_3$) were only partially successful. The oxygen flux through the coated tube was reduced to 30 % of its original value, and the resulting C_2 yield was only 7 %. A membrane tube was then fabricated from a non-combustion oxygen-permeating material, $BaCe_{0.8}Gd_{0.2}O_3$, and was utilized. C_2 yields as high as 16.5 % were obtained, higher than those observed in comparable fixed-bed studies. Changes in surface morphology were observed, however, for the side of the membrane in contact with the reducing atmosphere. In an interesting study Stoukides and coworkers (Athanasiou *et al.* [2.186], Tsiakaras *et al.* [2.187]) have studied the OCM reaction over a La-Sr-Co-Fe (LSCF) perovskite. They carried out three different types of experiments. In the first experiment the LSCF perovskite was used as a conventional catalyst, CH_4 and O_2 being the gaseous reactants. In the second experiment the LSCF perovskite was used as the anodic electrode of an electrochemical cell in which the O^{2-} was provided electrochemically. In the third experiment the LSCF was used as a dense mixed oxide membrane in a MR. The best OCM results were observed in the MR case. When the LSCF acted as an anodic electrode the dominant products were CO and H_2. When acting as a conventional catalyst the complete CH_4 oxidation was favored.

Dense membranes have good permselectivity towards oxygen, which allows for the use of air during the partial oxidation reaction. On the other hand, their permeability is generally low, which places an upper limit on the reactor's efficiency. This, in turn, has motivated the use of porous membranes. The first studies, using porous membranes, were in the CMR configuration (Chanaud *et al.* [2.188], Borges *et al.* [2.189]). They utilized catalytically active LaOCl membranes, which, unfortunately, were not very selective towards oxygen transport, and also did not exhibit the necessary stability at the reaction conditions. Efforts using the PBMR configuration have been more encouraging. The group at Zaragoza, Spain (Lafarga *et al.* [2.137] Coronas *et al.* [2.138, 2.139], Herguido *et al.* [2.140]) using non-permselective, commercially available alumina membranes and conventional oxidative coupling catalysts have attained promising yields (~25 %). In the work of Coronas *et al.* [2.138] the membrane was packed with a Li/MgO catalyst, the CH_4 was fed on the tubeside, while oxygen was fed on the shellside. The membrane reactor was shown to give better C_2 selectivity than the corresponding conventional reactor, but mostly in the region of low to moderate conversions. The authors also reported that the membrane reactor, with the oxidant and the hydrocarbon fed in separate compartments is intrinsically safer than the fixed bed reactor.

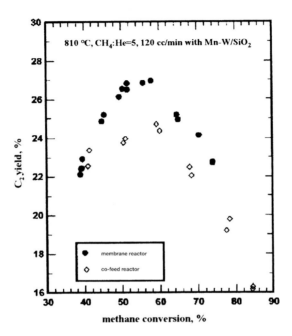

Figure 2.7. C_2 yields in a membrane and a co-feed reactor as function of methane conversion. From Ma *et al.* [2.190], with permission from Elsevier Science.

In the subsequent papers by the same group (Coronas *et al.* [2.139]; Herguido *et al.* [2.140]) they modified the structure of the commercial membranes in order to provide a nonuniform permeability pattern, and they also impregnated the membrane with LiCl. Further improvements in C_2 yield resulted. The group at WPI has also been very active in this area. In their earlier studies (Ramachandra *et al.* [2.191]; Lu *et al.* [2.192]) they utilized a Vycor® glass membrane, and a Sm_2O_3 catalyst. Compared to the conventional reactor the membrane reactor, for the same level of conversion, gave a better C_2 selectivity. The same group (Ma *et al.* [2.190], Lu *et al.* [2.193]) has reported a study of the oxidative coupling of methane using a lanthanum stabilized porous γ-alumina membrane reactor in a PBMR configuration. The catalyst used was a packed-bed of Mn-W-Na/SiO$_2$. The experimental results indicate that it was beneficial to distribute the feed of oxygen along the reactor length, with an enhancement of the C_2 yield obtained with the PBMR configuration, when compared to the yield obtained with the more conventional co-feed configuration (Figure 2.7). C_2 yields up to 27.5 % were obtained in the membrane reactor. Although the membrane reactor showed lower methane conversion than the conventional reactor under the same reaction conditions, for the same methane conversions it gave higher C_2

selectivity (~30 % higher) and yield (10 % higher). At similar C_2 yield and selectivity, the membrane reactor conversion was 15 % lower than that of a co-feed reactor.

A different approach to convert CH_4 into C_2 and higher hydrocarbons, utilizing a membrane reactor, was proposed by Garnier *et al.* [2.194]. They investigated a 2-step process using a PBMR with a Pd-Ag membrane and a 5 % Ru/Al_2O_3 catalyst. In the first step (which is favored at high temperatures) CH_4 is dehydrogenated into carbon and hydrogen. In the second step (which is favored at lower temperatures) the active carbon is re-hydrogenated into C_{2+} hydrocarbons. The two-step, $CH_4 \rightarrow C_{2+}$ process had been studied earlier by a number groups in a conventional reactor (Koerts *et al.* [2.195]). The advantage of the membrane reactor is in that it lowers the temperature required in the first step to effect complete conversion of CH_4, thus, favorably impacting on the economics. The robustness of the metal membrane in the reactive environment of step 1, and its ability to withstand thermal cycling between the two process steps, are the two key hurdles to be overcome.

Dehydrogenating hydrocarbons such as ethane and gasoline into hydrogen and carbon in a PBMR using a Pd-Ag membrane and Ni/Ca/graphite catalyst with the goal of simply producing hydrogen was recently proposed by Murata *et al.* [2.196, 2.197, 2.198]. They studied the decomposition reaction of ethane for hydrogen production at 550–750 K in a PBMR, and in a fixed-bed reactor (FBR). The use of the PBMR resulted in the formation of a smaller amount of methane (yield of less than 10 %) as a major by-product than the FBR (yield of 10–50 %). Murata *et al.* [2.196, 2.197] theorize that the beneficial effects of the PBMR relate to the reaction mechanism for the reaction. Ethane cracking gives methane, carbon, and hydrogen; the permeation of hydrogen through the membrane prevents the recombination of hydrogen with the carbon products to give methane. The PBMR system was also found to be effective for the formation of hydrogen by decomposition of commercially available gasoline at temperatures as low as 573 K. How one scales-up a process of this kind is somewhat unclear, though. A similar concept, intended initially to produce electricity in a fuel cell, but which also produces mixtures of H_2 and CO (otherwise known as synthesis gas – see further discussion below about the application of MR to synthesis gas production) was proposed in the early sixties by workers at General Electric (Tragert *et al.* [2.199]). Methane was fed in the anodic compartment, where it was broken down to carbon and hydrogen, in an electrochemical membrane reactor equipped with a YSZ solid electrolyte. The carbon that forms on the YSZ serves as the anodic electrode. The cathodic electrode was molten silver. Oxygen in the air fed in the cathodic compartment transported to the anode and oxidized the carbon to produce CO.

Another partial oxidation reaction that is attracting industrial attention for the application of reactive separations is the production of synthesis gas from methane [Stoukides, 2.127]. The earlier efforts made use of solid oxide solutions as electrolytes. Stoukides and coworkers (Eng and Stoukides [2.200, 2.126], Alqahtany *et al.* [2.201, 2.202]), for example, using a YSZ membrane in an electrochemical membrane reactor obtained a selectivity to CO and H_2 of up to 86 %. They found that a Fe anodic electrode was as active as Ni in producing synthesis gas from methane (Alqahtany *et al.* [2.201, 2.202]), and that electrochemically produced O^{2-} was more effective in producing CO than gaseous oxygen (no effect was found, though, on hydrogen selectivity), and more effective in eliminating carbon formation. Horita *et al*. [2.203] studied the same reaction in an electrochemical membrane reactor using a variety of anodes (Pt, Fe, Ni, and La-Ca-Cr perovskites). Takehira and co-workers (Sato *et al*. [2.204], Takehira *et al*. [2.205], Hamakawa *et al*. [2.206]) studied Rh anodes and found their performance during synthesis gas production to be dependent on the oxidation state of the metal; a reduced metal surface catalyzing the production of syn-thesis gas, with an oxidized metal surface favoring total oxidation. Hamakawa *et al*. [2.207], for example, utilized a Sm-doped ceria solid electrolyte (SDC) as an oxide ionic conductor in their study of the partial oxidation of methane to synthesis gas at 400–500 °C using a Rh anode and Ag as a cathode. On applying the direct current to the reaction cell, with a dilute mixture of CH_4 flowing in the anodic compartment and pure O_2 in the cath-ode, CH_4 oxidation was enhanced, and the synthesis gas formation rates increased linearly with increasing current. Selectivity to CO was 70 % under the oxygen pumping conditions at 0.6 mA/cm^2 at 500 °C. Hamakawa *et al*. [2.206] report that synthesis gas production is performed by the oxygen species, which is generated electrochemically at the Rh-SDC-gas phase triple-phase boundary, and then migrates to the Rh surface. Pd anodic electrodes were also studied by the same group (Hamakawa *et al*. [2.208]). The Pd\YSZ\Ag system has shown a very good activity for CO production at 500 °C, with a high CO selectivity (96.3 %) under oxygen pumping conditions at 5 mA. The H_2 production strongly depended on the oxidation state of the Pd anode. H_2 treatment of the Pd anode at 500 °C for 1 h drastically reduced the rate of H_2 production, while air treatment enhanced it. The group again reports that the reaction site of the electrochemical oxidation of CH_4 to synthesis gas is the Pd-YSZ-gas-phase triple-phase boundary. In a more recent effort Hamakawa *et al*. [2.209] report using an EMR with a Ni metal anode supported on a perovskite-type oxide ($Ca_{0.8}Sr_{0.2}Ti_{1-x}Fe_xO_{3-y}$) for CH_4, conversion into synthesis gas. Synthesis gas production has been accelerated by electrochemical oxygen pumping, and selectivity to CO at 900 °C was almost 100 %. The EMR exhibited good stability to carbon deposition, attributed to the oxidation of carbon deposits by the lattice oxygen species migrating to the Ni/$Ca_{0.8}Sr_{0.2}Ti_{1-x}Fe_xO_{3-y}$ boundary. High electrocatalytic activity towards formation of syngas from methane in a solid oxide fuel cell (SOFC) reactor using Pt and Ni electrodes was also recently reported by Sobyanin and Belyaev [2.210].

Synthesis gas production together with energy co-generation was studied by Ishihara and coworkers (Ishihara *et al*. [2.211], Hiei *et al*. [2.212]) using a La-Sr-Ga-Mg perovskite at 1000 °C. Good synthesis gas yields and power output was maintained for over 30 h. Hibino *et al*. [2.157] studied the effect of using an AC (rather than a DC volt-age) in a Pd/YSZ/Pd electrochemical cell. The AC voltage gave better performance than the DC voltage during synthesis gas production. Sobyanin and coworkers (Sobyanin *et al*. [2.213], Semin *et al*. [2.214]) studied the electrocatalytic conversion of methane to syn-thesis gas in an EMR using a YSZ membrane and a number of different anodic electrodes (Ag, Pt, Ni, and Pt+CeO_2) at 660–850 °C, and ratios of methane and electrochemically pumped oxygen flows of 0.8–2.0. Only the Ni and Pt were found to be active electrode-catalysts for partial oxidation of methane, producing syngas with a H_2/CO concentration ratio of 2. For the Pt electrode the methane conversion was 97 %, the CO selectivity was

95 %, and the EMR operation was stable without coking. Mazanec *et al.* [2.215] integrated the exothermic CH_4 partial oxidation with endothermic CH_4 steam reforming to obtain high syngas yields (97 %) in an electrochemical membrane reactor system that is potentially thermoneutral. They also extended the natural gas upgrading to oxygen transport membranes that do not need external circuitry, by the introduction of internally shorted, metal/YSZ composites, and single phase mixed conductors [2.216]. This group also reported [2.217] that their materials show stable microstructures after more than 500 h of operation generating synthesis gas. The concept of using mixed ionic/electronic conductors to produce synthesis gas in a membrane reactor is shown schematically in Figure 2.8.

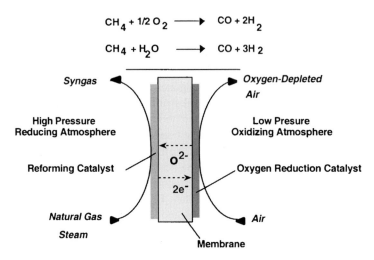

Figure 2.8. Synthesis gas production in a membrane reactor. From Dyer *et al.* [2.232], with permission from Elsevier Science.

Collaborative research efforts in this area by Amoco and the Argonne National Laboratory (Balachandran *et al.* [2.218, 2.219, 2.220, 2.221], Maiya *et al.* [2.222]) have made use of perovskite-like $Sr(Co,Fe)O_x$ materials. They have reported methane conversion efficiencies >99 %, and stable membrane operation for over 900 h. In a recent paper the same workers (Maiya *et al.* [2.222]), in order to enhance H_2 production, coupled their ceramic membrane reactor to a second catalytic reactor, in which, through the water gas shift reaction, CO reacts with steam to produce CO_2 and H_2. Experiments and thermodynamic calculations were used to establish the optimal temperature and steam-to-CO ratio to achieve thermodynamic efficiency while maximizing H_2 production. They observed no unusual synergisms by the combination of the two processes. Xu and coworkers (Li *et al.* [2.223], Jin *et al.* [2.224]) used a tubular $La_{0.6}Sr_{0.4}Co_{0.2}Fe_{0.8}O_{3-\delta}$ membrane to study the partial oxidation of CH_4 to synthesis gas over a pre-reduced Ni/Al_2O_3 catalyst. Conversions higher than 97 % with a selectivity to CO > 98 % were observed. The conversion decreased with increasing CH_4/O_2 ratio in the feed, but the selectivity remained constant.

Analysis of the used membrane by SEM, XRD and EDS showed metal ion reduction/sublimation and segregation of the elements on the membrane surfaces. This led to a redistribution of the metal ions across the membrane after being operated in the membrane reactor for 3–7 h. Xiong and coworkers (Dong *et al*. [2.225, 2.226], Shao *et al*. [2.227, 2.228]) have used a $Ba_{0.5}Sr_{0.5}Co_{0.8}Fe_{0.2}O_{3-\delta}$ membrane to study synthesis gas production from CH_4 utilizing a $LiLaNiO_x/\gamma\text{-}Al_2O_3$ catalyst. The MR gave at 875 °C a conversion of 98 % and a CO selectivity of 96 %, and oxygen permeation rates much higher than those previously reported in the literature (Balachandran *et al*. [2.218, 2.219], Tsai *et al*. [2.229]). The conversion and selectivity were reported to be constant over a 500 h run. However, evaluation of the membrane at the end of the run indicated Sr segregation and carbonate deposition on the exposed membrane surfaces, but this did not seem to affect the membrane performance during the partial oxidation to methane. Richie *et al*. [2.230] have developed a membrane by spray-deposition of a dense film of $La_{0.5}Sr_{0.5}Fe_{0.8}Ga_{0.2}O_{3-\delta}$ on a porous α-alumina tube. The membrane was placed inside a quartz tube and the shell-side was filled with a Rh catalyst. Air was fed into the tubeside and methane was fed to the shellside. At 850 °C the membrane reactor conversion was 97 %, and the selectivity to CO was close to 100%. Though the membrane was found to decomposes at 780 °C in pure CH_4, it remained stable up to 970 °C in a mixture of 90 % CH_4 and 10 % CO_2. Workers at Eltron Research have made use of brownmillerite membranes (Schwartz *et al*. [2.121], Sammels *et al*. [2.109]) with a reported chemical composition of $A_{2-x}A'_xB_{2-y}O_{5+z}$. They have also reported high yields to synthesis gas and a stable operation for over one year (Sammels *et al*. [2.122]). Eltron Research is one of the industrial partners of the Air Products/Ceramatec industrial consortium (see below).

Two industrial consortia have formed in the U.S.A. (Saracco *et al*. [2.7], Mazanec [2.231]) to commercialize solid-oxide MR-based synthesis gas production. One consortium is an industrial alliance among BP/Amoco, Praxair, Statoil, SASOL, and Phillips Petroleum, also enlisting the help of a number of Universities including MIT, and the Universities of Alaska Fairbanks, Houston, Missouri-Rolla, and Illinois-Chicago. A second group is headed by Air Products and Chemicals and Ceramatec, and a number of other industrial partners. A third consortium with the same purpose seems to be in the works in Europe [2.7]. The reason for the renewed industrial interest in synthesis gas is that gas conversion to liquids (GTL) is becoming recently more important for both economic and environmental reasons. The quantity of natural gas reserves is large, approximately the energy equivalent of the known crude oil reserves. Much of this gas, however, is in remote locations, where it is currently uneconomic to deliver to the market, and it is either vented or flared; conversion to high-grade, high-purity liquid fuels provides economic alternatives. The U.S. government appears to have a strong interest in remote natural gas conversion, since the United States has substantial remote gas reserves (in the Gulf of Mexico, along the Pacific coast, and on the Alaskan North Slope). In the Alaskan North Slope alone the known natural gas resources total approximately 38×10^{12} cubic feet, and estimates are that there is nearly double that amount in undiscovered fields. Recent studies indicate that natural gas conversion to liquids has the potential to keep the trans-Alaska pipeline system operational for a decade or more beyond its currently projected shutdown point; this would allow the recovery of the equivalent of an additional 1 billion barrels of crude oil from the North Slope. The major portion of capital costs in the GTL process is required for the generation of synthesis gas. Synthesis gas for liquid fuels production can be generated by a number of processes, including partial oxidation (POX), autothermal reforming (ATR), combinations of steam/methane reforming (SMR) and ATR, and gas heated reforming (GHR). Of all these processes the use of pure oxygen for POX results in the best economics. A significant GTL cost reduction can only come from a significant breakthrough in synthesis gas production. Such breakthrough is possible if the cryogenic air separation and the high-temperature syngas generation processes are combined into a single process step, through the use, for example, of membrane reactor technology. The estimated reduction in the capital investment for syngas by 25–50 % would have a major impact on overall GTL economics, and could be critical for the full commercialization of GTL production technology.

The efforts of the Air Products group are reviewed in a recent article by Dyer *et al.* [2.232]. Air Products has been developing mixed ionic conductor technology (which they define as ITM technology) for over a decade, and have joined forces with Ceramatec, Inc. in the development of appropriate membrane fabrication technology. They own more than 47 U.S. patents in the area [2.232]. The ITM materials and processes that are being developed are categorized in Figure 2.9. Membranes fabricated from pure oxygen ionic conductors are used (with a voltage applied across them) to separate and compress oxygen from air, to a high-pressure product oxygen stream. Air Products/ Ceramatec Inc. have developed stand-alone separator devices utilizing such membranes, designated as SEOS$^{(TM)}$ Oxygen Generators. Mixed ionic/electronic conductors are utilized in two types of processes: (i) Processes (designated ITM Oxygen), which impose an oxygen partial pressure differential across the membrane at high temperatures; these have potential application in large tonnage oxygen plants, in particular applications where the co-generation of electricity is required; (ii) and processes for which the driving force for oxygen transport is developed by depleting the oxygen partial pressure on one side of the membrane through chemical reaction, thus, allowing oxygen to be transported from a relatively low-pressure air feed to a higher pressure reaction product stream, like in the case of syngas production (designated as the ITM Syngas process). Recognizing the need for enhanced oxygen fluxes the Air Products/Ceramatec team has developed (Dyer *et al.* [2.232]) an asymmetric membrane, the multi-layer structure of which is shown schematically in Figure 2.10. The dense, thin membrane film is supported on a porous sub-layer, in turn, supported on layers with larger pore dimensions also made from an ITM Oxygen material, in order to minimize differential expansion effects. To minimize the mechanical stresses on the membranes, in the ITM Oxygen process a medium-pressure air feed stream (100–300 psia), and a low-pressure oxygen permeate stream (< 1 atm) are utilized.

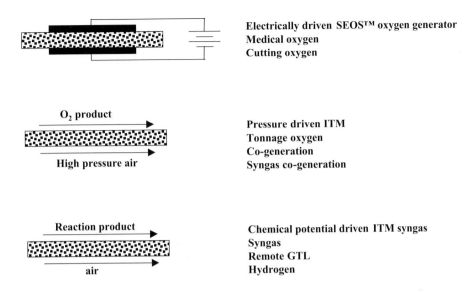

Figure 2.9. Applications of oxygen conducting membranes under development at Air Products. Adapted from Dyer *et al*. [2.232].

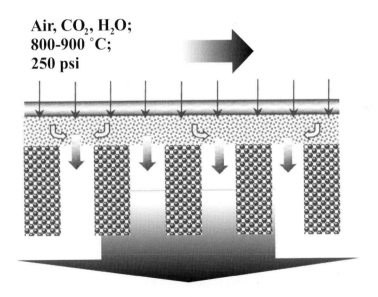

Pure oxygen product

Figure 2.10. An asymmetric oxygen conducting membrane. Adapted from Dyer *et al*. [2.232].

Since the ionic transport mechanism is activated, the ITM Oxygen process cycle must include a means to heat the feed to high temperatures, either by indirect heat exchange, direct firing with an inexpensive fuel source, or a combination of both. To achieve acceptable cycle efficiency, the energy associated with the hot, pressurized non-permeate stream can be recovered by integrating the ITM Oxygen membrane with a gas turbine power generation system. The ITM Oxygen technology is, thus, ideally suited for integration with power generation processes that require oxygen as a feedstock for combustion or gasification, or in any oxygen-based application with a need for power or an export power market. According to Dyer *et al.* [2.232], the Integrated Gasification Combined Cycle (IGCC) process is an ideal application for ITM Oxygen co-production technology. Figure 2.11 depicts the proposed flow-sheet for the integration of the ITM Oxygen process in an IGCC facility.

In the proposed process of Figure 2.11 air is extracted from the compressor section of a gas turbine and heated to the ITM Oxygen operating temperature by direct combustion with a slipstream of clean coal gas. The high-purity oxygen permeate exiting the membrane is cooled prior to compression for use in the coal gasifier, while the hot non-permeate stream is further heated by direct combustion with coal gas prior to introduction into the turbine section. A supplemental air compressor adds sufficient air to replace the oxygen removed by this process cycle, maintaining the gas turbine near its peak power output. The turbine exhaust then provides another heat source for the steam bottoming cycle. A small, stand-alone cryogenic nitrogen plant cost-effectively generates the inert gas required for coal handling and conveying to the gasifier. This nitrogen plant is approximately 35 times smaller than the cryogenic oxygen plant found in the conventional IGCC facility, which serves as the base case in the economic evaluation of the ITM Oxygen IGCC system. Table 2.1 shows the cost and performance comparison for the conventional, cryogenic base case and the ITM Oxygen-integrated IGCC facility (Dyer *et al.* [2.232]). The base case plant consumes 3180 tons per day (TPD) of Illinois #6 coal and 2565 TPD of oxygen (95 %), while the ITM Oxygen-integrated plant consumes 3176 TPD of Illinois #6 coal and 2420 TPD of oxygen (99+ %). The ITM Oxygen plant, including the supplemental air compressor, the additional combustor, ITM Oxygen modules, oxygen coolers, oxygen compressors, and the cryogenic nitrogen plant, saves 31 % of the installed cost for air separation equipment. The proposed economic benefits of ITM Oxygen/IGCC technology are reported to be a 2.9 % improvement in thermal efficiency, with a 6.5 % decrease in the generated electric power costs. The efficiency increase also produces a concomitant reduction in carbon dioxide and sulfur emissions. Currently Air Products/ Ceramatec and their partners (U.S. Department of Energy, Eltron Research Inc., McDermott Technology Inc., Northern Research and Engineering Corporation, Texaco Inc., the Pennsylvania State University, and the University of Pennsylvania) are in the midst of a multi-phase program (started in October 1998) for the development and commercialization of the ITM Oxygen Process/IGCC.

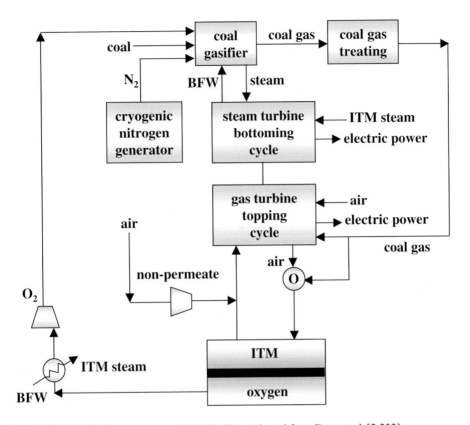

Figure 2.11. An ITM Oxygen process/IGCC facility. Adapted from Dyer *et al.* [2.232].

The Air Products/Ceramatec team, as noted previously, has currently under development the ITM Syngas process for synthesis gas production utilizing solid oxide membrane reactors. The process is conceptually illustrated in Figure 2.8. In the Air Products/ Ceramatec process the membrane structure incorporates the non-porous ITM and reduction and reforming catalyst layers. Intimate contact of the reforming catalyst with the ceramic membrane is claimed to be "critical" to accomplish depletion of the oxygen transported through the membrane by reaction at the ceramic membrane/catalyst interface, a necessary requirement to control the partial oxidation reaction exotherm and to provide a stable thermal operating mode for the membrane reactor. Tubular membranes have been tested for over 100 days of continuous operation in a laboratory apparatus, shown schematically in Figure 2.12, by contacting natural gas mixtures (250 psig) with the exterior surface of the membrane tubes at elevated pressure. The ITM Syngas process is currently being studied as part of a multi-phase development program by a team, which, in addition to Air Products/Ceramatec, Inc., also includes Chevron, Eltron Research Inc., McDermott Technology Inc., Norsk Hydro, Pacific Northwest National Laboratory, the Pennsylvania State University, the University of Alaska, and the University of Pennsylvania.

Despite these reported successes, concerns still appear to remain about the long-term stability of such materials in high temperature, reactive environments. Hendriksen and coworkers (Hendriksen [2.233], Hendriksen *et al.* [2.234]), for example, report that the material with the composition $SrFeCo_{0.5}O_x$ used by Balanchandran *et al.* [2.218, 2.219] is not stable under reducing conditions. It has been reported to be a multiphase system consisting of $Sr_4Fe_4Co_yO_{13+\delta}$ isostructural with $Sr_4Fe_6O_{13+\delta}$ and CoO. Neither of these phases is stable under reducing conditions (Fjellvag *et al.* [2.235]).

Table 2.1. Comparison between the cryogenic and ITM cases. Adapted from Dyer *et al.* [2.232].

	Cryogenic O_2 case	ITM O_2 case	Change	% Change
IGCC facility capital investment (US$)	$6.41\ 10^8$	$6.10\ 10^8$	$31\ 10^6$	$4.8\ 10^6$
IGCC facility capital investment (US$/kW)	1567	1453	(114)	(7.3)
Power production (Mw)	409	420	11	2.7
Thermal efficency (% HHV)	45.2	46.5	1.3	2.9
Electricity cost (mills/kW)	55.5	51.9	3.6	6.5

Wolf and coworkers [2.236, 2.237] have studied the production of synthesis gas in a PBMR configuration using a non-permselective porous alumina membrane. In their study they studied the partial oxidation of methane over a 3 % Rh/TiO_2 catalyst both in a fixed-bed and in the PBMR under autothermal conditions using O_2 as oxidant. The membrane reactor operates at millisecond residence rime. Methane conversions of up to 65 % with CO and H_2 selectivities of 90 and 82 % respectively have been achieved. The low methane/oxygen ratio and the high flow rates are key factors in attaining autothermal behavior. Basile *et al.* [2.238] and Basile and Paturzo [2.239] reported studying the partial oxidation of methane with a Ni-based catalyst in a laboratory size PBMR utilizing a Pd membrane tube. They compared the PBMR performance with that of a conventional reactor. The dimensions of both laboratory reactors were the same (length 25 cm, I.D. 0.67 cm). Under the same experimental conditions (T=550 °C, inlet pressure of 2.15 bar, and a feed gas composition of $CH_4/O_2/N_2$ of 2/1/14) the membrane reactor gave a CH_4 conversion of 96 % vs. 48 % for the conventional reactor. The membrane reactor, on the other hand, was shown to generate more carbon than the conventional reactor. Galuszka *et al.* [2.240] studied the same reaction (and also methane dry reforming) in a PBMR using Pd/Al_2O_3 membranes prepared by electroless-plating. The CH_4 conversion, and CO and H_2 yields were considerably enhanced in the membrane reactor for both processes, with CH_4 conversion increases ranging from 4–20 %, and CO and H_2 yield increases ranging from 2–20 % and 8–18 %, respectively. However, during reaction filamentous carbon formed on the palladium membrane, leading to swelling and eventual membrane failure. Based on this observation Galuszka *et al.* [2.240] provide a pessimistic assessment of the applicability of palladium membranes for hydrogen separation from process streams containing methane or carbon monoxide. Ostrowski *et al.* [2.241] studied the catalytic partial oxidation of methane to synthesis gas in a PBMR and in a fluidized-bed membrane reactor (FBMR) over a Ni/α-Al$_2$O$_3$ catalyst. The membranes were prepared by depositing a silicalite layer on alumina or porous stainless-steel supports. They also tested a membrane obtained by Pd deposition on the stainless-steel/silicalite membrane. The zeolite membranes were stable in the whole temperature range (700–750 °C) that they investigated, but the Pd-membrane proved unstable for temperatures higher than 650 °C. Neither the PBMR nor the FBMR showed an improvement in the syngas yield. The reason for the poor performance in the PBMR was the loss of methane. In the FBMR reactor the amount of permeated hydrogen was not sufficient to shift significantly the equilibrium towards a higher syngas yield. However, in the fluidized-bed membrane reactor significantly better selectivities of separation were achieved.

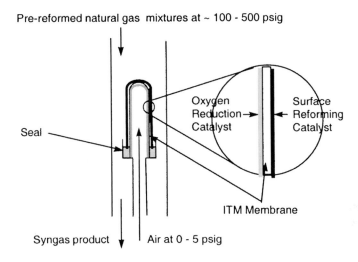

Figure 2.12. A schematic of the solid oxide membrane reactor for synthesis gas production. From Dyer *et al.* [2.232], with permission from Elsevier Science.

A number of other partial hydrocarbon oxidation reactions have also been studied using catalytic membrane reactors. Using zirconia-calcia-alumina porous membranes, Zhong-Tao and Ru-Xuan [2.242] studied the partial methane oxidation towards methanol and formaldehyde. They report a selectivity of over 96 % at 573 K. Oxidation of methane to formaldehyde using a zirconia-alumina macroporous membrane was studied by Yang *et al.* [2.243]. The membrane reactor gave better selectivities (for the same level of conversion) than the conventional packed-bed reactor, but the reported yields for both were rather small < 1 %. Direct catalytic oxidation of methane to methanol using catalytic membrane reactors was also reported by Lu *et al.* [2.244] with rather modest improvements in yield. An electrochemical reactor using a YBa$_2$Cu$_3$O$_x$ anodic electrode was used by Gür and Huggins [2.245] to study the same reaction. Again modest yields towards CH$_3$OH and HCHO were reported, the primary product being CO. The use of membranes for improving the yield of the homogeneous partial oxidation of methane to methanol was reported by Liu *et al.* [2.246]. In their set-up, shown in Figure 2.13, the non-permselective membrane provides a means for intimate contact for the reactants, and a way to separate the external hot wall from an internal, water-cooled tube, whose purpose is to remove CH$_3$OH by quenching. The concept of combining a membrane barrier with a quenching step (first proposed by Halloin and Wajc [2.247] for the catalytic hydrogenation of toluene to methylcyclohexane) seemed to work well for methanol production resulting in improved selectivity over the case in the absence of quenching.

Diakov *et al.* [2.248] have studied the partial oxidation of methanol to formaldehyde using a macroporous stainless steel membrane. The reaction is carried out industrially over an Fe-Mo oxide catalyst in a fixed-bed reactor in the temperature range of 300–400 °C. Under such conditions the reaction can be represented as

$$CH_3OH + 0.5O_2 \leftrightarrow HCHO + H_2O \tag{2.6}$$

$$HCHO + 0.5O_2 \leftrightarrow CO + H_2O \tag{2.7}$$

Kinetic investigations indicate that the selectivity to formaldehyde increases with decreasing oxygen concentration making this a candidate reaction for the application of a segregated feed MR. Diakov *et al.* [2.248] have carried theoretical investigations, which seem to indicate that the MR with a distributed oxygen feed provides better performance, selectivity-wise, than the conventional fixed-bed reactor. Experimental results seem to validate the results of the theoretical investigations. The same reaction was studied in a Ag/YSZ/Ag EMR by Neophytides and Vayenas [2.167] with the goal of co-producing electricity. CH$_3$OH conversions over 30 %, with selectivities of 90 % were achieved. Significant NEMCA effects were also reported.

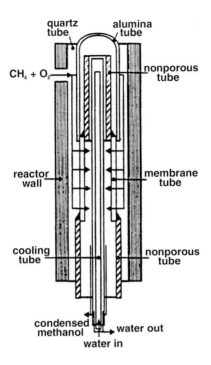

Figure 2.13. Schematic of the membrane reactor for methane partial oxidation. From Liu *et al* [2.246], with permission from Elsevier Science.

Grigoropoulou *et al.* [2.249] have studied the selective oxidation of benzyl alcohols to the corresponding benzyl aldehydes by hypochlorite and a phase transfer catalyst (tetrabutylammonium hydrogen sulfate) in a laboratory membrane reactor. They utilized a porous PTFE membrane to separate the aqueous (containing the hypochlorite) and organic (containing the alcohol and the phase transfer catalyst) phases. Phase transfer catalysis (PTC) has been used to transfer inorganic anions into organic media by using catalytic amounts of lipophilic quaternary ammonium or phosphonium salts. Most PTC reactions are carried out industrially in stirred-tank reactors, which require the subsequent separation of products. Key criteria for optimal PTC reactor design are a high interfacial area with little emulsification to enable easy phase separation. The use of membrane reactors for PTC reactions (first suggested by Stanley and Quinn [2.250]) potentially fulfills these criteria. The membrane acts as a stable interface contactor, and no emulsification problems are encountered. Grigoropoulou *et al.* [2.249] report that the PTC membrane reactor was successful in selectively oxidizing a series of substituted benzyl alcohols to the corresponding benzaldehydes with very high yields and selectivities. Phase transfer catalysis in the presence of a membrane has been studied by Luthra *et al.* [2.251]. They utilized NF membranes for separation and recycling of tetraoctylammonium bromide PT catalysts during the conversion of bromoheptane into iodoheptane. During three reaction cycles a 99+ % catalyst recycle was achieved, and no loss in PT catalyst activity was observed.

Hazburn [2.129] in a U.S. patent, reported ethane and propane partial oxidation to ethylene and propylene oxides using a TiYSZ mixed O^{2-}/electron conducting membrane with silver deposited on the hydrocarbon side of the membrane as a catalyst. For the ethane partial oxidation reaction catalyst at 250–400 °C, the ethylene oxide selectivity was (>75 %) but the ethane conversion was low (<10 %), limited by the low membrane oxygen flux at these temperatures. For the partial oxidation reaction to propylene oxide yields close to 5 % were reported. Using porous membranes Santamaria and coworkers (Mallada *et al.* [2.252, 2.253, 2.254, 2.255]), Mota *et al.* [2.256], and Xue and Ross [2.257] recently studied the oxidation of butane into maleic anhydride.

Varma and coworkers [2.258, 2.259] have studied the ethylene epoxidation reaction into ethylene oxide. The potential advantage of the membrane reactor (thought to be intrinsically a safer device) would be to be able to operate for higher feed concentrations of the hydrocarbon. Generally, a beneficial effect on the selectivity was observed with the membrane reactor, as compared with the classical packed-bed reactor. Varma and coworkers [2.258, 2.259] observed, that for the ethylene epoxidation reaction in the conventional reactor selectivity increases as the O_2/C_2H_4 molar feed ratio increases (in contrast to most other partial oxidation reactions). As a result, it is more beneficial to use the membrane for distributing the feed of the C_2H_4. They investigated two membrane reactor configurations, and compared their performance to that of a conventional fixed-bed reactor (FBR), under identical overall reaction conditions, using an alumina supported, cesium-doped silver catalyst. In one membrane reactor configuration (PBMR-O) oxygen was allowed to permeate through the membrane with ethylene flowing over the catalyst bed, while in the other configuaration (PBMR-E) ethylene permeated through the membrane with oxygen flowing over the catalyst bed. The two reactor configurations had different behavior, with the PBMR-O exhibiting smaller and the PBMR-E larger ethylene oxide selectivity, when compared to the conventional reactor. While maintaining the overall reaction conditions fixed, ethylene oxide production in the membrane reactor could be increased further by manipulating the residence time of the reactants over the catalyst bed; and the maximum value was obtained for the largest possible residence time. A mathematical model was also developed (Al-Juaied *et al.* [2.260]). The dusty gas model was used to describe transport through the porous stainless-steel membrane. The model results were in satisfactory agreement with the experimental data. The simulations showed that the imposed pressure gradient resulted in predominantly convective flow through the membrane, which inhibited backdiffusion from the catalyst bed. The model results confirmed that the PBMR-E is the best configuration, followed by the FBR, and the PBMR-O, respectively. Lafarga and Varma [2.259] also studied the addition in the feed of different levels (0–3.3 ppm) of 1,2-dichloroethane (DCE), which promotes selective ethylene epoxidation. Increasing the DCE concentration in the stream entering the catalyst bed increased the selectivity to values exceeding 80 % for low ethylene conversions. High levels of DCE, on the other hand, showed detectable catalyst deactivation after a few hours under reaction conditions; hence an optimum level exists (1–2 ppm) that maximizes ethylene oxide yield, while maintaining a reasonably high stable catalyst activity. Cheng *et al.* [2.261] for the same reaction have investigated, in detail, the performance of a PBMR with separated hydrocarbon and oxidant feeds. They investigated experimentally the effect of systematically varying both the length and the position along the catalyst bed of the membrane, through which oxygen is allowed to feed. Nonuniform feedthroughs were shown, in some instances, to be more effective than the uniform ones along the whole bed length.

For the reaction of butane to maleic anhydride Mallada *et al.* [2.252, 2.253, 2.254, 2.255] used a fixed bed of VPO catalysts and an inert membrane to distribute oxygen. The study was carried out using relatively high butane/oxygen ratios (up to seven times higher than those currently used in industrial practice), some of which would have been in the explosive range, if fed to a fixed-bed reactor. They studied two different reactor configurations, one in which the oxygen permeates inward from the shellside and the catalyst is packed on the tubeside (IMR), and another, in which the catalyst bed is packed on the shell side, and oxygen flows outwards from the tubeside (OFIMR). The radial oxygen and hydrocarbon concentration and temperature are different for the two types of reactors investigated; a considerably better performance is obtained for the OFIMR.

Mota *et al.* [2.256] studied the same reaction using a MFI zeolite membrane, which is created by plugging the pores of a commercial mesoporous alumina membrane. The catalyst, prepared by reduction of vanadyl phosphodihydrate by the isobutanol route, was packed in the interior membrane volume. Mota *et al.* [2.256] investigated three different reactor configurations. In the first configuration the oxygen and the butane were co-fed into the reactor. In the second configuration the butane was fed into the tubeside and the oxygen was fed through the shellside. And in the third configuration 20 % of the oxygen was co-fed with the butane, while 80 % of it was distributed through the membrane. Configuration 2 resulted in the reduction of the catalyst at the entrance into the reactor leading to an inferior performance. Configuration 3 also resulted into a heterogeneous catalytic activity. Reversing the flow into the interior volume resulted in a performance, which was superior to that of the conventional reactor. Xue and Ross [2.257] also studied the selective oxidation of *n*-butane to maleic anhydride using a tubular ceramic membrane reactor loaded with a commercial VPO catalyst. The study was carried out using a configuration, in which butane was fed to the membrane tubeside, where the catalyst was loaded, and O_2 was fed either entirely to the outer tube (shellside) or distributed in a certain ratio between the tube- and shellsides. Xue and Ross [2.257] report that the use of a butane-rich feed under this configuration gives a higher maleic anhydride yield than the conventional reactor, and that the way O_2 was distributed in the feed can have a significant effect on the reaction rate. Alonso *et al.* [2.262] in a modeling study proposed the use of an externally fluidized bed-membrane reactor for partial oxidation reaction of butane to maleic anhydride. Catalyst is loaded inside a porous stainless steel membrane tube into which the hydrocarbon is fed. An oxygen-rich gas fluidizes a powder on the shellside, where it crosses the membrane wall and reacts with butane. Alonso *et al.* [2.262] report with this reactor maleic anhydride yields, which are potentially 50 % higher to those attained in a conventional fixed-bed reactor, and a much broader operating range of reactant concentration.

The selective oxidation of C_2–C_4 alkenes and alkanes into oxygenates has been studied by a number of investigators. Oxidation of ethylene to acetaldehyde using porous V_2O_5/γ-Al_2O_3 membranes has been studied, for example, by Harold et al. [2.132]. The selectivity towards the intermediate oxidation product reported in the CMR was higher than that of a conventional reactor. Kolsch et al. [2.263] have studied the partial oxidation of propane to acrolein in a segregated feed MR (propane fed in the tubeside and oxygen fed through the membrane from the shellside), and compared the selectivity and yield with the co-feed situation, in which both propane and oxygen were fed through the tubeside of the same membrane. The membrane was a porous alumina membrane from the Hermsdorfer Institute for Technical Ceramics on the inside of which they deposited a precipitated $Mg_{10}Bi_{12}Mo_{12}O_x$ layer as a partial oxidation catalyst. The segregated feed configuration gave a higher acrolein selectivity, albeit at the cost of a reduced propane conversion (63 % vs. 18 % difference in selectivity, and 2.3 % vs. 1.9 % difference in yield). *In situ* hydrolysis of tetraethylsilicate in the pores of the γ-Al_2O_3 resulted in a hydrophilic membrane, which showed enhanced permeation towards the reaction products (acrolein and water). Using this membrane in the two different reactor configurations resulted in the segregated feed having a yield, which was nine times higher than that of the co-feed.

The selective oxidation of C_2–C_4 alkanes into oxygenates has been studied in an electrochemical membrane reactor using YSZ and ceria-based membranes by Hamakawa and coworkers [2.264, 2.265, 2.266]. Hamakawa et al. [2.267], for example, studied the partial oxidation of ethane to acetaldehyde using a dual chamber Au/YSZ/Ag EMR at 475 °C. Upon electrochemically pumping oxygen to the Au electrode, acetaldehyde and CO_2 were produced with corresponding selectivities of 45 and 55 %. In open circuit studies CO and CO_2 were formed, but no acetaldehyde. York et al. [2.264, 2.265] in the same reactor studied the partial oxidation of propane to acetone and acrylaldehyde, butane to methylethyl and methyl-vinyl ketones, and of isobutane to methacrolein by pumping electrochemically oxygen to the Au electrode. In the open circuit operation only CO and CO_2 were produced. In the studies of Hamakawa and coworkers the selectivities are good, but generally very dilute hydrocarbon mixtures were used, and the reported conversions are rather low. As previously noted, the same group has also studied the production of synthesis gas from CH_4 using a similar reactor system and membranes ([2.206, 2.207]).

Takehira and coworkers (Hayakawa et al. [2.268, 2.269], Tsunoda et al. [2.270, 2.271, 2.272, 2.273], Hamakawa et al. [2.274, 2.275, 2.276]) have also studied the selective oxidation of C_2–C_4 alkenes into oxygenates. Using a YSZ solid oxide electrolyte and a Au electrode Takehira and coworkers (Hayakawa et al. [2.268, 2.269]) first reported the production of acrolein from propylene back in 1986. They subsequently studied the reactivity of a number of catalysts deposited on the Au anode including Bi_2O_3, MoO_3, as well as bismuth molybdate. During the EMR operation it was shown that the oxygen atoms bound to the Mo participate in the reaction by inserting into the propylene to form acrolein. In a later study (Hamakawa et al. [2.274, 2.275]) the YSZ solid oxide electrolyte was replaced by $(CeO_2)_{0.8}(SmO_{1.5})_{0.2}$ in order to achieve higher fluxes, and to be able to operate the reaction at lower temperatures (350–450 °C). More recently the same group [2.276] used the same electrolyte, but coated with YSZ on the anode side. At 450 °C the acrolein selectivity was 36 %, which is four times higher than what was obtained in the absence of YSZ. Tsunoda *et al.* [2.271, 2.272, 2.273] also studied the partial oxidation of ethylene to acetaldehyde and of but-1-ene to methyl-vinyl-ketone and crotonaldehyde.

A number of studies using membrane reactors to control oxygen addition have investigated the oxidative dehydrogenation of hydrocarbon and other organic compounds, including ethane to ethylene (Coronas *et al.* [2.277]; Tonkovich *et al.* [2.278, 2.279]) ethylene to acetylene [2.280], propane to propylene [2.281, 2.282, 2.283, 2.284], butane to butene [2.285], butene to butadiene [2.129], and isobutane to isobutylene [2.286]. Iwahara *et al.* [2.287, 2.62] used a proton ionic conductor in a dual-chamber EMR to study ethane dehydrogenation to ethylene with the simultaneous production of electrical energy (steam was added to the ethane feed to improve performance). The EMR operated at 800–1000 °C. Both Ni and Pt cathodes were tested with stable performance. Mazanec *et al.* [2.217] in a U.S. patent describe the use of a mixed oxygen/electron conductor in an EMR to produce ethylene and acetylene from ethane, and propene and propylene from propane. Yang *et al.* [2.288] have recently studied the oxidative dehydrogenation of ethane to ethylene using a MR with a disk-shaped $Ba_{0.5}Sr_{0.5}Co_{0.6}Fe_{0.4}O_{3-\delta}$ solid oxide membrane. The MR utilized is shown schematically in Figure 2.14 with C_2H_6 passing on the one side of the membrane and oxygen passing on the other side. These authors also performed the experiments in a fixed-bed reactor, in which the membrane was used only as a catalyst using the same approximate flows of C_2H_6, O_2, and He as those used in the MR experiments. The MR performed significantly better than the fixed-bed reactor, as can be seen in Figure 2.14.

Michaels and Vayenas [2.168, 2.169] studied the oxidative dehydrogenation of ethylbenzene to styrene in a Pt/YSZ/Pt EMR at 575–600 °C, with simultaneous oxygen pumping. The best selectivity to styrene was achieved at modest current densities. Raybold and Huff [2.286] have studied the oxidative dehydrogenation of isobutane over Pt/α-Al_2O_3 and Rh/α-Al_2O_3 monoliths and Pt/γ-Al_2O_3 pellets in a Pd membrane reactor. Though the presence of the membrane tends to alleviate the thermodynamic limitations of the dehydrogenation reaction, these processes, however, can still be limited by the slow kinetics. Adding O_2 can suppress the kinetic limitations, while the removal of H_2 through the Pd membrane can alleviate the equilibrium limitation imposed on the system by the dehydrogenation route. Raybold and Huff have investigated i-C_4H_{10}:O_2 ratios of 1.0 to 2.0, contact times ranging from 0.04 to 0.25 s, and temperatures ranging from 400 to 700 °C. Continuously removing the H_2 produced increased isobutylene yields. Yield improvements depended strongly on the balance of reaction time with H_2 removal time, and the importance of the dehydrogenation in the overall reaction scheme.

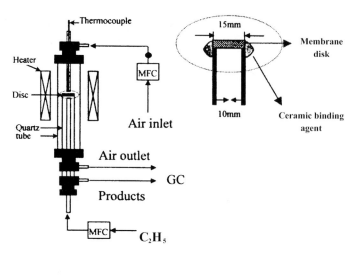

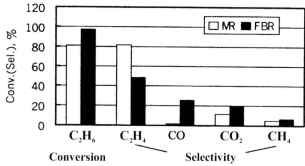

Figure 2.14. The membrane reactor and some typical experimental results for the ethane oxidative dehydrogenation. From Yang *et al.* [2.288].

Alfonso *et al.* (2.283, 2.284] studied the oxidative dehydrogenation of propane to propylene. They investigated two types of catalytic membranes using alumina membrane supports, (a) V/MgO membranes, where MgO was deposited in the support pores before impregnation with V, and (b) V/Al_2O_3 membranes, where the alumina supports were impregnated with a boehmite sol before V deposition. For both membranes the introduction of V was carried from the internal side of the porous tube, using short impregnation times and a radial distribution of active material was obtained. The two membranes were tested giving yields to propene of up to 15 %. It was found that the V/MgO membranes are more selective, but less active than the V/Al_2O_3 membranes. The same group [2.289] proposed an interesting MR, shown schematically in Figure 2.15, for the oxidative dehydrogenation of butane. The membrane consists of two catalytic layers: An inside layer containing a V/MgO catalyst, which is active towards the oxidative dehydrogenation of butane and an outside layer containing a Pt/Sn/Al_2O_3 catalyst prepared by a sol-gel technique, which is thought to be active towards the dehydrogenation of butene. The presence of both layers allows one to couple the two reactions both energetically, but also chemically, since the product of the oxidative dehydrogenation reaction is the reactant for the dehydrogenation reaction. An additional beneficial effect of this reaction coupling appears to be that the steam and CO_2 that are formed in the oxidative dehydrogenation seem to help the dehydrogenation catalyst maintain its catalytic activity. The authors report that the presence of the $Pt/Sn/Al_2O_3$ layer increases the butane conversion by 2–13 %.

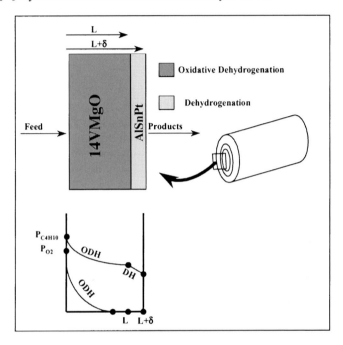

Figure 2.15. The two-layer membrane for butane dehydrogenation. From Alfonso *et al.* [2.289], with permission from the American Chemical Society.

The propylene oxidative dimerization reaction to benzene and hexadiene using a $(Bi_2O_3)_{0.85}(La_2O_3)_{0.15}$ oxygen conducting membrane disk was studied by Di Cosimo *et al.* [2.290, 2.291]. The membrane was exposed to propylene on one side, and oxygen on the other side. The oxidative dimerization utilized the lattice oxygen of the membrane, which was replenished by the oxygen flowing on the oxidant side. The selectivity to C_3 dimers was significantly higher than in the case when propylene and oxygen were co-fed on the same membrane side. The same reaction was later studied by Bordes and coworkers [2.292, 2.293] in a CMR using again a catalytic membrane made of metal doped bismuth oxide. A 36 % reactor yield was obtained at 773 K. Courson *et al.* [2.293] studied the structural and textural properties, catalytic and non-catalytic reactivities of M-doped Bi_2O_3 (M = La, Ce, Eu, Er, V, Nb). The dopants were shown to increase the rates of reduction and reoxidation of Bi_2O_3; the reoxidation of the M-doped Bi_2O_3 was shown to be rate limiting. During the CMR operation, the best yields in 1,5-hexadiene and benzene were obtained with a cerium-doped Bi_2O_3.

In a related study Piera *et al.* [2.294] studied the dimerization of isobutene to produce branched octenes using MFI zeolite (silicalite, ZSM5) membranes in a catalytic membrane reactor. This dimerization reaction is typically carried out in the liquid phase utilizing an acid catalyst, but the reaction, in addition to the desired branched octenes, also produces undesired C_{12} and C_{16} species. The idea behind using the membrane reactor for this reaction is that the membrane removes the C_8 product, so that it does not undergo further undesired reactions. Indeed the membrane was shown to preferably permeate *i*-octene from *i*-butene (selectivity 23 at room temperature), but the selectivity declined rapidly at higher temperatures (1 at 70 °C). The reactor experiments were carried out with Amberlyst®15 as a catalyst packed in the reactor tubeside (W/F=69 kg catalyst·s/kg *i*-butene) at 15 bar, and a N_2 sweep of 1500 cm^3(STP)/min. Consistent with the permeation experiments, best results were obtained at room temperature (20 °C). The MR gave similar conversions with a PFR operated under the same conditions but a higher yield to different *i*-octene species (66 % vs. 40 % for the conventional reactor).

Recently Brinkmann *et al.* [2.295] reported a theoretical and experimental study of the oxidative dehydrogenation of methanol to CO_2 and H_2 by using a CMR with a Pt impregnated tubular alumina membrane. The authors studied the influence on reactor performance of the transmembrane pressure or the purge stream to feed ratio. Conversion increased as either one of these increased, the reactor conversion reaching as high as 84 %, while the hydrogen selectivity remained near 50 %.

Using membranes as ozone distributors has been recently discussed in a number of applications involving the environmental remediation of polluted wastewaters. The concept here is very similar to that previously discussed, i.e., having the membrane act as an effective distributor of the oxidant (in this case ozone). Typically ozone treatment of polluted waters is achieved by bringing the water in direct contact through bubble columns, and specialized ozone injectors. Sometimes these devices suffer from excessive formation of foam and a high energy consumption for pumping the gas. The membrane offers a more controlled way of bringing the two phases together. Ceramic membranes have been proposed as contactors to bring together ozone and polluted water by Janknecht *et al.* [2.296, 2.297, 2.298]. Inorganic membranes are better suited, since they are resistant to ozone. The ozone mass transfer per contactor volume through hydrophilic membranes was in the same order of magnitude as in conventional bubble contacting, with the pressure differential between the gaseous and aqueous phases as well as the membrane microstructure determining the transfer rate. A hydrophobic coating of the membrane surface led to a considerable increase in transfer by eliminating water infiltration in the membrane structure that creates the added membrane phase resistance. Polymeric membranes have been used by other groups. Hashino *et al.* [2.299], for example, have utilized an ozone resistant hollow-fiber membrane module made of polyvinylidenefluoride (PVDF). A new filtration system using this membrane module together with ozone dosing provided three to four times higher permeate flux compared with the filtration without ozone. The reaction of ozone with organic materials in feed water was necessary to occur on the surface of the membrane to have higher permeate flux. Wydeven *et al.* [2.300] incorporated a polymeric membrane into a commercial UV/ozone stripper/cleaner system in an effort to improve UV/ozone technology; the membrane was found to be an effective diffuser for ozone/oxygen mixtures, and there was no apparent degradation of the membrane due to exposure to UV/ozone.

A novel recent application reported by Vankelecom *et al.* [2.301] involves the use of polymeric membranes, in which one incorporates active catalysts, as interfacial contactors to carry out partial oxidation (but also hydrogenation and hydroformylation) of various organic compounds. This concept is similar, in principle, to that previously applied to hydrogenation reactions. In addition to the potential advantages that were previously noted, in this case the membrane also allows carrying out the reaction without the need of a co-solvent, and prevents the loss of catalyst. In the study by Vankelecom and coworkers [2.301], an organometallic complex of manganese ([5,10,15,20-tetrakis(2,6-dichloro-phenyl)porphyrinato]manganese(III)chloride), which is catalytically active for epoxidation reactions in the homogeneous phase, was incorporated into a polydimethylsiloxane (PDMS) membrane. The catalytic membrane was tested for the 3-penten-2-ol epoxidation using hydrogen peroxide. The experimental results showed that the selectivity is dramatically increased (from ~20 to 100 %), when using the catalytic membrane instead of the free catalyst. This behavior was attributed to the reduced mobility of the ligand in the membrane, when compared with its free form in solution. Another interesting observation was that the hydrogen peroxide activity was reduced in the membrane because the accessibility of metal active sites was reduced. The feasibility of this concept has also been demonstrated by Wu *et al.* [2.302] for the oxyfunctionalization of *n*-hexane into a mixture of hexanols and hexanones by hydrogen peroxide, using titanium silicalite (TS1) as the catalyst imbedded in a modified PDMS membrane. It was shown that the oxyfunctionalization products are formed in, and separated by the catalytic membrane. Langhendries *et al.* [2.303] have demonstrated the feasibility of the same concept with the oxidation of cyclohexane, cyclodecane, and *n*-dodecane using *t*-butylhydroperoxide as the oxidant, and zeolite-encaged iron phthalocyanine as the catalyst embedded in a hydrophobic polymeric membrane matrix. During *n*-dodecane oxidation in the catalytic membrane reactor, the alcohol and ketone products are exclusively recovered in the organic *n*-dodecane phase, demonstrating an added advantage of the CMR configuration.

Frusteri and coworkers (Espro *et al.* [2.304], Frusteri *et al.* [2.305]) have studied the selective oxidation of light alkanes (CH_4, C_2H_6, C_3H_8) under mild conditions (80–120 °C, 140 kPa) in a three-phase catalytic membrane reactor containing superacid catalytic membranes. In this reactor Frusteri and coworkers study the functionalisation of the C_1–C_3 alkanes to the corresponding oxygenates on superacid catalytic membranes mediated by the Fe^{2+}/H_2O_2 Fenton oxidation system. The membrane is sandwiched between two porous Teflon plates with the catalytic side towards the liquid phase, which consists of a solution of H_2O_2 and $[Fe^{2+}]$ or $[Fe^{3+}]$. The gas phase consists of a mixture of the hydrocarbon and N_2. The two half-cells of the three-phase MR operate as batch reactors with separate recirculation for the two phases. Among the various superacid catalytic membranes, the Nafion®-based ones showed the best performance. Frusteri and coworkers present a general pathway for the reaction, which involves the activation of the C-H bond of the alkane molecule on the superacid membrane sites, and the subsequent reaction of the activated alkane with primary reactive intermediates, generated from the Fe^{2+}/H_2O_2 system.

Carbon monoxide oxidation coupled to oxygen transport through a $SrFe_{1.125}Co_{0.375}O_y$ oxide membrane was investigated by Ran *et al.* [2.306] in the temperature range of 850–900 °C. An oxygen flux as large as 1.3×10^{-6} mol/cm^2.s was observed through a 1.2 mm thick membrane at 900 °C by exposing one side of the membrane to air, and the other side to mixture of helium and carbon monoxide with P_{CO} at 0.16 atm. This flux, which is two orders of magnitude higher than the value measured in the absence of CO, remained almost unchanged as the membrane thickness varied. It showed a weak temperature dependence with an activation energy of 28 +/– 3 kJ/mol, and increased with the CO partial pressure. Hasegawa *et al.* [2.307] also studied the CO to CO_2 oxidation reaction in a membrane reactor using CO_2 permselective, ion-exchanged (K, Rb, Cs) NaY zeolite membranes. The motivation for the study is fuel cell applications. The ion-exchange seemed to increase the CO_2 separation characteristics of the membranes.

A number of groups have discussed the application of membrane reactors in the context of coal gasification. Nakagawa and Ishida [2.308], for example, used a Pt/YSZ/Pt EMR in order to produce CO and electricity from charcoal in the temperature range of 800–1000 °C. The charcoal was gasified by electrochemically supplied O^{2-} to produce CO with a high selectivity (95 % at 1000 °C). Yentekakis *et al.* [2.309] used an imaginative dual-chamber EMR design, which utilized Pt as the cathode exposed to air together with a molten Fe bath, which served as both the anode, but also as the carbon reservoir. In the temperature range of 1000–1500 °C the O^{2-} that was transported through the membrane would react with the carbon (which was supplied in the form of fine particles to the molten Fe bath) to produce CO and electricity. Horita *et al.* [2.203] tested in a similar EMR the applicability of a number of other anodes including ZrC, WC, VC, and TiC. VC was proven to be the most effective. Sammels *et al.* [2.310] also reported the use of a membrane reactor using brownmillerite membranes for gasifying coal. The membrane-based gasifier was reported to operate below the slagging temperature, and coal gasification at practical rates was claimed.

To summarize the above discussion, membrane-based reactive separations have been applied to partial oxidation reactions and some interesting results have been obtained. However, for the technical concept to have a significant beneficial impact it will be necessary to develop mechanically and chemically stable membrane materials, which prevent the desired partially oxidized products from coming into prolonged contact with the oxidant. This is because, generally, the reactivity of the product towards oxygen or the oxide catalysts is higher than that of the hydrocarbon. Non-permeable dense membranes are successful in that regard. However, the oxygen transport through the oxide lattice is low, when compared to its transport rate through porous materials. This requires the use of high temperatures of operation limiting the use of the membrane reactor concept to only a handful of reactions. Preparing dense, solid-oxide membranes with improved oxygen permeabilities is the subject of ongoing research efforts. An alternative approach is to synthesize porous ceramics with good oxygen permeability and selectivity. This is not a simple task and it is also the subject of considerable ongoing research. The development of microporous zeolite membranes capable of providing a high separation factor between oxygen and hydrocarbons offers hope in this area.

Another problem with the use of separated feed MR for partial oxidation reactions is that the molar ratio of hydrocarbon to oxygen reduces along the length of the reactor as the hydrocarbon gets consumed and the oxygen flux increases, because of a decrease in the feed-side pressure as a result of a pressure drop through the catalyst bed. One idea to overcome this problem is to prepare a membrane, whose permeance is sensitive to the chemical environment, allowing more oxygen to go through in a reducing environment, and less so in an oxidizing environment (the concept is shown schematically in Figure 2.16). The preparation of such a "chemical valve" membrane was recently reported by Farrusseng *et al.* [2.311] and Julbe *et al.* [2.312]. They prepared this membrane by filling the internal structure of an asymmetric α-alumina tubular support with V_2O_5 and $AlPO_4$ crystallites. The redox properties of the V_2O_5, which converts to V_2O_3 under a reducing atmosphere, were found to significantly influence membrane properties (grain morphology, pore size and pore size distribution, and surface area). Farrusseng *et al.* [2.311] carried out an experiment, in which a C_3H_8/He mixture was flown (20 ml/min) past one side of the membrane, with a He/O_2 mixture passing on the other. The experimental results, shown in Figure 2.16, indicate clearly that the $(He + O_2)$ permeance through the membrane is a function of the C_3H_8/O_2 ratio, i.e., that the membrane, indeed, acts as a chemical valve.

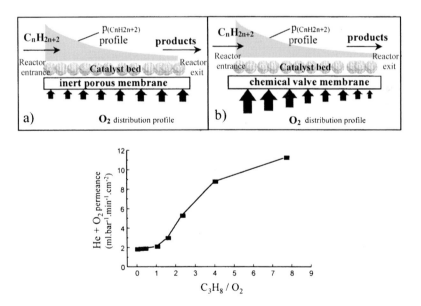

Figure 2.16. Top, O_2 profiles generated in a membrane reactor for the partial oxidation of alkanes. (a) conventional MR (b) a chemical valve MR. Bottom, Evolution of permeance of the chemical valve membrane as a function of the ratio C_3H_8/O_2 without any catalyst.

2.4   Other Catalytic Reactions

Ohashi *et al*. [2.313] studied the thermal decomposition of hydrogen sulfide in membrane reactor with a ZrO_2-SiO_2 porous membrane, which has Knudsen diffusion characteristics. A modest enhancement in conversion was observed. A mathematical model was also developed, which was found to be in good agreement with the experimental results in the range of low temperatures, but not in such good agreement at higher reactor temperatures. The same group theorized that an additional separation step would be necessary to increase the hydrogen yield [2.314]. They have proposed a membrane reactor consisting of three chambers (one chamber for reaction, and two others in series for separation). The analysis showed that this membrane reactor gives a better performance than the single stage membrane reactor.

Ma *et al*. [2.315] studied the H_2S decomposition reaction in a PBMR configuration using a Vycor® porous glass membrane and a molybdenum sulfide catalyst. No sweep gas was used in the experiments in order to avoid the complications in determining the conversion enhancement that is only due to dilution effects. They concluded that in this case, in order to provide a significant shift in the reaction equilibrium, highly selective membranes may be required. Such a selective, composite Pt-V membrane was used for the study of this reaction by Edlund and Pledger [2.316]. Pd membranes, tested by this group, had previously proved unstable in the same reaction environment. Their Pt-V membranes, though stable, were, unfortunately, significantly less permeable than the Pd membranes, and were, themselves, not free of some of the other problems associated with dense metallic membranes. The chemical stability of Pt and Pd composite membranes towards hydrogen sulfide was more recently evaluated by Kajiwara *et al*. [2.317]. Their Pt membranes consisted of a platinum layer created by CVD and supported on a porous alumina tube. The palladium composite membranes were prepared by an electroless-plating technique. The CVD Pt composite membranes gave hydrogen fluxes comparable to that of the Pd composite membranes, as well as high ideal permselectivity for hydrogen over nitrogen (240 for Pd and 210 for platinum at 773 K). When these composite membranes were brought in contact with a gas stream containing hydrogen sulfide, their hydrogen permeability decreased rapidly. However, for the Pd composite membranes many cracks were formed on the metallic surface, and, as a result, other gases besides hydrogen permeated through the cracks formed. On the other hand, very few cracks were formed in the Pt composite membrane. The difference in behavior was attributed to the fact that the lattice constant of Pd was expanded from 0.39 to 0.65 nm upon sulfidation of the metallic layer, while that of Pt was only slightly changed from 0.39 to 0.35 nm. Upon treating the sulfurized Pt composite membranes in pure oxygen, their hydrogen permeability recovered to 50 % of that of the fresh membrane. A nonisothermal 1-D model for H_2S catalytic decomposition in a PBMR using a Pt-coated Nb tubular membrane was presented by Chan *et al*. [2.318]. The conversion in the PBMR rose monotonically along the length of the reactor exceeding the equilibrium conversion by as much as eight times under some conditions.

The H_2S decomposition reaction has also been studied utilizing electrochemical membrane reactors. Alqahtany *et al*. [2.319], for example, studied the reaction in a Pt/YSZ/Pt

EMR. The reaction exhibited a strong NEMCA effect with oxygen electrochemical pumping increasing the open circuit decomposition rate by 11-fold. Winnick and coworkers [2.320, 2.321] studied the electrochemical decomposition of H_2S using both H^+ and O^{2-} conducting membranes. A proton conductor (Li_2SO_4) gave the best performance, with a maximum current density at 725 °C of 12 mA/cm^2, and a stable operation without any apparent degradation.

The decomposition of dilute mixtures of NH_3 in a PBMR using Pd-alloy membranes was studied by Collins and Way [2.322], and by Gobina *et al.* [2.323]. This application is of potential interest in the treatment of coal gasification streams, and the laboratory results showed promise. It would be interesting to see, whether the same membranes prove robust in the real coal-gas environment. The use of a PBMR to study the hydrodechlorination of dichloroethane was reported by Chang *et al.* [2.324]. The reported potential advantage of the membrane would be in preferentially removing the by-product HCl, which deactivates the catalyst. The authors attribute the observed improved performance, however, to a dilution effect.

Electrochemical membrane reactors have also been applied to a number of environmentally relevant reactions. The decomposition of NO to nitrogen and oxygen was among the first such reactions that were studied. Huggins and coworkers [2.325, 2.326] studied the reaction using Pt and Au electrodes and a scandia stabilized zirconia electrolyte for a broad range of temperatures (600–900 °C). Electrochemically pumping the oxygen away from the NO decomposition catalyst was shown to enhance the open circuit rate by up to a six orders of magnitude factor. The decomposition of SO_2 to S_8 and oxygen was studied by Cicero and Jarr [2.327] using an oxygen conducting membrane and metal oxides as cathodic electrodes in the temperature range of 650–1950 °C. Oxygen was electrochemically pumped from the cathode, in order to decompose the SO_2; a conversion of about 40 % was reported.

The reaction between CO-NO-O_2 was studied by Hibino and coworkers [2.328, 2.329, 2.330] in a single chamber electrochemical reactor, in which the Ce-based solid oxide electrolyte and Pd electrodes were all immersed in the same gas atmosphere. The reaction was studied in the temperature range of 400–800 °C, and the effect of gaseous species like CO_2 and H_2O was investigated. Oxygen pumping was shown to have a considerable effect on the reaction rate. The same group also studied the reaction using Au electrodes in the temperature range of 800–900 °C. In the presence of an excess amount of CO_2/H_2O no effect of the electrochemical oxygen pumping was found for this case. The reduction of NO by CO as well as by propylene has also been studied by Vayenas and coworkers [2.331, 2.332, 2.333] in a single chamber EMR. Both Na^+ conducting and more conventional YSZ electrolytes were utilized [2.334]. The reaction was shown to exhibit a strong NEMCA effect. The reaction of NO with H_2 was studied by the same group [2.335] using a single chamber EMR containing again a Na^+ conducting solid oxide electrolyte and Pt and Au electrodes. At 300–430 °C upon electrochemically pumping the Na^+ to the Pt electrode a strong NEMCA effect was reported.

Hibino and coworkers [2.336, 2.337, 2.338, 2.339, 2.340, 2.341, 2.342, 2.343, 2.344] also studied the reaction between NO and CH_4 both in a single chamber, and in a dual-chamber EMR. In a Pd/YSZ/Pd dual-chamber EMR the NO was reduced to N_2 at the cathode, and CH_4 was oxidized to CO and CO_2 both at the anode and cathode. In a single-chamber $Pd/Ce_{0.8}Sm_{0.2}O_{1.9}/Pd$ EMR the reaction was carried out successfully at temperatures as low as 400 °C. The same group [2.345] has recently studied the Pt-catalyzed reduction of nitric oxide by ethane in the temperature range between 600 and 700 °C. Their dual-chamber EMR can be represented as (NO, C_2H_6, O_2), Pt or (Pt + M_xOy)\YSZ\Pt, air. When the reactant gas stream entering the anodic compartment contains no oxygen, almost all of the oxygen species transported to the Pt catalyst through the YSZ membrane is consumed in the reaction with NO and ethane; under open-circuit conditions the same reaction could not proceed when gaseous oxygen was supplied in an amount equimolar to the transported oxygen species. At O_2 concentrations of 2–12% in the reactant gas stream for the anodic compartment, the oxygen species transported to the Pt catalyst showed a promotional effect on the reduction of NO by ethane; this promotional effect is further enhanced by the addition of 15 wt % Mn_2O_3 or 15 wt % MoO_3 to the Pt catalyst. At potentials of 2 V the reaction rates are about four times higher than those measured under open-circuit conditions; the reaction over the MoO_3-containing Pt catalyst is accelerated by the application of negative potentials, which was opposite to the behavior observed with Pt or the (Pt + Mn_2O_3).

The catalytic reaction of NO_x with various hydrocarbons in the presence of oxygen has also been studied in a membrane reactor by Xue *et al.* [2.346]. The two different membrane reactor configurations proposed by these authors are shown in Figure 2.17. The reaction of NO_x with various hydrocarbons seems to be facilitated by the presence of NO_2 in the feed. One approach that may be tried is the use of a two-reactor configuration. In the first reactor NO is catalytically oxidized to NO_2 using the oxygen in the exhaust, while in the second reactor the effluent from the first reactor is mixed with the hydrocarbon to be reduced to N_2 and water. Xue *et al.* [2.346] have proposed, instead, the use of a single membrane reactor to accomplish the same task carried out by the proposed two-reactor configuration. In the first reactor configuration the NO oxidation and reduction catalysts are put into different compartments in the membrane reactor. In the second reactor configuration either both catalysts are put in the same compartment B, or one of them is impregnated on the membrane. The membrane reactor system offers a more compact design than the two-reactor system.

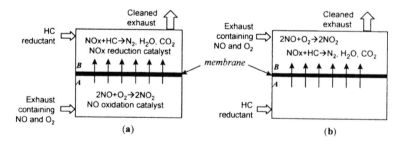

Figure 2.17. Different membrane reactors for NO reduction. From Xue *et al.* [2.346].

The coupling of membranes with a chemical reaction has also been applied to a number of other reactions of relevance to environmental applications. For these applications, discussed below, the membrane simply plays the role of providing the reactive interface between reactants flowing on either membrane side (as is the case with some of the two-phase hydrogenation reactions previously discussed). For these reactions the goal is to attain full conversion of the pollutant, and to avoid its slippage out of the reactor. The concept was first demonstrated by Zaspalis *et al.* [2.347] for the selective catalytic reduction (SCR) of nitric oxide with ammonia to nitrogen and water. The role of the membrane in this case is to create a localized reaction interface, in which the nitric oxide completely reacts with ammonia. More recently, Saracco *et al.* [2.7] extended the concept by depositing a catalyst in the pores of a fly-ash filter to combine NO SCR by NH_3 with particulate removal. NO reduction by H_2 under an excess O_2 concentration was studied in a membrane reactor using a Ni impregnated porous Vycor® glass membrane by Yamamoto *et al.* [2.348]. The NO and O_2 gas mixture was fed in the shellside, while the H_2 containing mixture was fed in the tubeside. Maximum conversion of NO to N_2 was 45 %. As one of advantages of using a membrane reactor Yamamoto *et al.* [2.348] report the potential of longer catalyst lifetime.

Sloot *et al.* [2.349, 2.350] studied the Claus reaction between hydrogen sulfide and sulfur dioxide in a CMR configuration using a catalytically active, but non-permselective membrane. They fed the hydrogen sulfide and sulfur dioxide on the opposite sides of the membrane, and the reaction took place within the membrane itself. They demonstrated that if the reaction rate is sufficiently fast, a sharp reaction front is formed within the membrane, thus, preventing slippage from either side. One then does not need to use a permselective membrane. They concluded, further, that this kind of system is very attractive, because by properly adjusting the operating conditions, undesired side reactions can be avoided. The same principle has been demonstrated with experiments and model calculations with CO [2.351, 2.352] and H_2S oxidation [2.353]. Neomagus *et al.* [2.353] studied the oxidation of H_2S in a membrane reactor utilizing a sintered stainless steel membrane; this type of membrane is easier to integrate into a membrane reactor. In addition, it exhibits catalytic properties towards H_2S oxidation.

The same group [2.354] has also recently reported on the performance of a membrane reactor with separate feed of reactants for the catalytic combustion of methane. In this membrane reactor methane and air streams are fed at opposite sides of a Pt/γ-Al_2O_3-activated porous membrane, which also acts as catalyst for their reaction. In their study Neomagus *et al.* [2.354] assessed the effect of a number of operating parameters (temperature, methane feed concentration, pressure difference applied over the membrane, type and amount of catalyst, time of operation) on the attainable conversion and possible slip of unconverted methane to the air-feed side. The maximum specific heat power load, which could be attained with the most active membrane, in the absence of methane slip, was approximately 15 kW m^2 with virtually no NO_x emissions. These authors report that this performance will likely be exceeded with a properly designed membrane, tailored for the purpose of energy production.

Saracco *et al.* [2.355] and Saracco and Specchia [2.356] used a very similar membrane reactor with separate reactant feeds to study the catalytic combustion of propane. In their early study they investigated in a pilot-plant the importance of the amount of catalyst (1% Pt on γ-Al$_2$O$_3$), deposited on the pore walls of the membrane (an α-Al$_2$O$_3$ porous tube with a pore diameter of 0.7 μm, length of 100 mm, I.D. of 1.4 mm, and thickness of 3 mm). The experiments involved feeding air on one membrane side, and a propane/N$_2$ mixture on the other side. The kinetic- and transport-limited operating regimes were both investigated; in the transport-limited region studies were carried out both in the presence and in the absence of trans-membrane pressure gradients. The studies indicate that the catalyst loading on the membrane affects significantly the reactor performance. The higher the amount of catalyst deposited in the membrane pores, the higher is the conversion attainable in the kinetically controlled regime (low temperature operation), and the easier it is to reach the transport-controlling condition at higher temperatures. However, this last feature is achieved at the price of relatively low overall propane conversions due to the lowering of membrane permeability as a result of the catalyst deposition in the pores. Saracco *et al.* [2.355] conclude that, provided the propane feed concentration and the eventual trans-membrane pressure gradient are not too high, a limited amount of catalyst should be employed, thus, allowing higher membrane permeability and propane conversions, with minor slip of reactants through the membrane. In their most recent study with this reaction Saracco and Specchia [2.356] developed a model to interpret the experimental data. The model takes into account the mass and heat transfer inside and outside the membrane (see further discussion in Chapter 5). The external heat transfer towards the cooling surfaces of the reactor, and the uneven catalyst distribution over the membrane were identified as the most important features governing the behaviour of the reactor.

A different concept was recently suggested by Pina *et al.* [2.357, 2.358] and Irusta *et al.* [2.359]. They have studied the use of CMR for the destruction of volatile organic compounds (VOC) in contaminated air streams. The idea is to create a more effective contact of the VOC with the catalyst by forcing, in a flow-through mode, the VOC containing air stream through the membrane. The membrane was prepared by *in situ* crystallization of different La-based perovskites inside a porous alpha-alumina matrix, with perovskite loads of 2 wt% and higher. For VOC (toluene and methyl ethyl ketone) containing air streams, at concentrations between 875 and 3450 ppmv and space velocities of up to 27,200 h^{-1}, the membrane reactors exhibited total VOC combustion at moderate temperatures. Sarraco and Specchia [2.360] used catalytic ceramic membranes (filters) prepared by depositing a γ-Al$_2$O$_3$ layer on the pore walls of α-Al$_2$O$_3$ filters by *in situ* precipitation, followed by Pt dispersion by wet impregnation. They were tested, in the catalytic combustion of selected volatile organic compounds (naphthalene, propylene, propane and methane), fed alone or in a mixture. The VOC conversion results, obtained by varying the operating temperature, the superficial velocity and the catalyst loading in the filter, showed that catalytic filters might outperform conventional technologies in the treatment of flue gases, where simultaneous dust removal and catalytic abatement of VOC have to be accomplished.

Jacoby *et al.* [2.361] have applied the same idea to the photocatalytic oxidation of carbonyl compounds and bioaerosols using a porous film of TiO_2 (acting both as an exclusion filter and a photocatalytic layer) cast on an underlying support. Tsuru *et al.* [2.362] also combined photocatalytic reaction under conditions of blacklight irradiation, and filtration through a porous TiO2 membrane in a laboratory study for the destruction of of a model pollutant, i.e., trichloroethylene (TCE) and a model foulant, i.e., polyethyleneimine (PEI). In addition to enhanced conversion, the combined system showed less sensitivity to membrane fouling by PEI.

The use of a photocatalytic MR for the destruction of organics dissolved in water has also been reported by Makhmotov *et al.* [2.363, 2.364, 2.365]. They have investigated three different process configurations: (i) Filtration of a solution through a TiO_2 semiconductor membrane, whose surface was simultaneously UV-irradiated; (ii) premixing the water with an anatase suspension, followed by its filtration through an inert membrane under UV irradiation, and (iii) an alternative of technique (ii) above involving *in situ* TiO_2 sol formation by the addition of a specially prepared titanium alkoxide solution, followed by membrane filtration of the sol solution with continuous or periodic illumination of the membrane surface. Makhmotov *et al.* [2.365] recently reported testing the third concept above successfully in a laboratory scale MR with a model wastewater containing 2,3,7,8,-tetrachlorodibenzo-*p*-dioxin.

Coupling a photocatalytic reactor to a membrane unit, which allows the purified water to permeate through, while retaining the catalyst for recycle back into the reactor has also been proposed by Drioli and coworkers (Molinari *et al.* [2.366, 2.367, 2.368], Artale *et al.* [2.369, 2.370]). In their studies they have investigated a model pollutant 4-nitrophenol (4-NP) using TiO_2 as the photocatalyst. A variety of polymeric membranes were tested. Some of these, in addition to retaining the catalyst, showed the ability to also adsorb the pollutant. This provides an additional advantage for the proposed process, since the presence of the membrane, itself, could provide a buffering action towards spikes in the concentration of the pollutants. Drioli and coworkers [2.366, 2.367, 2.368, 2.369] report that the concentration of 4-NP in the permeate stream were as low as 6–7 % of the initial 4-NP concentration. A similar technical concept was studied by Sopajaree *et al.* [2.371, 2.372]. In their study they combined a batch-recirculated photoreactor with a hollow-fiber membrane ultrafiltration unit. Methylene Blue (MB) and TiO_2 (Degussa, P-25) were used as the model substrate and photocatalyst respectively. These authors studied also the influence on the photocatalytic reaction rate of the MB concentration, recirculation flow rate, TiO_2 dose, and solution volume and pH in the batch reactor over ten repeat cycles of operation of the integrated photoreactor-UF assembly. The photocatalyst separation was complete, as determined by turbidity measurements. However, a continuous degradation in the photocatalyst performance was noted with each repeat cycle, which Sopajaree *et al.* [2.371, 2.372] attributed to the agglomeration of the TiO_2 particles as a result of the UF process. They have proposed a number of possible solutions to this problem.

The CMR concept involving the forced flow of the reactant stream through the membrane, was also applied by Binkerd *et al.* [2.373] for the oxidative coupling of methane, and by Lange *et al.* [2.374] and Lambert and Gonzalez [2.375] for selective hydrogenations. For these reactions the benefits of utilizing the CMR result from diminished side reactions between the desired intermediate products and the reactants (H_2, O_2), due to the short contact times. Lambert and Gonzalez [2.375], for example, used a Pd/γ-Al_2O_3 membrane with an average pore diameter of 3.6 nm supported on a macroporous α-Al_2O_3 tube, prepared by sol-gel processing, in order to study the partial hydrogenation of acetylene and 1,3-butadiene. The highest selectivity to the partially hydrogenated products occurred, when the reactant was premixed with H_2 and passed through the membrane wall.

Steam reforming of methane is one of the most interesting reactions, which catalytic membrane reactors have been applied to. Uemiya *et al.* [2.376] studied this reaction using a membrane prepared by coating a thin Pd film on a Vycor® glass tube utilizing an electroless plating technique. They observed significant enhancements in methane conversion over the calculated equilibrium values. Interestingly, the effect of reactor pressure on the membrane reactor conversion is the opposite of what would be expected for the same reaction being carried-out in a conventional reactor. According to Kikuchi [2.377], this behavior is indicative of the fact that in the membrane reactor the conversion is limited by the hydrogen diffusion through the metallic membrane. Kikuchi [2.378] recently reported that these composite membranes are being evaluated by Tokyo Gas and Mitsubishi Heavy Industries in membrane reformers in polymer electrolyte fuel cell systems. Kikuchi [2.379] have studied the methane steam reforming in a membrane reactor at 773 K using two types of supported metal membranes. These are palladium membranes made by electroless-plating, and palladium, platinum, and ruthenium membranes prepared using CVD. A CVD platinum membrane with a H_2/N_2 separation factor as high as 280 had performance comparable to the palladium membrane prepared by electroless-plating. The advantages of the CVD platinum membrane versus the electroless-plating palladium membrane are reported to be lower metal loading and cost, as well as the lower tendency toward hydrogen embrittlement.

Adris *et al.* [2.380, 2.381], Adris and Grace [2.382], and Mleczko *et al.* [2.383] have studied the same reaction in a FBMR. Adris and coworkers have studied the reaction in a pilot-plant scale FBMR system, shown in Figure 2.18, which combines a fluidized catalytic bed (using a Ni/Al_2O_3 catalyst) with a bundle of twelve tubular Pd membranes. Reactor performance was shown to be sensitive to the way the membrane area is distributed between the dense and dilute fluidized-bed phases. Methane and steam conversion enhancements were observed, as high as 1.3 times that of the same reactor without the membranes being present. Despite the high installed membrane permeation capacity it was observed that the low permeation of palladium membranes still limited the conversion enhancement. The same group has (Grace *et al.* [2.384]) recently studied the feasibility of operating the FBMR under autothermal conditions by adding some oxygen into the feed, in order to provide some of the required heat of the endothermic reforming reaction. They have utilized a simple equilibrium model (which reportedly is in agreement with their experimental data in the FBMR), for which the product distribution in the non-permeate stream is set equal to the shifted chemical equilibrium composition after allowance is made for the removal of hydrogen through the membrane. The model results indicate that the FBMR should be able to operate autothermally, and free of coke formation over a considerable range of temperatures, pressures, and steam-to-methane ratio. Sream reforming of methane has also been studied in a PBMR configuration utilizing mesoporous membranes, but with somewhat more modest gains in conversion by Minet *et al.* [2.385]. Inagaki *et al.* [2.386] have proposed coupling the steam reformer to a membrane module using polyimide membrane, which separates the unreacted methane for direct recycling. They investigated the gas separation characteristics of the membrane separator, and experimentally tested the combined system. The results showed that the recycling system improves methane conversion by 60 % over the conventional system without recycling.

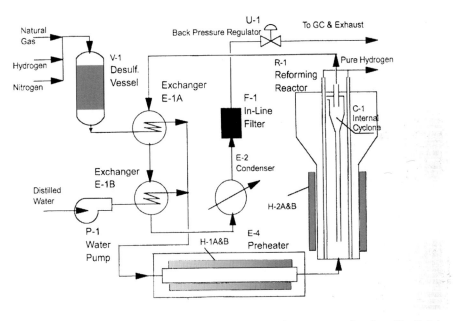

Figure 2.18. Schematic representation of the pilot-plant for methane steam reforming. The R-1 is the reforming reactor, which contains a bundle of tweve palladium tubular membranes and a fluidized-bed of catalysts. From Adris *et al.*[2.381], with permission from Elsevier Science.

The dry (i.e. using CO_2 rather than steam) reforming of methane is also recently attracting attention. Hibino *et al.* [2.337] studied this reaction in an electrochemical reactor utilizing a solid oxide electrolyte, which is conductive to H^+ ions. Hydrogen pumping accelerated the rate of the reforming reaction. Tagawa *et al.* [2.387] proposed using partial oxidation together with carbon dioxide reforming of methane in a molten carbonate fuel cell (MCFC) system operating as a membrane reactor. This provides a means to utilize CO_2 to co-generate electric power together with syn-gas. They found that the selection of the anode catalyst is important for optimizing the reforming reaction. They tested several transition, rare earth, and noble metals and chose Ni as the anode catalyst. The reaction has also been studied in a PBMR configuration using dense Pd membranes [2.240, 2.388, 2.389], dense SiO_2 [2.390, 2.391], and modified mesoporous Al_2O_3 membranes [2.392].

Despite the significant attention that methane reforming has received in the membrane reactor area, questions still remain about the potential effect that hydrogen removal through the membrane may have on accelerated catalyst deactivation due to coking and poisoning by H_2S [2.393, 2.394]. Hou *et al.* [2.393], for example, have carried out simulations of the steam reforming of methane in a membrane reactor for pressures up to 6 bar, and temperatures in the range of 773 to 873 K, with emphasis on the effect of hydrogen removal on poisoning due to traces of H_2S and carbon deposition. They compared membrane reactor performance with that of fixed-bed reactor, and concluded that removal of hydrogen increases the tendency to poisoning from H_2S and from carbon deposition. These effects can, however, be reduced by operating at higher temperatures, where H_2S poisoning is less dominant, or by increasing pressure, in order to lessen the extent of carbon deposition. The overall economic benefits of methane reforming MR, furthermore, still remain rather debatable, at least in the context of large-scale applications [2.395, 2.396, 2.392], see further discussion in Chapter 6.

As was described in Section 2.3, Sammels *et al.* [2.310] have recently reported the use of solid oxide membranes (brownmillerite) for gasifying coal. In the same study they have reported the use of such membranes in a catalytic MR for the autothermal reforming of logistic fuels (JP-8 and DF-2), in order to deliver a gaseous feedstock compatible for subsequent use in a SOFC. This is an important application, which makes it feasible for directly using such fuels in a SOFC.

Methanol reforming for the production of H_2 using a PBMR and Pd-type membranes has also been investigated recently by a number of groups [2.397, 2.398] in the context of mobile fuel cell applications. A double-jacketed membrane reactor, in order to carry out the methanol steam reforming reaction for the production of hydrogen, has been proposed by Lin and Rei [2.399, 2.400]. The reactor consists of a concentric module, which included a supported Pd membrane at its center, and two separate stainless steel tubes separately assembled as the two outer jackets. A Cu-based catalyst was used for the methanol reforming reaction, and it was placed in the annulus between the membrane and the inner stainless steel tube. A Pd/Al_2O_3 catalytic oxidation catalyst was placed in the annulus created by the two stainless steel tubes to oxidize the retentate mixture exiting the membrane reactor, thus, generating heat, which is used *in situ* to drive the endothermic reforming reaction. A mixture of methanol and water was fed to the MR at 350 °C and in the pressure range of 6–15 atm. The hydrogen produced by the reforming reaction is separated by the membrane, and removed through its interior volume. Electrical heating is used to provide energy that may be needed, in addition to that resulting from burning the reject stream. The amount of electrical heating required depends on the amount of hydrogen recovered through the membrane. An autothermal operation is attained for hydrogen recoveries in the range of 70–80 %. A key operating parameter is the load (in terms of mole CH_3OH/h) to membrane surface area ratio. As expected, with increasing load to surface area ratios the hydrogen flux through the membrane increases, but the hydrogen recovery rate decreases. In contrast, both the flux and the recovery rate increase with membrane reactor operating pressure. At 350 °C and reactor pressures of above 12 atm, a recovery yield of

97 % and a hydrogen flux of 3.7 m^3/m^2.h are attained for a load to surface ratio of 50 mole/m^2.h, which is much higher than the conventional yield of 75 %.

Itoh *et al.* [2.401] have also recently reported on the use of a Pd membrane reactor in a PEM fuel cell utilizing liquid methanol as a fuel. They have experimentally evaluated the transport characteristics of a Pd alloy membrane (Pd$_{0.91}$Ru$_6$In$_3$). They have also utilized a computer code to evaluate the performance of the overall system (Membrane Reactor + PEM Fuel Cell). Using methanol as a liquid fuel to deliver pure hydrogen, requires that the reactor is operated at an elevated pressure. Some of the produced hydrogen remains on the feedside and is utilized as a fuel to provide the energy for endothermic methanol re-forming reaction. Running the reactor at higher pressures increases the reactor yield and the hydrogen recovery, but, on the other hand, reduces the amount of hydrogen on the feedside, which is available as a fuel for the the reaction. Itoh *et al.* report, as a result, that an optimum operating pressure exists. Frustreri *et al.* [2.402] have reported the use of a Ru impregnated zeolite membrane in a MR for methanol steam reforming. A Cu-based methanol reforming catalyst was utilized and packed in the annular space between the membrane and the reactor external cell. The presence of the Ru-zeolite membrane helped reduce the CO concentration down to 50 ppm. Since membrane stability issues dictate operating at temperatures > 300 °C [2.403], for methanol reforming there is no beneficial effect due to equilibrium conversion displacement. Suggested potential benefits, as previously noted, include reduced reactor volumes, and a more compact unit design.

Direct H$_2$O decomposition in order to produce hydrogen has been investigated by a number of groups. The earlier investigations were by Baumard and coworkers [2.404, 2.405], who studied the thermal H$_2$O decomposition in a MR using a calcia-stabilized zirconia membrane at high temperatures (1400–1800 °C). Naito and Arashi [2.406] and Nigara *et al.* [2.407] studied the same reaction using a mixed conducting electrolyte, in order to attain higher oxygen fluxes through the membrane and, therefore, also higher hydrogen production rates. Operation of an EMR for hydrogen production from steam utilizing a CaTi$_{0.8}$Fe$_{0.2}$O$_y$ membrane having mixed oxygen ion and electronic conductivity has been reported by Demin *et al.* [2.408]. They also presented a model to obtain the dependence of the EMR productivity on the anode and cathode and membrane properties. Guan *et al.* [2.409] used a mixed protonic-electronic conductor (doped Ba-cerate) as the membrane to study the same reaction at lower temperatures (500–800 °C). Protonic conductors were also used to study the same reaction in a dual-chamber EMR by Iwahara and coworkers [2.410, 2.411, 2.412] at 700–900 °C. The process of hydrogen production by solar thermal water splitting (HSTWS) has been studied at the Weizmann Institute of Science [2.413, 2.414, 2.415]. In the process under development water vapor is partially dissociated in a solar reactor at temperatures approaching 2500 K, and the product hydrogen is separated from the hot mixture by a porous ceramic membrane. The reactor thermodynamic efficiency in this process rises with increasing reaction temperature, and with decreasing pressure at the downstream side of the gas-separating membrane. Kogan and coworkers utilized zirconia porous membranes, but these membranes, when exposed to the high temperatures in the solar reactor, lost their gas permeability due to pore closure by sintering. Efforts to inhibit the membrane sintering process by the use of special stabilized zirconia powders, consisting of particles with a rounded shape, proved unsuccessful. The HSTWS process, though potentially of significance, faces a number of technical challenges including the need (a) to achieve very high solar hydrogen reactor temperatures by secondary concentration of solar energy, (b) to overcome the materials problems encountered in the manufacture of the solar reactor, and (c) to develop special porous ceramic membranes that resist clogging by sintering at the very high temperatures of reactor operation.

There have also been a number of membrane reactor applications to the Sabatier reaction, i.e. the catalytic hydrogenation of CO_2 to produce CH_4 and H_2O, of current interest in NASA's human exploration program to planet Mars for CO_2 capture and beneficiation for the CO_2 rich Mars atmosphere. One of the earliest studies is by Gryaznov and coworkers [2.416, 2.417]. They utilized a thin Pd-Ru foil as a hydrogen permselective membrane. They fed CO_2 on one side of the membrane, and hydrogen on the other side. On the CO_2 membrane side they electrolytically deposited a porous film of Ni to act as the catalyst for the methanation reaction. In a more recent study Ohya *et al.* [2.418] used a SiO_2 mesoporous membrane to carry out the reaction at somewhat elevated pressures (0.2 Mpa). Higher pressures favor the equilibrium of this reaction. More importantly, for the reactive separation application the authors envision the SiO_2 membranes become water permselective under these conditions. In the experiments, H_2 and CO_2 mixtures (H_2/CO_2 feed ratio in the range of 1–5) are fed on the one side of the membrane over a commercial 0.5 % Ru on alumina catalyst. The water produced during the reaction permeates selectively through the membrane. Removing the product from the reaction zone increases the reactor conversion. In the range of the space velocities investigated (0.03–0.123 s^{-1}) and temperatures (480–719 K), a maximum 18 % increase in conversion over the reactor conversion attained in the absence of the membrane was observed.

The direct thermal CO_2 decomposition reaction to CO and O_2 has also been investigated in the context of NASA's human exploration program to planet Mars. The current approach involves the use of a Pt/YSZ/Pt EMR [2.419] utilizing electric energy to produce O_2 from CO_2 at 700–1000 °C. Studies utilizing mixed ionic conductors, for which no need exist for the use of electrodes and electric energy have also appeared [2.420, 2.421]. Another interesting application of membrane reactor technology to CO_2 capture was recently reported by Nishiguchi *et al.* [2.422]. In this application, CO_2 was catalytically hydrogenated to produce methane and water. The methane, thus, produced was then fed into a membrane reactor utilizing a Pd membrane, operating at 500 °C, to be converted over a Ni/SiO_2 catalyst into graphitic carbon and hydrogen. In this two-stage reactor system over 70 % of the CO_2 was, thus, converted into graphitic carbon. CO_2 hydrogenation to produce synthesis gas was studied by Uemura *et al.* [2.423]. In their reactor the membrane was made of a $LaNi_5$ alloy, typically utilized in hydrogen storage devices. The hydrogen was provided by the cyclohexane dehydrogenation reaction coupled to CO_2 hydrogenation through the membrane, which also acted as catalyst for both reactions.

CO_2 hydrogenation to produce methanol in a membrane reactor at low temperatures (<200 °C) has been studied by Struis *et al.* [2.424]. Their study utilized Li-exchanged perfluorinated hydrocarbon membranes, which selectively permeated both products of the reaction (H_2O and CH_3OH). Polymeric membranes (PDMS, polyamide) impregnated with catalytically active components (active carbons, USY and β-zeolites) were also used by Vital *et al.* [2.425, 2.426] to study the catalytic hydration of *a*-pinene into monoterpenes, alcohols and hydrocarbons at 50 °C, using aqueous acetone as solvent. The use of the membranes appears to have a favorable influence on selectivity. By controlling the reaction variables it is possible to make it selective to the terpenic alcohols, namely α-terpineol. A simple diffusion-kinetic model was also developed which fitted the experimental concentration data well

Van de Graaf *et al.* [2.427, 2.428] reported the use of a silicalite-1 membrane in a reactor to study the metathesis reaction of propene into ethylene and 2-butene, and the isomerization of *cis*-2-butene into *trans*-2-butene. Both reactions, interestingly enough, take place at room temperature. The membrane at these conditions shows a modestly preferential permeation of *trans*-2-butene over propene (a selectivity of ~5) and *cis*-2-butene (a selectivity of ~1.6). Its use in the reactor results in noticeable improvements in conversion over the calculated equilibrium values. Under optimal operating conditions the propene conversion was 13 % higher than the thermodynamic equilibrium conversion, and a 34 % increase in the ratio, in which *trans*-2-butene and *cis*-2-butene are formed, was observed. The *trans*-2-butene yield from the geometrical isomerization of *cis*-2-butene increased ~4 %,when compared to the thermodynamic equilibrium.

Hydrogen production will certainly play a key role in the coal-based power plant of the future. All Integrated Gasification Combined Cycle (IGCC) plants are envisioned to produce hydrogen, either as a product for power production in a turbine or a fuel cell, or as a reactant to produce fuels and chemicals. Hydrogen utilizing fuel cells, in particular, are significantly more efficient and produce lower SO_x and NO_x emissions than conventional combustion technology for power generation. For the coal-based power plant co-production concept to be successful, it requires the development of better hydrogen separation technologies. Such technologies must have significantly improved efficiencies to impact on the plant economics, and must be robust to process conditions, which involve high temperatures and pressures, and the handling of complex mixtures, containing steam and reactive compounds like SO_2 and NH_3. High temperature membrane-based technologies show the greatest promise for non-incremental technology leaps in this area. Applying membrane-based reactive separation technology in the context of the IGCC concept has been discussed by a number of investigators. Pruschek *et al.* [2.429], for example, have investigated the concept of CO_2 removal from coal-fired power stations. Their study indicates that net efficiencies of 40 % are feasible. About 88 % of the total CO_2 produced can be recovered using the IGCC concept employing a water gas shift (WGS) reaction step, and CO_2 separation either by physical absorption or, potentially, H_2/CO_2 separation with membranes.

The concept of combining the WGS reaction and the hydrogen separation step in a single unit has been investigated by a consortium of European academic and industrial entities [2.430]. They investigated the concept of the WGS membrane reactor (WGS-MR) for CO_2 removal in the IGCC system. In order to establish full insight into the possibilities of the application of such a reactor, a multidisciplinary feasibility study was carried out, comprising system integration studies, catalyst research, membrane research, membrane reactor modelling, and bench-scale membrane reactor experiments [2.431]. The study concluded that the application of the WGS-MR concept in IGCC systems is an attractive future option for CO_2 removal as compared to conventional options. The net efficiency of the IGCC process with integrated WGS-MR is 42.8 % (LHV) with CO_2 recovery (80 % based on coal input). This figure has to be compared with 46.7 % (LHV) of an IGCC without CO_2 recovery and based on the same components, and with 40.5 % (LHV) of an IGCC with conventional CO_2 removal. Moreover, an economic analysis indicates favorable investment and operational costs. The study concluded that development of the process is considered to be, in principle, technically feasible. However, the authors report that the currently available high temperature selective gas separation membranes (noble metal and SiO_2) are not capable of withstanding the harsh environments found in the WGS reaction step and, therefore, further development in this area remains essential.

Though the idea by the European consortium to apply the WGS-MR concept to IGCC was novel, the WGS-MR, concept, itself was not. The WGS reaction, $CO+H_2O \leftrightarrow CO_2+H_2$, ($\Delta H^0$ = -41.1 kJ/mol), is among the oldest catalytic reactions used to produce hydrogen for ammonia synthesis, and numerous other applications. The reaction is exothermic, so lower temperatures favor higher conversions, but kinetic considerations favor higher temperatures. In practice water gas shift reactor systems consist of two reactors. A first reactor, which is operated at high temperatures (623–673 K) using a high temperature shift catalyst (HTS), typically an Fe-Cr-based catalyst with high thermal stability; and a second reactor, which is operating at low temperature (ranging from 473 to 523 K), using an active Cu-Zn-based low temperature shift (LTS) catalyst. To attain good conversions, typically H_2O/CO ratios significantly higher than 1 are utilized. The use of a membrane reactor (WGS-MR) provides the opportunity to attain high conversions in a single stage, with diminished needs for steam (see further discussion), and to simultaneously provide high purity hydrogen. The first to study the WGS-MR concept was the group of Kikuchi and co-workers [2.432, 2.433]. They studied the water gas shift reaction at 673 K in a membrane reactor equipped with a Pd membrane deposited on a porous glass tube. They used a commercial Fe-Cr catalyst (Girdler G-3). Ar was used as the inert sweep gas and the reactor provided levels of conversion higher than equilibrium. Also, the amount of steam required to achieve reasonable levels of conversion was reduced. Esaka *et al.* [2.434] studied the same reaction in an EMR utilizing a mixed (O^{2-} and e^-) conducting $CaTi_{1-x}Fe_xO_{3-a}$ membrane. Steam was passed on one side of the membrane with CO in the other. Hydrogen and O^{2-} formed on the steam side of the membrane; the oxygen ions migrated to the other side, where they reacted with CO to form CO_2. The advantage of this reactor configuration is in that the hydrogen produced is separated from the CO_2 product. The disadvantage is in the need to maintain high temperatures in the reactor (800–1000 °C).

Another emerging application of the membrane assisted WGS concept (WGS-MR) is a process to recover tritium from tritiated water from breeder-blanket fluids in fusion reactor systems. A conceptual process model to accomplish this has been proposed by Violante *et al.* [2.435]. It uses two membrane reactor units. The first membrane reactor unit removes the hydrogen isotopes from the purge gas (He) via oxidation. The second unit uses the tritiated water to recover tritium using the WGS reaction. In a companion study, Basile *et al.* [2.436] studied the WGS reaction in a membrane reactor using a composite Pd membrane created by a film of Pd deposited on the interior of a commercial alumina membrane by the so-called co-condensation technique. Their membrane was imperfect with an ideal H_2/N_2 separation factor of 8.2 at 595 K and 11.2 at 625 K. A Halder-Topsoe commercial catalyst was utilized. They determined the optimal conditions for reactor performance as a function of H_2O/CO ratio, temperature, reactor pressure, and gas feed flow rate, with and without N_2 sweep gas. They observed that for the conventional reactor the maximum conversion attained increases with temperature. For the membrane reactor, on the other hand, an optimum temperature exists. Reaction studies at three temperatures (595, 615 and 633 K) show that reactor conversion increases with W/F. In a subsequent study, Tosti *et al.* [2.437] developed a model for the overall process, which accounts for the isotopic competition effect in the permeation process. The model, supported by a computer code, has been used to study the hydrogen isotopes extraction from the gas stream arising from a tritiated water gas shift reactor under conditions relevant to a solid breeding blanket of ITER (International Thermonuclear Experimental Reactor) size. Tosti *et al.* [2.438] have also tested in a WGS MR-based, pilot-plant blanket tritium recovery system Pd and Pd-Ag membranes prepared by coating porous ceramic tubes with a thin metal layer using different techniques (electroless deposition, ion sputtering, and rolling of thin metal sheets). The reported conversion was ~100 %, above the corresponding equilibrium of 80 % at 350 °C. The same concept was tested experimentally by Birdsell and Willms [2.439] in a 2-stage PBMR using a tubular Pd membrane. Tritium was recovered from a molecular sieve container loaded with 2050 g of water and 4.5 g of tritium. The maximum water processing rate for the PBMR was determined to be 0.5 slpm. A method was developed to automatically liberate steam from the molecular sieve and add the amount of CO needed for the WGS reaction before injection into the PBMR. The maximum decontamination factor (DF) achieved in the first stage ranged from 100 to 260, depending on the inlet flow rate. Performance of the second stage could not be measured because the outlet tritium concentration was below the background of the ion chamber used for analysis. Although the DF could not be measured, it is presumed that it was high, because no tritium was detected, except during start-up, in the tritium waste treatment system that was downstream from the PBMR.

An interesting potential application of membrane reactors involves their use in the Fischer-Tropsch (FT) reaction for the production of higher hydrocarbons from synthesis gas. The FT reaction represented as follows

$$CO + (1+x) H_2 \leftrightarrow CH_{2x} + H_2O \tag{2.8}$$

produces, in addition to hydrocarbons, water, which is thought to deactivate the FT catalysts. Simultaneous removal of water is potentially a way for providing a more favorable kinetic regime, and for improving the per pass conversion. Hydrophilic zeolite membranes may provide the means for *in situ* removal of water under reactive conditions. Recently Espinoza *et al.* [2.440] presented results of a transport investigation of mordenite and ZSM5 zeolite membranes under a simulated FT environment. The long-term future of this

application will depend on the robustness of the membrane under the real FT environment, and, of course, on the comparison of economics of the membrane process, with the
simpler alternative process involving condensation of water out of the product stream and
recycle.

The CO and CO_2 hydrogenation to produce CH_4 had been studied earlier on by Gür and
Huggins [2.245, 2.441, 2.442] and by Gür *et al.* [2.443] in an EMR. They utilized a
Ni/YSZ/Pt or a Pt/YSZ/Pt EMR, and their idea was to electrochemically pump O^{2-} away
from the electrode in contact with the CO/CO_2-H_2 mixture. It was suggested that the CO
hydrogenation to produce CH_4 proceeds by abstraction of oxygen from the adsorbed CO
to produce surface carbon, which is then hydrogenated to form methane. Electrochemically pumping the oxygen away from the electrode surface tends to increase the rate by a
couple of orders of magnitude over the open circuit case. For the CO_2 hydrogenation the
first step is the subtraction of oxygen from CO_2 to form adsorbed CO, which is then hydrogenated following the mechanism previously outlined.

Membrane reactors for removing dissolved oxygen (DO) from water have been studied
at the laboratory scale by Li and coworkers [2.444, 2.445, 2.446], and their performance
was compared to conventional ion-exchange columns. The membrane reactors have a
shell-and-tube type construction with polypropylene hollow-fiber membranes or a tubular
membrane acting as the tubes. A catalyst (Pd doped anion-exchange resin based on styrene and divinyl benzene) was packed in the void space of the reactor made from the hollow-fiber membranes, while for the tubular membrane reactor, the catalyst was packed
within the membrane tube. Hydrogen gas was employed as a reducing agent. Water-
containing saturated DO was fed into the reaction compartment containing the catalyst
particles, while hydrogen was introduced countercurrently into the gas compartment of the
reactor, so that the normal physical and chemical processes for the DO removal take place
simultaneously. The ion-exchange column was fabricated from glass and packed with an
anionic resin (Purolite A-310) regenerated in the sulfite form. The performance of the two
methods was analyzed; it was found that the membrane reactor, which reduced the dissolved
oxygen in water by both physical stripping and chemical reduction, gave the best results in
terms of the outlet dissolved oxygen level. In their most recent study Li and Tan [2.446]
used a system which combines the membrane reactor with UV irradiation. The reactor consisted of a UV light surrounded by silicone rubber hollow-fiber membranes. Water saturated
with oxygen was fed into the lumen of the hollow fibers, while hydrogen was introduced
into the reactor shellside. The membranes served as gas distributors for the hydrogen to react
with the dissolved oxygen under the UV irradiation. Experimental data obtained at various
operating conditions were compared with the results of a theoretical model.

A interesting new process for the removal of metallic (Cu(II)) compounds in the treatment of industrial waste waters has been proposed by Flores and Cabassud [2.447, 2.448].
It consists of a membrane reactor with a moving-bed of porous ion-exchange resin particles. These authors have carried out laboratory experiments with this novel membrane reactor process. Prior to the initiation of the experiments a number of cationic resins were
screened, and IRP 69, a resin whith a small particle diameter and high exchange rate, was
chosen for the study. For these resin particles, the influence of stirring velocity and tem-

perature on the apparent reaction rate was analyzed in a stirred batch reactor, and the rate constants were independently determined. A separate particle filtration study defined the optimal operating conditions (circulation velocity and pressure) needed for one to determine a compromise between power consumption, and the number of modules to be set up. The behavior of the membrane reactor system was compared with that of the conventional packed-bed reactor. For the same flow rate and a conversion of 95 %, the membrane reactor running under optimum conditions provides reductions by factors of 2.9 for the volume of adsorbing particles, 26 for the pressure loss, and 4.75 for the power supplied to the system. Based on these results, the process shows significant promise; its performance still remains, however, to be validated by a pilot-plant study.

Drogui *et al.* [2.449] have recently proposed the coupling of an electrolysis unit with a microfiltration device for wastewater treatment. The electrolysis system had two electrodes, a carbon felt cathode and an anode consisting of titanium coated with RuO_2. Hydrogen peroxide was formed in this system from an electroperoxidation reaction involving anodic oxygen formed from air. The hydrogen peroxide formed was then used for the treatment of a real municipal wastewater through the coupling of the oxidation reaction simultaneously with the microfiltration process. Drogui *et al.* [2.449] report that the combined process was able to decrease the chemical oxygen demand by a factor of 2, whereas the turbidity was decreased by more than 99 %. In addition, as shown in Figure 2.19, the steady-state flux of the treated water increased, when compared with the results of the microfiltration unit alone. The latter result was attributed to the fact that the oxidation reaction was able to degrade the species fouling the membrane.

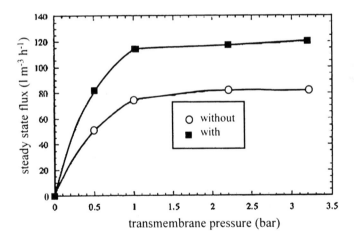

Figure 2.19. Steady state flux vs transmembrane pressure for the wastewater treatment with the membrane device alone and coupled with the electroperoxidation unit. From Drogui *et al.* [2.449], with permission from Elsevier Science.

The vapor-phase decomposition of methyl-*t*-butyl ether (MTBE) into methanol and isobutene was examined in a shell- and tube-type catalytic membrane reactor by Choi and coworkers [2.450, 2.451, 2.452]. They studied both PBMR and CMR systems. In the PBMR systems $H_3PW_{12}O_{40}$ supported on silica was used as an active catalyst for the reaction. Polyphenylene oxide (PPO), polysulfone (PSF), or cellulose acetate (CA) composite membranes, prepared by coating the polymeric layer on a porous alumina tube, were utilized. The permeability of methanol was higher than the permeability of isobutene and MTBE through all three polymeric membranes. The selective removal of methanol through the polymeric membranes helped to shift the chemical equilibrium in the PBMR, and MTBE conversions exceeded equilibrium, as shown in Figure 2.20. The CA membrane reactor showed the best performance at high contact times, whereas the PPO membrane reactor showed the best performance at low contact times, in terms of MTBE conversion, isobutene yield, and isobutene selectivity. The PBMR performance improved with a decrease in the thickness of the polymeric membrane and reaction temperature, and with an increase of the partial pressure of MTBE. For the CMR studies 12-tungstophosphoric acid (PW) was again used as a catalyst, and poly-2,6-dimethyl-1,4-phenylene oxide (PPO) was used as a polymer material, in which the PW was immobilized. Two different types of membranes were tested. A membrane, in which a single-phase (PW-PPO) was immobilized on a porous Al_2O_3 membrane (PW-PPO/Al_2O_3), and a composite membrane, in which the (PW-PPO) was deposited on a PPO/Al_2O_3 membrane (PW-PPO/PPO/Al_2O_3). The single-phase PW-PPO/Al_2O_3 membrane was shown to be permselective towards the reaction products. As a result the selective removal of methanol through the catalytic membrane shifted the chemical equilibrium toward the favorable direction in the MTBE decomposition. The PW-PPO/PPO/Al_2O_3 showed better performance than PW-PPO/Al_2O_3. The enhanced performance of the PW-PPO/ PPO/Al_2O_3 CMR was due to the additional separation capability of the sub-layered PPO membrane.

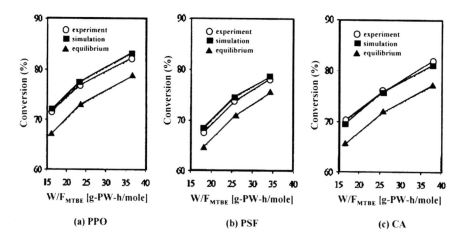

Figure 2.20. Comparison between the simulation, experimental and equilibrium MTBE conversion values for different polymers deposited on the alumina as function of the contact time. At 2.3 atm and 80 °C. (a) polyphenylene oxide; (b) polysulfone; (c) cellulose acetate. From Choi *et al.* [2.452], with permission from Elsevier Science.

MTBE synthesis from *t*-butanol and methanol in a membrane reactor has been reported by Salomon *et al.* [2.453]. Hydrophilic zeolite membranes (mordenite or NaA) were employed to selectively remove water from the reaction atmosphere during the gas-phase synthesis of MTBE. This reaction was carried out over a bed of Amberlyst® 15 catalyst packed in the inside of a zeolite tubular membrane. Prior to reaction, the zeolite membranes were characterized by measuring their performance in the separation of the equilibrium mixture containing water, methanol, *t*-butanol, MTBE, and isobutene. The results obtained with zeolite membrane reactors were compared with those of a fixed-bed reactor (FBR) under the same operating conditions. MTBE yields obtained with the PBMR at 334 K reached 67.6 %, under conditions, where the equilibrium value without product removal (FBR) would be 60.9 %.

A polymer electrolyte membrane (PEM) reactor is described by Hicks and Fedkiw [2.454] for use during Kolbe electrolysis, which involves the anodic oxidation of an alkyl carboxylic acid, and its subsequent decarboxylation and coupling to produce a dimer.

$$2RCOOH \rightarrow R\text{-}R + 2CO_2 + 2e^- + 2H^+ \tag{2.9}$$

The electrochemical membrane reactor utilizes platinized Nafion® 117 as the PEM. It functions both as the electrolyte and separator. Hicks and Fedkiw [2.454] present results for the oxidation, at atmospheric pressures, of acetic acid (in the vapor phase with a nitrogen diluent) to ethane and carbon dioxide, accompanied with hydrogen evolution at the counter electrode. For cell voltages ranging from 4 to 10 V, current densities from 0.06 to 0.4 A/cm^2, Kolbe current efficiencies of 10 to 90 % were reported. The best results were obtained using PEM's plantinized by a non-equilibrium impregnation-reduction method. The results obtained by these authors are encouraging for Kolbe electrolysis in a PEM cell. It would be interesting to see, whether similar results can be attained using a liquid-phase reactant.

Polymeric membrane supported catalysts are also finding application in hydroformylation reactions. Though no membrane reactor applications for this reaction have been published as yet a number of recent studies point out the potential benefits of immobilizing the catalysts in polymeric membranes. Feldman *et al.* [2.455], for example, have reported the synthesis of a cellulose acetate membrane containing HRh(CO)(PPH$_3$)$_3$, a rhodium hydroformylation catalyst. The HRh(CO)(PPH$_3$)$_3$ catalyst was incorporated in the membrane, by introducing it into the cellulose acetate solution prior to the casting procedure. The membranes obtained were tested in the hydroformylation reaction of ethylene and propylene [2.456], and were shown to be very active (turnover rate numbers with the film supported catalyst were ~5 times higher than the ones observed with the corresponding homogeneous catalysts). The *normal/iso* aldehyde concentration ratio was also reported to be better. More recently Naughton and Drago [2.457] reported using a water soluble rhodium complex incorporated in a PEG thin film for the hydroformylation of 1-hexene. The catalytic activity of the film supported catalyst was better than that observed with the unsupported rhodium complex. The experimental results with the thin film supported catalysts indicate the promise that catalytic membrane reactors hold for this application.

Hydroformylation of ethylene has been studied in a membrane reactor by Kim and Datta [2.458]. These authors utilized a three-layer membrane, which consisted of a porous matrix supporting a liquid-phase catalyst dissolved in a high boiling point organic solvent, which was sandwiched between two different hydrophobic polymeric membranes. The reaction was studied using a flat-disk membrane. In this configuration the reactants (C_2H_4, CO, H_2) were fed into one compartment of the membrane reactor, and a sweep gas was used in the second compartment, in order to effect a continuous elimination of the product. The experimental results and the theoretical model presented were in good agreement, and showed that very high separation factors and good conversion were possible for this type of CMR. The authors also proposed a scaled-up design, utilizing hollow-fiber membranes, shown in Figure 2.21. In this design the space in between the fibers is filled with the porous matrix, which is again used to immobilize the hydroformylation catalyst.

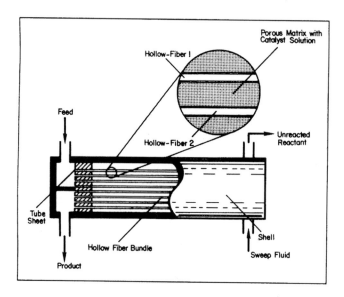

Figure 2.21. Design of a hollow fiber membrane reactor for a homogeneous catalytic reaction. From Kim and Datta [2.458], with permission from the American Institute of Chemicals Engineers.

To conclude this chapter, as the preceeding discussion has, hopefully, indicated there is considerable interest today, and significant emerging potential for the application of catalytic membrane reactors, particularly, those making use of hydrogen and oxygen permeable high temperature membranes. The main challenge that remains is in the development of membranes, which are sufficiently robust under the harsh temperature and pressure conditions typically encountered in such applications. Though significant progress has been made recently to overcome some of these deficiencies, major breakthroughs in materials science are still needed, before such membranes become viable in high temperature/pressure separations of relevance to the aforementioned applications. The low and moderate temperature applications are also attracting recent interest; here polymeric membranes may turn out to be the option of choice. Mixed-matrix membranes incorporating the catalyst into the membrane bulk have shown good promise for a variety of applications. As with the high temperature membranes, however, the key challenge that remains here, before commercial success is realized, is to prove long-term membrane robustness.

2.5 References

[2.1]   V.M. Gryaznov, *Platinum Met. Rev.* **1986**, 30, 68.

[2.2]   V.M. Gryaznov, *Platinum Met. Rev.* **1992**, 36, 70.

[2.3]   J. Shu, B.P.A. Grandjean, A. van Neste, and S. Kaliaguine, *Can. J. Chem. Engng.* **1991**, 69, 1036.

[2.4]   O.Wolfrath, L. Kiwi-Minsker, and A. Renken, *Proc. AIChE Annual Meeting*, Los Angeles, CA, November 2000.

[2.5]   H.Weyten, J. Luyten, K. Keizer, L. Willems, and R. Leysen, *Catal. Today* **2000**, 56, 3.

[2.6]   J. Ali and A. Baiker, *Appl. Catal.* A. **1996**, 140, 99.

[2.7]   G.H.W. Saracco, J.P. Neomagus, G.F. Versteeg, and W.P.M. van Swaaij, *Chem. Engng. Sci.* **1999**, 54, 1997.

[2.8]   T. Matsuda, I. Koike, N. Kubo, and E. Kikuchi, *Appl. Catal.* A. **1993**, 96, 3.

[2.9]   E. Gobina and R. Hughes, *J. Membr. Sci.* **1994**, 90, 11.

[2.10] A.M. Champagnie, T.T. Tsotsis, R.G. Minet, and I.A. Webster, *Chem. Engng. Sci.* **1990**, 45, 2423.

[2.11] A.M. Champagnie, T.T. Tsotsis, R.G. Minet, and E. Wagner, *J. Catal.* **1992**, 134, 713.

[2.12] J. Bitter, Br. Patent 2,201,159, **1988**.

[2.13] Z.D. Ziaka, R.G. Minet, and T.T. Tsotsis, *AIChE J.* **1993**, 39, 526.

[2.14] R.S.A. de Lange, J.H.A. Hekkink, K. Keizer, and A.J. Burggraaf, *J. Membr. Sci.* **1995**, 99, 57.

[2.15] N.K. Raman and C.J. Brinker, *J. Membr. Sci.* **1995**, 105, 273.

[2.16] A. Ayrall, C. Balzer, T. Dabadie, C. Guizard, and A. Julbe, *Catal. Today* **1995**, 25, 219.

[2.17] R.M. de Voss and H. Verweij, *Science* **1998**, 279, 1710.

[2.18] M. Tsaspatis and G.R. Gavalas, *J. Membr. Sci.* **1994**, 87, 281.

[2.19] S. Kim and G. R. Gavalas, *Ind. Engng. Chem. Res.* **1995**, 34, 168.

[2.20] S. Morooka, S. Yan, K. Kusakabe, and Y. Akiyama, *J. Membr. Sci.* **1995**, 101, 89.

[2.21] H. Weyten, K. Keizer, A. Kinoo, J. Luyten, and R. Leysen, *AIChE J.* **1997**, 43, 1819.

[2.22] J.P. Collins, R.W. Schwartz, R. Seghal, T.L. Ward, C.J. Brinker, G.P. Hagen, and C.A Udovich, *Ind. Engng. Chem. Res.* **1996**, 35, 43.

[2.23] T. Ioannides and G.R. Gavalas, *J. Membr. Sci.* **1993**, 77, 207.

[2.24] R. Schaefer, M. Noack, P. Kolsch, S. Thomas, A. Seidel-Morgensten, and J. Caro, *Proc. 6th International Conference on Inorganic Membranes*, Montpellier, France, June 26-30, 2000, 118.

[2.25] N. Itoh and K. Haraya, *Catal. Today* **2000**, 56, 103.

[2.26] M.S. Strano and H.C. Foley, *AICHE J.* **2001**, 47, 66.

[2.27] W.O. Haag and J.G. Tsikoyiannis, U.S. Patent 5,069,794, **1991**.

[2.28] J.N. van de Graaf, F. Kapteijn, and J.A. Moulijn, "Catalytic Membranes", in *Structured Catalysts and Reactors,* A. Cybulski and J.A.Moulijn, eds., Marcel Dekker Inc., New York, 1998.

[2.29] D. Casanave, A.G. Fendler, J. Sanchez, R. Loutaty, and J.A. Dalmon, *Catal. Today* **1995,** 25, 309.

[2.30] D. Casanave, P. Ciavarella, K. Fiaty, and J.A. Dalmon, *Chem. Engng. Sci.* **1999**, 54, 2807.

[2.31] P. Ciavarella, D. Casanave, H. Moueddeb, S. Miachon, K. Fiaty, and J.A. Dalmon, *Catal. Today,* **2001,** 67, 177.

[2.32] L.T. Yin Au, J.L. Hang Chau, C. Tellez Ariso, and K.L.Yeung, *J. Membr. Sci.* **2001**, 183, 269.

[2.33] G.R. Gallaher Jr., T.E. Gerdes, and P.T.K. Liu, *Sepn. Sci. and Technol.* **1993**, 28, 309.

[2.34] F. Tiscareno-Lechuga, C.G. Hill Jr., and M.A. Anderson, *Appl. Catal. A: General* **1993***, 96, 33.

[2.35] F. Tiscareno-Lechuga, C.G. Hill Jr., and M.A. Anderson, *J. Membr. Sci.* **1996***, 118, 65.

[2.36] T. Tagawa, H. Itoh, and S. Goto, *Proc. 5th International Congress on Inorganic Membranes,* Nagoya, June 22-26, 1998.

[2.37] Y. She and Y.H. Ma, in, "Dehydrogenation of Ethylbenzene to Styrene in a Pd Membrane Reactor", *Proc. 6th International Congress on Inorganic Membranes*, Montpellier, France*, June 26-30, 2000, 130.

[2.38] Y. She, J. Han, and YH. Ma, *Catal. Today* **2001**, 67, 43.

[2.39] Z.D. Jiang and J.Q. Wang, *Sep. Sci. Technol.* **1998**, 33,1379.

[2.40] Z. Xiongfu, L. Yongsh, W. Jinqu, and T. Huairong, *Proc. 6th International Congress on Inorganic Membranes*, Montpellier, France*, June 26-30, 2000, 59.

[2.41] A.M. Mondal and S. Ilias, *Sep. Sci. Technol.* **2001**, 36, 1101.

[2.42] P.A. Terry, M. Anderson, and I. Tejedor, *J. Porous Mat.* **1999**, 6, 267.

[2.43] H.L. Frisch, S. Maaref, and H. D. Nemer, *J. Membr. Sci.* **1999**, 154, 33.

[2.44] W.J. Koros and D.G. Woods, *J. Membr. Sci.* **2001**, 181, 157.

[2.45] M.E. Rezac and B. Schoberl, *J. Membr. Sci.* **1999**, 156, 211.

[2.46] D.W. Mouton, J.N. Keuler, and L.L. Lorenzen, *Proc. 6th International Congress on Inorganic Membranes*, Montpellier, France*, June 26-30, 2000, 137.

[2.47] A. Trianto, L.Q. Wang, and T. Kokugan, *J. Chem. Eng. Jpn.* **2001**, 34, 1065.

[2.48] B.A. Raich and H.C. Foley, *Ind. Engng. Chem. Res.* **1998**, 37, 3888.

[2.49] B.S. Liu, W.L. Dai, G.H. Wu, and J.F. Deng, *Catal. Lett.* **1997***, 49, 181.

[2.50] B.S. Liu Liu, G.H. Wu, G.X. Niu, and J.F. Deng, *Appl. Catal. A-Gen.* **1999**, 185, 1.

[2.51] D. Xue, H. Chen, G.H. Wu, and J.F. Deng, *Appl. Catal. A-Gen.* **2001**, 214, 87.

[2.52] W. Lefu, J. Hongbing, and Z. Jie, *Proc. 4ᵗʰ International Conference on Catalysis in Membrane Reactors*, 2000, 63.

[2.53] S.V. Gorshkov, G.I. Lin, A.Y. Rozovskii, Y.M. Serov, and S.J. Uhm, *Kinet. Catal.* **1999***, 40, 92.

[2.54] H. Amandusson, L.G. Ekedahl, and H. Dannetun, *Surf. Sci.* **1999**, 442, 199.

[2.55] S. Hara, K. Sakaki, and N. Itoh, *Ind. Engng. Chem. Res.* **2000**, 38, 4913.

[2.56] H. Amandusson, L.G. Ekedahl, and H. Dannetun, *Proc. International Congress on Catalysis with Membrane Reactors-2000*, Zaragoza, Spain, July 3-5, 2000, 177.

[2.57] S.H. Jung, K. Kusakabe, S. Morooka, and S.D. Kim, *J. Membr. Sci.* **2000***, 170, 53.

[2.58] S.M. Liu, X.Y. Tan, K. Li, and R. Hughes, *Catal. Rev.-Sci. Eng.* **2001**, 43, 147.

[2.59]   A. Anderson, I.M. Dahl, K. Jens, E. Rytter, A. Slagtern, and A. Solbakken, *Catal. Today* **1989**, 4, 389.
[2.60]   H. Iwahara, T. Esaka, H. Uchida, and N. Maeda, *Solid State Ionics* **1981**, 3/4, 359.
[2.61]   H. Iwahara, T. Esaka, H. Uchida, T. Yamauchi, and K. Ogaki, *Solid State Ionics* **1986**, 18/19, 1003.
[2.62]   H. Iwahara, H. Uchida, and S. Tanaka, *J. Appl. Electrochemistry* **1986**, 16, 663.
[2.63]   H. Iwahara, *Solid State Ionics* **1995**, **77**, 289.
[2.64]   H. Iwahara, H. Uchida, K. Ono, and K. Ogaki, *J. Electrochem. Soc.* **1988**, 135, 529.
[2.65]   H.Iwahara, H. Uchida, K. Morimoto, and S. Hosogi, *J. Appl. Electrochemistry* **1989**, 19, 448.
[2.66]   H. Hamakawa, S.T. Hibino, and H. Iwahara, *J. Electrochem. Soc.* **1993**, 140, **459**.
[2.67]   P.H. Chiang, P.H., D. Eng, and M. Stoukides, *J. Electrochem. Soc.* **1991**, 138, L11.
[2.68]   P.H. Chiang, P.H., D. Eng, and M. Stoukides, *Solid State Ionics* **1993**, 61, **99**.
[2.69]   P.H. Chiang, P.H., D. Eng, and M. Stoukides, *J. Catal.* **1993**, 139, 683.
[2.70]   P.H. Chiang, P.H., D. Eng, and M. Stoukides, *Solid State Ionics* **1994**, 67, 179.
[2.71]   P.H. Chiang, P.H., D. Eng, T. Panagiotis, and M. Stoukides, *Solid State Ionics* **1995**, 77, 305.
[2.72]   C.G. Vayenas, S. Bebelis, I.V. Yentekakis, and H.G. Lintz, *Catal. Today* **1992**, 11, 303.
[2.73]   S. Hamakawa, T. Hibino, and H. Iwahara, *J. Electrochem. Soc.* **1994**, 141, **1720**.
[2.74]   J. Langguth, R. Dittmeyer, H. Hofmann, and G. Tomandl, *Appl. Catal. A: General* **1997**, 158, 287.
[2.75]   T. Terai, X.H. Li, F. Tomishige, and K. Fujimoto, *Chem. Lett.* **1999**, 323.
[2.76]   L. Li, E.C. Lu, A. Li, R.W. Borry, and E. Iglesia, 1, **2000**. *Proc. International Congress on Catalysis with Membrane Reactors-2000*, Zaragoza, Spain, July 3-5, 2000, 1.
[2.77]   V.M. Gryaznov, V.S. Smirnov, and M.G. Slin'ko, in: G.C. Bond, P.B. Wells, and F.C. Tompkins, Eds., "Binary Palladium Alloys as Selective Membrane Catalysts", *Proc 6th Intl. Cong. Catal.*, 1976, 2, 894.
[2.78]   V.M. Gryaznov, and A.N. Karavanov, *Khim.-Farm.Zh.* **1979**, 13, 74.
[2.79]   N. Itoh, W.C. Xu, S. Hara, and K. Sakaki, *Catal. Today* **2000**, 56, 307.
[2.80]   N. Itoh, S. Hara, K. Sazaki, Y. Kaneko, and A. Igarashi, *Proc. International Congress on Catalysis with Membrane Reactors-2000*, Zaragoza, Spain, July 3-5, 2000, 111.
[2.81]   V.M. Gryaznov and V.S. Smirnov, *Kinet. and Catal.* **1977**, 18, 485.
[2.82]   V.M. Gryaznov, A.N. Karavanov, T.M. Belosljudova, A.M. Ermolaev, A.P. Maganjunk, and I.K. Sarycheva, U.S. Patent 4,388,479, 1983.
[2.83]   M.M. Ermilova, N.V. Orekhova, and V.M. Gryaznov, in: R. Bredesen, Ed., "Optimization of the Selective Hydrogenation Process by Membrane Catalysts", *Proc. Fourth Workshop: Optimisation of Catalytic Membrane Reactors Systems*, Oslo, Norway, May, 1997, 187.
[2.84]   M. Shirai, Y. Pu, M. Arai, and Y. Nishiyama, *Appl. Surf. Sci.* **1998**,126, 99.
[2.83]   N. Itoh, M. Machida, and K. Adachi, *Chem. Lett.* **2000**, 1162.
[2.86]   K. Otsuka and T. Yagi, *J. Catal.* **1994**, 145, 289.
[2.87]   E.L. Panagos, Y. Voudouris, and M. Stoukides, *Chem. Engng. Sci.* **1996**, 51, 3175.
[2.88]   G. Marnellos and M. Stoukides, *Science* **1998,** 282, 95.
[2.89]   P. Cini and M.P. Harold, *AIChE J.* **1991**, 37, 997.
[2.90]   M. Torres, J. Sanchez, J.A. Dalmon, B. Bernauer, and J. Lieto, *Ind. Engng. Chem. Res.* **1994**, 33, 2421.

[2.91]   O. Monticelli, A. Bezzi, A. Bottino, G. Capannelli, and A. Servida, in: "Hydrogenation of Cinnamaldehyde: the Use of Three Phase Catalytic Membrane Reactors", *Proc. Fourth Workshop: Optimisation of Catalytic Membrane Reactors Systems*, Oslo, Norway, May, 1997, 109.

[2.92]   X.L. Pan, B.J. Liu, S.S. Sheng, G.X. Xiong, J. Liu, and W.S. Yang, in, "Preparation of Catalytic Membranes and the Applications in the Selective Hydrogenation of Liquid Cinamaldehyde", *Proc. 6th International Congress on Inorganic Membranes*, Montpellier, France, June 26-30, 2000, 130.

[2.93]   J.M. Veldsink, *J. Am. Oil Chem. Soc.* **2001**, 78, 443.

[2.94]   C.Q. Liu, Y. Xu, S.J. Liao, D.R. Yu, Y.K. Zhao, and Y.H. Fan, YH, *J. Membr. Sci.* **1997**, 137, 139.

[2.95]   C.Q. Liu, Y. Xu, and D. Yu, *Appl. Catal. A: General* **1998**, 172, 23.

[2.96]   C.Q. Liu, Y. Xu, S.J. Liao, and D.R. Yu, *J. Mol. Catal. A-Chem.* **2000**, 157, 253.

[2.97]   S. Ziegler, J. Theis, and D. Fritsch, *J. Membr. Sci.* **2001**, 187, 71.

[2.98]   K.D. Smet, I.F.J. Vankelecom, and P.A. Jacobs, *Proc. International Congress on Catalysis with Membrane Reactors-2000*, Zaragoza, Spain, July 3-5, 2000, 119.

[2.99]   T. Dwars, J. Haberland, I. Grassert, G. Oehme, and U. Kragl, *J. Mol. Catal. A-Chem.* **2001**, 168, 81.

[2.100]   S. Laue, L. Greiner, J. Woltinger, and A. Liese, *Adv. Synth. Catal.* **2001**, 342, 711.

[2.101]   W.D. An, J.K. Hong, and P.N. Pintauro, *J. Appl. Electrochem.* **1998**, 28, 947.

[2.102]   W.D. An, J.K. Hong, P.N. Pintauro, K. Warner, and W. Neff, *J. Am. Oil Chem. Soc.* **1998**, 75, 917.

[2.103]   W.D. An, J.K. Hong, P.N. Pintauro, K. Warner, and W. Neff, *J. Am. Oil Chem. Soc.* **1999**, 76,215.

[2.104]   K.D. Vorlop and U. Prusse, *Environmental Catalysis* **1999,** 10, 195.

[2.105]   U. Prusse and K.D. Vorlop, *J. Mol. Catal. A-Chem.* **2001**, 173, 313.

[2.106]   A. Pintar, J. Batista, and J. Levec, *Wat. Sci. Tech.* **1998**, 37, 177.

[2.107]   H. Berndt, I. Monnich, B. Lucke, and M. Menzel, *Appl. Catal. B-Environ.* **2001**, 30, 111.

[2.108]   K. Daub, G. Emig, M.J. Chollier, M. Callant, and R. Dittmeyer, *Chem. Engng. Sci.* **1999**, 54, 1577.

[2.109]   R. Dittmeyer, V. Hollein, and K. Daub, *J. Mol. Catal. A-Chem.* **2001**, 173, 135.

[2.110]   K. Daub, V. Wunder, and R. Dittmeyer, *Catal.Today* **2001**, 67, 257.

[2.111]   O.M. Ilinitch and F.P. Cuperus, *Proc. International Congress on Catalysis with Membrane Reactors-2000*, Zaragoza, Spain, July 3-5, 2000, 41.

[2.112]   O.M. Ilinitch, F.P. Cuperus, L.V. Nosova, and E.N. Gribov, *Catal. Today* **2000**, 56, 137.

[2.113]   O.M. Ilinitch, F.P. Cuperus, R.W. van Gemert, E.N. Gribov, and L.V. Nosova, *Sepn. Pur. Technol.* **2000,** 21, 55.

[2.114]   O.M. Ilinitch, A.A. Kirchanov, N.A. Kulikovskaya, and V.I. Vereschagin, *Proc. 6th International Congress on Inorganic Membranes*, Montpellier, France, June 26-30, 2000, 107.

[2.115]   G. Strukul, R. Gavagnin, F. Pinna, E. Modaferri, S. Perathoner, G. Centi, M. Marella, and M. Tomaselli, *Catal. Today* **1999**, 55, 139.

[2.116]   J. Coronas, M. Menendez, and J. Santamaria, *J. Loss Prev. Process Ind.* **1995**, 8, 97.

[2.117]   A.G. Dixon, "Innovations in Catalytic Inorganic Membrane Reactors", *Specialist Periodical Reports: Catalysis*, Vol 14, The Royal Society of Chemistry, 1999.

[2.118]   B.C.H. Steele, *Mater. Sci. Engng.* **1992**, B13, 79.

[2.119]   M. Stoukides, *Ind. Engng. Chem. Res.* **1988**, 27, 1745.

[2.120] Y. Teraoka, H.M. Zhang, S. Furukawa, and N. Yamazoe, *Mater. Res. Bull.* **1988**, 33, 51.
[2.121] M. Schwartz, J.H. White, M.G. Myers, S. Deych, and A.F. Sammels, in: "The Use of Ce-
         ramic Membrane Reactors for the Partial Oxidation of Methane to Synthesis Gas", *Pre-
         prints 213th ACS National Meeting, San Francisco, CA, April 13-17,* 42, 1997.
[2.122] A.F. Sammels, M. Schwartz, R. A. Mackay, T.F. Barton, and D.R. Peterson, *Catal. Today*
         2000, 56, 325.
[2.123] R.L. Mieville, U.S. Patent 5,276,237, 1994.
[2.124] G.E. Keller and M.M. Bhasin, *J. Catal.* **1982**, 73, 9.
[2.125] E.E. Wolf, *Methane Conversion by Oxidative Processes*, Reinhold, New York, 1992.
[2.126] D. Eng and M. Stoukides, *Catal. Rev. Sci. Engng.* **1991**, 33, 375.
[2.127] M. Stoukides, *Catal. Rev-Sci. Eng.* **2000**, 42,1.
[2.128] K. Otsuka, S. Yokoyama, and A. Morikawa, *Chem. Lett.* **1985**, 319.
[2.129] E.A. Hazburn, U.S. Patent 4,791,079, 1988.
[2.130] H. Nagamoto, K. Hayashi, and H. Inoue, *J. Catal.* **1990**, 126, 671.
[2.131] T. Nozaki, O. Yamazaki, K. Omata, and K. Fujimoto, *Chem. Engng. Sci.* **1992**, 47, 2945.
[2.132] M.P. Harold, C. Lee, A.J. Burggraaf, K. Keizer, V.T. Zaspalis, and R.S.A. de Lange, *Mat.
         Res. Soc. Bull.* **1994**, April, 34.
[2.133] T. Hibino, T. Sato, K. Ushiki, and Y. Kuwahara, *J. Chem. Soc. Faraday Trans.* **1995**, 91,
         4419.
[2.134] Y. Zeng and Y.S. Lin, *Ind. Engng. Chem. Res.* **1997**, 36, 277.
[2.135] Y. Zeng and Y.S. Lin, *Appl. Catal. A: General* **1997**, 159, 101.
[2.136] Y. Zeng and Y.S. Lin "Experimental Study of Oxidative Coupling of Methane in Dense
         Ceramic Membrane Reactor", *Proc. 5th International Congress on Inorganic Membranes,*
         Nagoya, June 22-26, 1998, 354.
[2.137] D. Lafarga, J. Santamaria, and M. Menendez, *Chem. Engng. Sci.* **1994**, 49, 2005.
[2.138] J. Coronas, M. Menendez, and J. Santamaria, *Chem. Engng. Sci.* **1994**, 49, 2015.
[2.139] J. Coronas, M. Menendez, and J. Santamaria, *Chem. Engng. Sci.* **1994**, 49, 4749.
[2.140] J. Herguido, D. Lafarga, M. Menendez, J. Santamaria, and C. Guimon, *Catal. Today* **1995**,
         25, 263.
[2.141] T. Kanno, J. Horiuchi, and M. Kobayashi, *React. Kinet. Catal. Lett.* **2001**, 72, 195.
[2.142] M.P. Harold, V.T. Zaspalis, K. Keizer, and A.J. Burggraaf, *Chem. Engng. Sci.* **1993**, 48,
         2705.
[2.143] J. Santamaria, E. Miro, and E.E. Wolf, *Ind. Engng. Chem. Res.* **30**, 1157, **1991**.
[2.144] J. Santamaria, M. Menendez, and J.I. Barahona, *Catal. Today* **1992**, 13, 353.
[2.145] S.C. Reyes, E. Iglesia, and C.P. Kelkar, *Chem. Engng. Sci.* **1993**, 48, 2643.
[2.146] S.C. Reyes, C.P. Kelkar, and E. Iglesia, *Catal. Lett.* **1993**, 19, 167.
[2.147] S. Cheng and X. Shuai, *AIChE J.* **1995**, 41, 1589.
[2.148] W. Wang and Y. S. Lin, *J. Membr. Sci.* **1995**, 103, 219.
[2.149] C. Finol, M. Menendez, A. Monzon, and J. Santamaria, *Chem. Engng. Commun.* **1995**,
         135, 175.
[2.150] K. Omata, H. Hashimoto, H. Tominaga, and K. Fujimoto, *Appl. Catal.* **1989**, 52, L1.
[2.151] T. Nozaki, Y. Osamu, O. Kohji, and K. Fujimoto, *Ind. Engng. Chem. Res.* **1993**, 32, 1174.
[2.152] T. Nozaki and K. Fujimoto, *AIChE J.* **1994**, 40, 870.
[2.153] A.G. Anshits, A.N. Shigapov, S.N. Veresshchagin, and V.N. Shevin, *Catal. Today* **1990**, 6,
         593.

[2.154]   X.M. Guo, K. Hidajat, C.B. Ching, and H.F. Chen, *Ind. Engng. Chem. Res.* **1997**, 36, 3576.
[2.155]   C.G. Vayenas, S. Bebelis, and S. Ladas, *Nature* **1990**, 343, 625.
[2.156]   C.G. Vayenas, S. Bebelis, I.V. Yentekakis, P. Tsiakaras, and H. Karasali, *Platinum Metals Rev.* **1990**, 34 3.
[2.157]   T. Hibino, A. Masegi, and H. Iwahara, *J. Electrochem. Soc.* **1995**, 142, L72.
[2.158]   T. Hibino, K. Ushiki, Y. Kuwahara, A. Masegi, and H. Iwahara, *J. Chem. Soc. Faraday Trans.* **1996**, 92, 2393.
[2.159]   N.U. Pujare and A.F. Sammels, *J. Electrochem. Soc.* **1988**, 135, 2544.
[2.160]   K. Tagawa, K.K. Moe, M. Ito, and S. Goto, *Chem. Engng. Sci.* **1999**, 54, 1553.
[2.161]   X.M. Guo, K. Hidajat, and C.B. Ching, *Catal. Today* **1999**, 50, 109.
[2.162]   C.G. Vayenas and R.D. Farr, *Science* **1980**, 208, 593.
[2.163]   R.D. Farr and C.G. Vayenas, *J. Electrochem. Soc.* **1980**, 127, 1478.
[2.164]   C.T. Sigal and C. G. Vayenas, *Solid State Ionics* **1981**, 5, 567.
[2.165]   M.R.S. Manton, H. Sawin, R.M. Scharfman, C.T. Sigal, C.G. Vayenas, and J.Wei, *Energy Technol.* **1984**, 11, 794.
[2.166]   N.M. Sammes and B.C.H. Steele, *J. Catal.* **1994**, 145, 187.
[2.167]   S. Neophytides and C. G. Vayenas, *J. Electrochem. Soc.* **1990**, 137, 839.
[2.168]   J. N. Michaels and C.G. Vayenas, *J. Electrochem. Soc.* **1984**, 131, 2544.
[2.169]   J. N. Michaels and C.G. Vayenas, *J. Catal.* **1984**, 85, 477.
[2.170]   H. Iwahara, H. Uchida, and S. Tanaka, *J. Appl. Electrochem.* **1986**, 16, 663.
[2.171]   N. Kiratzis and M. Stoukides, *J. Electrochem. Soc.* **1987**, 134, 1925.
[2.172]   E. McKenna and M. Stoukides, *Chem. Engng. Sci.* **1992**, 47, 2951.
[2.173]   Y. Teraoka, H.M. Zhang, S. Furukawa, and N. Yamazoe, *Chem. Lett.* **1985**, 1367.
[2.174]   Y. Teraoka, H.M. Zhang, and N. Yamazoe, *Chem. Lett.* **1985**, 1743.
[2.175]   Y.K. Kao, L. Lei, and Y.S. Lin, *Ind. Engng. Chem. Res.* **1997**, 36, 3583.
[2.176]   J.E. Ten Elshof, B.A. Vanhassel, and H.J.M. Bouwmeester, *Catal. Today* **1995**, 25, 397.
[2.177]   J.E. Ten Elshof, H.J.M. Bouwmeester, and H. Verweij, *Appl. Catal. A: General* **1995**, 25, 130.
[2.178]   Y.S. Lin and Y. Zeng, *J. Catal.* **1996**, 164, 20.
[2.179]   J.S. Xu and W.J. Thompson, *AIChE J.* **1997**, 43, 2731.
[2.180]   Y.S. Lin, W. Wang, and J. Han, *AIChE J.* **1994**, 40, 786.
[2.181]   Y. Zeng, Y.S. Lin, and S.L. Swartz, *J. Membr. Sci.* **1998**, 150, 87.
[2.182]   Y. Zeng and Y.S. Lin, *J. Catal.* **2000**, 193, 58.
[2.183]   Y. Zeng and Y.S. Lin, *AICHE J.* **2001**, 47, 436.
[2.184]   Y.P. Lu, A.G. Dixon, W.R. Moser, Y.H. Ma, and U. Balachandran, *Catal. Today* **2000**, 56, 297.
[2.185]   Y.P. Lu, A.G. Dixon, W.R. Moser, Y.H. Ma, and U. Balachandran, *J. Membr. Sci.* **2000**, 170, 27.
[2.186]   C. Athanasiou, G. Marnellos, J. E. ten Elshof, P. Tsiakaras, H. J. M. Bouwmeester, and M. Stoukides, *Ionics* **1997**, 3, 128.
[2.187]   P. Tsiakaras, G. Marnellos, C. Athanasiou, M. Stoukides, J. E. ten Elshof, and H.J.M. Bouwmeester, *Appl. Catal.* **1998**, 169, 247.
[2.188]   P. Chanaud, A. Julbe, A. Larbot, C. Guizard, L. Cot, H. Borges, A.G. Fendler, and C. Mirodatos, *Catal. Today* **1995**, 25, 225.

[2.189] H. Borges, A.G. Fendler, C. Mirodatos, P. Chanaud, and A. Julbe, *Catal. Today* **1995**, 25, 377.

[2.190] Y.H. Ma, Y. Lu, A.G. Dixon, and W.R. Moser, *Proc. 5th International Congress on Inorganic Membranes,* Nagoya, June 22-26, 1998, 330.

[2.191] A.M. Ramachandra, Y. Lu, Y.H. Ma, W.R. Moser, and A.G. Dixon, *J. Membr. Sci.* **1996**, 116, 253.

[2.192] Y. Lu, A.G. Dixon, W.R. Moser, and Y.H. Ma, *Ind. Engng. Chem. Res.* **1997**, 36, 359.

[2.193] Y. Lu, A. G. Dixon, W.R. Moser, and Y.H. Ma, *Chem. Engng. Sci.* **2000**, 55, 4901.

[2.194] O. Garnier, J. Shu and B.P.A. Grandjean, *Proc. 4th International Congress on Inorganic Membranes*, Gatlinburg, TN, July 14-18, 1996, 334.

[2.195] T. Koerts, M.J.A.G. Deelen, and R.A. van Santen, *J. Catal.* **1992**, 138, 101.

[2.196] K. Murata, N. Ito, T. Hayakawa, K. Suzuki, and S. Hamakawa, *Chem. Commun.* **1999**, 7, 573.

[2.197] K. Murata, T. Hayakawa, and K. Suzuki, *Sekiyu Gakkaishi* **2000**, 43, 162.

[2.198] K. Murata, T. Hayakawa, S. Hamakawa, and K. Suzuki, *React. Kinet. Catal. Lett* **2001**, 73, 297.

[2.199] W.E. Tragert, R.L. Fullman, and R.E. Carter, U.S. Patent 3,138,490, 1964.

[2.200] D. Eng and M. Stoukides, *Proc. 9th Int. Congr. Catalysis*, Chem. Inst. of Canada, Ottawa, 1988, 2, 974.

[2.201] H. Alqahtany, D. Eng, and M. Stoukides, *J. Electrochem. Soc.* **1993**, 140, 1677.

[2.202] H. Alqahtany, D. Eng, and M. Stoukides, *Energy & Fuels* **1993**, 7, 495.

[2.203] T. Horita, N. Sakai, T. Kawada, H. Yokokawa, and M. Dokiya, *J. Electrochem. Soc.* **1996**, 143, 1161.

[2.204] K. Sato, J. Nakamura, T. Uchijima, T. Hayakawa, S. Hamakawa, T. Tsunoda, and K. Takehira, *J. Chem. Soc. Faraday Trans.* **1995**, 91, 1655.

[2.205] K. Takehira, T. Hayakawa, S. Hamakawa, T. Tsunoda, K. Sato, J. Nakamura, and T. Uchijima, *Catal. Today* **1996**, 29, 397.

[2.206] S. Hamakawa, K. Sato, T. Hayakawa, A.P.E. York, T. Tsunoda, K. Suzuki, M. Shimizu, and K. Takehira, *J. Electrochem Soc.* **1997**, 144, 1.

[2.207] S. Hamakawa, T. Hayakawa, K. Suzuki, R. Shiozaki, and K. Takehira, *Denki Kagaku.* **1997**, 65, 1049.

[2.208] S. Hamakawa, M. Koizumi, K. Sato, J. Nakamura, T. Uchijima, K. Murata, T. Hayakawa, and K. Takehira, *Catal. Lett.* **1998**, 52, 191.

[2.209] S. Hamakawa, T. Hayakawa, K. Suzuki, K. Murata, K. Takehira, S. Yoshino, J. Nakamura, and T. Uchijima, *Solid State Ion.* **2000**, 136, 761.

[2.210] V.A. Sobyanin and V.D. Belyaev, *Solid State Ion.* **2000**, 136, 747.

[2.211] T. Ishihara, Y. Hiei, and Y. Takita, *Solid State Ion.* **1995**, 79, 371.

[2.212] Y. Hiei, T. Ishihara, and Y. Takita, *Solid State Ion.* **1996**, 86-88, 1267.

[2.213] V.A. Sobyanin, V.D. Belyaev, and V.V. Gal'vita, *Catal. Today.* **1998**, 42, 337.

[2.214] G.L. Semin, V.D. Belyaev, A.K. Demin, and V.A. Sobyanin, *Appl. Catal. A-Gen.* **1999**, 181, 131.

[2.215] T.J. Mazanec, T. L. Cable, and J. G. Frye, Jr., *Solid State Ion.* **1992**, 53, 111.

[2.216] T.L. Cable, T.J. Mazanec, and J.G. Frye, *Europ. Pat. Appl.* 0,399,833, Nov. 28, 1990.

[2.217] T.J. Mazanec, T.L. Cable, J.G. Frye, and W.R. Kliewer, U.S. Patent 5,591,315, 1997.

[2.218] U. Balachandran, J.T. Dusek, S.M. Sweeney, R.B. Poeppel, R.L. Mieville, P.S. Maiya, M.S. Kleefisch, S. Pei, T.P. Kobylinsky, and C.A. Udovich, *Ceram. Bull.* **1995**, 74, 71.

[2.219]   U. Balachandran, J.T. Dusek, R.L. Mieville, R.B. Poeppel, M.S. Kleefisch, S. Pei, T.P. Kobylinsky, C.A. Udovich, and A.C. Bose, *Appl. Catal. A Gen.* **1995**, 133, 19.

[2.220]   U. Balachandran, J.T. Dusek, P.S. Maiya, B. Ma, R.L. Mieville, M.S. Kleefisch, and C.A. Udovich, *Catal. Today* **1997**, 36, 265.

[2.221]   U. Balachandran, P.S. Maiya, B. Ma, J.T. Dusek, R.L. Mieville, and J.J Picciolo, *Proc. 5th International Congress on Inorganic Membranes,* Nagoya, June 22-26, 1998.

[2.222]   P.S. Maiya, T.J. Anderson, R.L. Mieville, J.T. Dusek, J.J. Picciolo, and U. Balachandran, *Appl. Catal. A-Gen.* **2000**, 196, 65.

[2.223]   S. Li, W. Jin, X. Gu, N. Xu, J. Shi, and Y.H. Ma, *Proc. International Congress on Catalysis with Membrane Reactors-2000*, Zaragoza, Spain, July 3-5, 2000, 23.

[2.224]   W.Q. Jin, S.G. Li, P. Huang, N.P. Xu, J. Shi, and Y.S. Lin, *J. Membr. Sci.* **2000**, 166, 13.

[2.225]   H. Dong, Z. Shao, G.T. Xiong, J. Tong, S. Sheng, and W.S. Yang, *Catal. Today* **2001**, 67, 3.

[2.226]   H. Dong, G.X. Xiong, Z.P. Shao, S.L. Liu, and W.S. Yang, *Chin. Sci. Bull.* **2000**, 45, 224.

[2.227]   Z. Shao, G. Xiong, Y. Cong, and W. Yang, *J. Membr. Sci.* **2000**, 172, 177.

[2.228]   Z. Shao, G. Xiong, H. Dong, Y. Cong, and W. Yang, *Proc. 6th International Congress on Inorganic Membranes*, Montpellier, France, June 26-30, 2000, 26.

[2.229]   C.Y. Tsai, A.G. Dixon, and Y.H. Ma, *J. Am. Ceram. Soc.* **1998**, 81, 1437.

[2.230]   J.T. Ritchie, J.T. Richardson, and D. Luss, *AICHE J.* **2001**, 47, 2092.

[2.231]   T.J. Mazanec, Private Communication, **1999**.

[2.232]   P.N. Dyer, R.E. Richards, S.L. Russek, and D.M. Taylor, *Solid State Ion.* **2000**, 134, 21.

[2.233]   P.V. Hendriksen, *Proc. International Congress on Catalysis with Membrane Reactors-2000*, Zaragoza, Spain, July 3-5, 2000, 19.

[2.234]   P.V. Hendriksen, P.H. Larsen, M. Mogensen, F.W. Poulsen, and K. Wiik, *Catal. Today* **2000**, 56, 283.

[2.235]   H. Fjellvag, B.C. Hauback, and R. Bredesen, *J. Materials Chem.* **1997**, 7, 2415.

[2.236]   M. Alibrando, H.S. Hahm, and E.E. Wolf, *Catal. Lett.* **1997**, 49, 1.

[2.237]   J. Yun, H. Hahm, M. Alibrando, and E.E. Wolf, *Proc. III International Conference on Catalysis in Membrane Reactors*, Copenhagen, Sept. 8-10, 1998, paper O4.

[2.238]   A. Basile, L. Paturzo, and V. Durante, *Proc. International Congress on Catalysis with Membrane Reactors-2000*, Zaragoza, Spain, July 3-5, 2000, 87.

[2.239]   A. Basile and L. Paturzo, *Catal. Today* **2001**, 67, 55.

[2.240]   J. Galuszka, R. Pandey, and S. Ahmed, *Catal. Today* **1998**, 46, 82.

[2.241]   T. Ostrowski, A. Giroir-Fendler, C. Mirodatos, and L. Mleczko, *Catal. Today* **1998**, 40, 191.

[2.242]   H. Zhong-Tao and L. Ru-Xuan, *Proc. 1st Int. Workshop on Catalytic Membranes*, Lyon-Villeurbanne, France, Sept. 26-28,1994.

[2.243]   C. Yang, N. Xu, and J. Shi, *Proc. 5th International Congress on Inorganic Membranes,* Nagoya, June 22-26, 1998, 616.

[2.244]   G. Lu, S. Shen, and R. Wang, *Catal. Today* **1996**, 30, 41.

[2.245]   T.M. Gür and R.A. Huggins, *J. Catal.* **1986**, 102, 443.

[2.246]   Q. Liu, J. Rogut, B. Chen, J.L. Falconer, and R.D. Noble, *Fuel* **1996**, 75, 1748.

[2.247]   V.L. Halloin and S.J. Wajc, *Chem. Engng. Sci.* **1994**, 49, 4691.

[2.248]   V. Diakov, D. Lafarga, and A. Varma, *Proc. International Congress on Catalysis with Membrane Reactors-2000*, Zaragoza, Spain, July 3-5, 2000, 33.

[2.249]   G. Grigoropoulou, J.H. Clark, D.W. Hall, and K. Scott, *Chem. Commun.* **2001**, 6, 547.

[2.250] T.J. Stanley and J.A. Quinn, *Chem. Eng. Sci.* **1987**, 42, 2313.

[2.251] S.S. Luthra, X.J. Yang, L.M.F dos Santos, L.S. White, and A.G. Livingston, *Chem. Commun.* **2001**, 16, 1468.

[2.252] R. Mallada, M. Menendez, and J. Santamaria, *Proc. 5th International Congress on Inorganic Membranes,* Nagoya, June 22-26, 1998, 612.

[2.253] R. Mallada, M. Menendez, and J. Santamaria, *Catal. Today* **2000**, 56, 191.

[2.254] R. Mallada, M. Pedernera, M. Menendez, and J. Santamaria, *Ind. Engng. Chem. Res.* **2000**, 39, 620.

[2.255] R. Mallada, M. Menendez, and J. Santamaria, *Catal. Today* **2000**, 56, 191.

[2.256] S. Mota, S. Miachon, J.C. Volta, and J.A. Dalmon, *Catal. Today* **2001**, 67, 169.

[2.257] E.Z. Xue and J.R.H. Ross, *Catal. Today* **2000**, 61, 3.

[2.258] M. Pena, D.M. Carr, K.L. Yeung, and A. Varma, *Chem. Engng. Sci.* **1998**, 53, 3821.

[2.259] D. Lafarga and A. Varma, *Chem. Engng. Sci.* **2000**, 55, 749.

[2.260] M.A. Al-Juaied, D. Lafarga, and A. Varma, *Chem. Eng. Sci.* **2001**, 395, 56.

[2.261] Y.S. Cheng, C. Tellez, K.L. Yeung, and M.A. Pena, *Proc. International Congress on Catalysis with Membrane Reactors-2000*, Zaragoza, Spain, July 3-5, 2000, 37.

[2.262] M. Alonso, M.J. Lorences, M.P. Pina, and G.S. Patience, *Catal. Today* **2001**, 67, 151.

[2.263] P. Kolsch, M. Noack, and J. Caro, *Proc. International Congress on Catalysis with Membrane Reactors-2000*, Zaragoza, Spain, July 3-5, 2000, 183.

[2.264] A.P.E. York, S. Hamakawa, T. Hayakawa, K. Sato, T. Tsunoda, and K. Takehira, *J. Chem. Soc., Faraday Trans.* **1996**, 92, 3579.

[2.265] A.P.E.S. York, S. Hamakawa, T. Hayakawa, T. Tsunoda, and K. Takehira, *J. Electrochem Soc.* **1996**, 143, L249.

[2.266] S. Hamakawa, T. Hayakawa, K. Suzuki, M. Shimizu, and K. Takehira, *Proc. 3rd Word Congress on Oxidation Catalysis*, 1997, 1323.

[2.267] S. Hamakawa, K. Sato, T. Hayakawa, A.P.E. York, T. Tsunoda, K. Suzuki, M. Shimizu, and K. Takehira, *J. Electrochem. Soc.* **1997**, 144, 1.

[2.268] T. Hayakawa, T. Tsunoda, H. Orita, T. Kameyama, H. Takahashi, K. Takehira, and K. Fukuda, *J. Chem. Soc. Chem. Commun.* **1986**, 961.

[2.269] T. Hayakawa, T. Tsunoda, H. Orita, T. Kameyama, H. Takahashi, K. Fukuda, and K. Takehira, *J. Chem. Soc. Chem. Commun.* **1987**, 780.

[2.270] T. Tsunoda, T. Hayakawa, Y. Imai, T. Kameyama, K. Takehira, and K. Fukuda, *Catal. Today* **1995**, 25, 371.

[2.271] T. Tsunoda, T. Hayakawa, K. Sato, T. Kameyama, K. Fukuda, and K. Takehira, *J. Chem. Soc. Faraday Trans.*, **1995**, 91, 1111.

[2.272] T. Tsunoda, T. Hayakawa, K. Sato, T. Kameyama, K. Fukuda, and K. Takehira, *J. Chem. Soc. Faraday Trans.* **1995**, 91, 1117.

[2.273] T. Tsunoda, T. Hayakawa, T. Kameyama, K. Fukuda, and K. Takehira, *J. Chem. Soc. Faraday Trans.* **1995**, 91, 1125.

[2.274] S. Hamakawa, T. Hayakawa, H. Yasuda, K. Suzuki, M. Shimizu, and K. Takehira, *J. Electrochem Soc.* **1995**, 142, L159.

[2.275] S. Hamakawa, T. Hayakawa, A.P.E. York, T. Tsunoda, Y.S. Yoon, K. Suzuki, M. Shimizu, and K. Takehira, *J. Electrochem. Soc.* **1996**, 143, 1264.

[2.276] S. Hamakawa, T. Hayakawa, T. Tsunoda, K. Suzuki, K. Murata, and K. Takehira, *Electrochem. Solid State Lett.* **1998**, 1, 220.

[2.277] J. Coronas, M. Menendez, and J. Santamaria, *Ind. Engng. Chem. Res.* **1995**, 34, 4229.

[2.278]   A.L.Y. Tonkovich, R. Secker, E. Reed, E. Roberts, and J. Cox, *Sepn. Sci. Technol.* **1995**, 30, 397.

[2.279]   A.L.Y. Tonkovich, J.L. Zilka, D.M. Jimenez, G.L. Roberts, and J.L. Cox, *Chem. Engng. Sci.* **1996**, 51, 89.

[2.280]   K. Keizer, V.T. Zaspalis, R.S.A. de Lange, M.P. Harold, and A.J. Burggraaf, p 415 in: *Membrane Processes in Separation and Purification*, J.G. Crespo and K.W. Boddeker, Eds., Kluwer Academic Pub., The Netherlands, 1994.

[2.281]   A. Pantazidis, J.A. Dalmon, and C. Mirodatos, *Catal. Today* **1995**, 25, 403.

[2.282]   G. Capannelli, E. Carosini, F. Cavani, O. Monticelli, and F. Trifiro, *Chem. Engng. Sci.* **1996**, 51, 1817.

[2.283]   M.J. Alfonso, A. Julbe, D. Farrusseng, M. Menendez, and J. Santamaria, *Chem. Engng. Sci.* **1999**, 54, 1265.

[2.284]   M.J. Alfonso, M. Menendez, and J. Santamaria, *Catal. Today* **2000**, 56, 247.

[2.285]   C. Tellez, M. Menendez, and J. Santamaria, *AIChE J.* **1997**, 43, 777.

[2.286]   T.M. Raybold and M.C. Huff, *Catal. Today* **2000**, 56, 35.

[2.287]   H. Iwahara, T. Yajima, T. Hibino, and H. Uchida, *J. Electrochem. Soc.* **1993**, 140, 1687.

[2.288]   W. Yang, Y. Ping, Y. Cong, Z. Shao, H. Li, G. Xiong, and L. Lin, *Proc. International Congress on Catalysis with Membrane Reactors-2000*, Zaragoza, Spain, July 3-5, 2000, 35.

[2.289]   M.J. Alfonso, M. Menendez, and J. Santamaria, *Ind. Engng. Chem. Res.* **2001**, 40, 1058.

[2.290]   R.D. Di Cosimo, J.D. Burrington, and R.K. Grasselli, *J. Catal.* **1986**, 102, 234.

[2.291]   R.D. Di Cosimo, J.D. Burrington, and R.K. Grasselli, U. S. Patent 4,571,443, 1986.

[2.292]   S. Azgui, F. Guillaume, B. Taouk, and E. Bordes, *Proc. of the 1st Int. Workshop on Catalytic Membranes*, Lyon-Villeurbanne, France, Sept. 26-28, 1994.

[2.293]   C. Courson, B. Taouk, and E. Bordes, *Catal. Lett.* **2000**, 66, 129.

[2.294]   E. Piera, C.Tellez, J. Coronas, M. Menendez, and J. Santamaria, *Catal.Today* **2001**, 67, 77.

[2.295]   T. Brinkmann, S.P. Perera, and W.J. Thomas, *Chem. Engng. Sci.* **2001**, 54, 2807.

[2.296]   P. Janknecht, P.A. Wilderer, C. Picard, and A. Larbot, *Proc. 6th International Congress on Inorganic Membranes*, Montpellier, France, June 26-30, 2000, 73.

[2.297]   P. Janknecht, P.A. Wilderer, C. Picard, A. Larbot, and J. Sarrazin, *Chem. Engng. Technol.* **2000**, 23, 674.

[2.298]   P. Janknecht, P.A. Wilderer, C. Picard, A. Larbot, and J. Sarrazin, *Ozone-Sci. Engng.* **2000,** 22: (4) 379-392.

[2.299]   M. Hashino, Y. Mori, Y. Fujii, N. Motoyama, N. Kadokawa, H. Hoshikawa, W. Nishijima, and M. Okada, *Water Sci. Technol.* **2000**, 41, 17.

[2.300]   T. Wydeven, P. Wood, and P. Tsuji, *Ozone-Sci. Engng.* **2000**, 22, 427.

[2.301]   I.F.J. Vankelecom and P.A. Jacobs, *Catal. Today* **2000**, 56, 147.

[2.302]   S.Q. Wu, J.E. Gallot, M. Bousmina, C. Bouchard, and S. Kaliaguine, *Catal. Today* **2000**, 56, 113.

[2.303]   G. Langhendries, G.V. Baron, I.F.J. Vankelcom, R.F. Parton, and P.A. Jacobs, *Catal. Today* **2000**, 56, 131.

[2.304]   C. Espro, F. Arena, F. Frusteri, and A. Parmaliana, *Catal. Today* **2001**, 67, 247.

[2.305]   F. Frusteri, C. Espro, F. Arena, E. Passalacqua, A. Patti, and A. Parmaliana, *Catal. Today* **2000**, 61, 37.

[2.306]   S. Ran, X. Zhang, P.H. Yang, M. Jiang, D.K. Peng, and C.S. Chen, *Solid State Ion* **2000**, 135, 681.

[2.307] H. Hasegawa, K. Kusakabe and S. Morooka, *J. Membr. Sci.* **2001**, 190, 1.
[2.308] N. Nakagawa and M. Ishida, *Ind. Engng. Chem. Res.* **1998**, 27, 1181.
[2.309] I.V. Yentekakis, P.G. Debenedetti, and B. Costa, *Ind. Engng. Chem. Res.* **1989**, 28, 1414.
[2.310] A.F. Sammels, D.R. Peterson, and S.T. Hasford, *Proc. 6th International Congress on Inorganic Membranes*, Montpellier, France, June 26-30, 2000, 33.
[2.311] D. Farrusseng, A. Julbe, D. Cot, and C. Guizard, *Proc. International Congress on Catalysis with Membrane Reactors-2000*, Zaragoza, Spain, July 3-5, 2000, 59.
[2.312] A. Julbe, D. Farrusseng, D. Cot, and C. Guizard, *Catal. Today* **2001**, 67, 139.
[2.313] H. Ohashi, H. Ohya, M. Aihara, Y. Negishi, and S.I. Semenova, *J. Membr. Sci.*, **1998**, 146, 39.
[2.314] J. Fan, H. Ohashi, H. Ohya, M. Aihara, T. Takeuchi, Y. Negishi, and S. Semenova, *J. Membr. Sci.* **2000**, 166, 239.
[2.315] Y.H. Ma, W.R. Moser, S. Pien, and A.B. Shelekhin, *Proc. 3rd International Congress on Inorganic Membranes*, Worcester, MA, USA., July 10-14, 1994, 281.
[2.316] D.J. Edlund and W.A. Pledger, *J. Membr. Sci.* **1994**, 94, 111.
[2.317] M. Kajiwara, S. Uemiya, and T. Kojima, *Int. J. Hydr. Energy* **1999**, 24, 839.
[2.318] P.P.Y Chan, K. Vanidjee, A.A. Adesina, and P.L. Rogers, PL, *Catal. Today* **2001**, 63, 379.
[2.319] H. Alqahtany, P.H. Chiang, D. Eng, A. Robbat, and M. Stoukides, *J. Catal. Lett.* **1992**, 13, 289.
[2.320] T.J. Kirk and J. Winnick, *J. Electrochem. Soc.* **1993**, 140, 3494.
[2.321] D. Peterson and J. Winnick, *J. Electrochem. Soc.* **1996**, 143, L55.
[2.322] J.P. Collins and J.D. Way, *J. Membr. Sci.* **1994**, 96, 259.
[2.323] E.N. Gobina, J.S. Oklany, and R. Hughes, *Ind. Engng. Chem. Res.* **1995**, 34, 3777.
[2.324] C.C. Chang, C.M. Reo, and C.R.F. Lund, *Appl. Catal. B – Envir.* **1999**, 20, 309.
[2.325] S. Pancharatnam, R.A. Huggins, and D.M. Mason, *J. Electrochem. Soc.* **1975**, 122, 869.
[2.326] T.M. Gür and R.A. Huggins, *J. Electrochem. Soc.* **1979**, 126, 1067.
[2.327] D.C. Cicero and L.A. Jarr, *Sepn. Sci. Tech.* **1990**, 25, 1455.
[2.328] T. Hibino, K.I. Ushiki, Y. Kuwahara, A. Masegi, and H. Iwahara, *J. Chem. Soc.Faraday Trans.* **1996**, 92, 2393.
[2.329] T. Hibino, K. Ushiki, and Y. Kuwahara, *Solid State Ion* **1997**, 93, 309.
[2.330] T. Hibino, K. Ushiki, and Y. Kuwahara, *Solid State Ion* **1996**, 89, 13.
[2.331] R.M. Lambert, M. Tikhov, A. Palermo, I.V. Yentekakis, and C.G. Vayenas, *Ionics* **1995**, 1, 366.
[2.332] R.M. Lambert, I.R. Harkness, I.V. Yentekakis, and C.G. Vayenas, *Ionics* **1995**, 1, 29.
[2.333] A. Palermo, M.S. Tikhov, N.C. Filkin, R.M. Lambert, I.V. Yentekakis, O. Marina, and C.G. Vayenas, *J. Catal.* **1996**, 161, 471.
[2.334] M. Marwood and C.G. Vayenas, *J. Catal.* **1997**, 170, 275.
[2.335] O.A. Mar'ina, I.V. Yentekakis, C.G. Vayenas, A. Palermo, and R.M. Lambert, *J. Catal.* **1997**, 166, 218.
[2.336] T. Hibino, *Chem. Lett.* **1994** , 927.
[2.337] T. Hibino, S. Hamakawa, T. Suzuki, and H. Iwahara, *J. Appl. Electrochem.* **1994**, 24, 126.
[2.338] T. Hibino, *J. Appl. Electrochem.* **1995**, 25, 203.
[2.339] T. Hibino, A. Masegi, and H. Iwahara, *J. Electrochem. Soc.* **1995**, 142, L72.
[2.340] T. Hibino, A. Masegi, and H. Iwahara, *J. Electrochem. Soc.* **1995**, 142, 3262.
[2.341] T. Hibino, T. Sato, K.I. Ushiki, and Y. Kuwahara. *J. Chem. Soc. Faraday Trans.* **1995**, 91, 4419.

[2.342] T. Hibino, K. Ushiki, T. Sato, and Y. Kuwahara, *Solid State Ion* **1995**, 81, 1.
[2.343] T. Hibino, K. Ushiki, Y. Kuwahara, and M. Mizuno, *J. Chem. Soc. Faraday Trans.* **1996**, 92, 4297.
[2.344] T. Hibino, K. Ushiki, Y. Kuwahara, and M. Mizuno, *Solid State Ion* **1996**, 89, 13.
[2.345] T. Hibino, T. Inoue, and M. Sano, *J. Electrochem. Soc.* **2000**, 147, 3745.
[2.346] E. Xue, E. O'Keefe, J.B. Breen, and J.R.H. Ross, *Proc. International Congress on Catalysis with Membrane Reactors-2000*, Zaragoza, Spain, July 3-5, 2000, 97.
[2.347] V.T. Zaspalis, W. van Praag, K. Keizer, J.G. van Ommen, J.R.H. Ross, and A.J. Burggraaf, *Appl. Catal.* **1991**, 74, 249.
[2.348] M. Yamamoto, S. Ona, S. Noda, and M. Sadakata, *J. Chem. Eng. Jpn.* **2001**, 34, 834.
[2.349] H.J. Sloot, C.A. Smolders, W.P.M. van Swaaij, and G.F. Versteeg, *AIChE J.* **1991**, 37, 997.
[2.350] H.J. Sloot, G.F. Versteeg, and W.P.M. van Swaaij, *Chem. Engng. Sci.* **1990**, 45, 2415.
[2.351] J.M. Veldsink, G.F. Versteeg, and W.P.M. van Swaaij, *Proc. 1st Int. Workshop on Catalytic Membranes*, Lyon-Villeurbanne, France, Sept. 26-28,1994.
[2.352] H.W.J.P. Neomagus, G. Saracco, and G.F. Versteeg, *Chem. Eng. Commun.* **2001**, 184, 49.
[2.353] H.W.J.P. Neomagus, W.P.M. van Swaaij, and G.F. Versteeg, *J. Membr. Sci.* **1998**, 148, 147.
[2.354] H.W.J.P. Neomagus, G. Saracco, H.F.W. Wessel, and G.F. Versteeg, *Chem. Engng. J.* **2000**, 77, 165.
[2.355] G. Saracco, J.W. Veldsink, G.F. Versteeg, and W.P.M. van Swaaij, *Chem. Engng. Commun.* **1996**, 147, 29.
[2.356] G. Saracco and V. Specchia, *Chem. Engng. Sci.* **2000**, 55, 3979.
[2.357] M.P. Pina, M. Menendez, and J. Santamaria, *Appl. Catal . B.* **1996**, 11, L19.
[2.358] M.P. Pina, S. Irusta, M. Menendez, J. Santamaria, R. Hughes, and N. Boag, *Ind. Engng. Chem. Res.* **1996**, 36, 4557.
[2.359] S. Irusta, M.P. Pina, M. Menedez, and J. Santamaria, *Catal. Lett.* **1998**, 54, 69.
[2.360] G. Saracco and V. Specchia, *Chem. Eng. Sci.*, **2000**, 55, 897.
[2.361] W.A. Jacoby, P.C. Maness, D.M. Blake, and E.J. Wolfrum, *Proc. III International Conference on Catalysis in Membrane Reactors*, Copenhagen, Sept. 8-10, 1998, paper O13.
[2.362] T. Tsuru, T. Toyosada, T. Yoshioka, and M. Asaeda, *J. Chem. Eng. Jpn.* **2001**, 34, 844.
[2.363] F.A. Makhmotov, R.N. Mishkin, and Y.I. Tzareva, *Proc. Euromembrane'97*, June 23-27, 1997, University of Twente, the Netherlands, 259.
[2.364] F.A. Makhmotov, R.N. Mishkin, *Proc. Euromembrane'99*, September 19-22, 1999, Leuven, Belgium, 547-548.
[2.365] F.A. Makhmotov and Y.I. Tzareva, *Proc. 6th International Congress on Inorganic Membranes*, Montpellier, France, June 26-30, 2000, 136.
[2.366] R. Molinari, M. Mungari, E. Drioli, A. Di Paola, V. Loddo, L. Palmisano, and M. Schiavello, *Catal. Today* **2000**, 55, 71.
[2.367] R. Molinari, C. Grande, E. Drioli, L. Palmisano, and M. Schiavello, *Catal.Today* **2001**, 67, 273.
[2.368] R. Molinari, M. Borgese, E. Drioli, L. Palmisano, and M. Schiavello, *Ann. Chim.* **2001**, 91, 197.
[2.369] M.A. Artale, V. Augugliaro, E. Drioli, G. Golemme, C. Grande, V.Loddo, R. Molinari, L. Palmisano, and M. Schiavello, *Ann. Chim.* **2001**, 91, 127.

[2.370] M.A. Artale, V. Augugliaro, A. Di Paola, E. Garcia-Lopez, V. Loddo, G. Marci, L. Palmisano, and M. Schiavello, 310 *Fresenius Environ. Bull.*, **2001**, 10, 310.

[2.371] K. Sopajaree, S.A. Qasim, S. Basak, K. Rajeshwar, *J. Appl. Electrochem.* **1999**, 29, 533.

[2.372] K. Sopajaree, S.A. Qasim, S. Basak, and K. Rajeshwar, *J. Appl. Electrochem.* **1999**, 29, 1111.

[2.373] C.R. Binkerd, Y.H. Ma, W.R. Moser, and A.G. Dixon,"An Experimental Study of the Oxidative Coupling of Methane in Porous Ceramic Radial-Flow Catalytic Membrane Reactors", *Proc. 4th International Congress on Inorganic Membranes*, Gatlinburg, TN, July 14-18, 1996, 441.

[2.374] C. Lange, S. Storck, B. Tesche, and W.F. Maier, *J. Catal.* **1998**, 175, 280.

[2.375] C.K. Lambert and R.D. Gonzalez, *Catal. Lett.* **1999**, 57, 1.

[2.376] S. Uemiya, N. Sato, H. Ando, T. Matsuda, and E. Kikuchi, *Appl. Catal.* **1991**, 67, 223.

[2.377] E. Kikuchi, *Proc. 1st Int. Workshop on Catalytic Membranes*, Lyon-Villeurbanne, France, Sept. 26-28, 1994.

[2.378] E. Kikuchi, *Catal. Today* **2000**, 56, 97.

[2.379] E. Kikuchi, Y. Nemoto, M. Kajiwara, S. Uemiya, and T. Kojima, *Catal. Today* **2000**, 56, 75.

[2.380] A.M. Adris, J.R. Grace, C.J. Lim, and S.S.E.H. Elnashaie, U.S. Patent Appl. 07965011. Canad. Patent Appl. 2081170, 1992.

[2.381] A.M. Adris, C.J. Lim, and J.R. Grace, *Chem. Engng. Sci.* **1994**, 49, 5833.

[2.382] A.M. Adris and J.R. Grace, *Ind. Engng. Chem. Res.* **1997**, 36, 45.

[2.383] L. Mleczko, T. Ostrowski, and T. Wurzel, *Chem. Engng. Sci.* **1996**, 51, 3187.

[2.384] J.R. Grace, X. Li, and C.J. Lim, *Catal. Today* **2001**, 64, 141.

[2.385] R.G. Minet, S.P. Vasileiadis, and T.T. Tsotsis, *Proc. of Symp. on Natural Gas Upgrading*, 1992, 37, 245.

[2.386] Y. Inagaki, K. Haga, H. Aita, K. Sekita, H. Koiso, and R. Hino, *J. At. Energy Soc. Jpn.* **1998**, 40, 59.

[2.387] T. Tagawa, M. Ito, and S. Goto, *Appl. Organomet. Chem.* **2001**, 15, 127.

[2.388] E. Kikuchi and A. Chen, *Stud. Surf. Sci. Catal.* **1997**, 107, 547.

[2.389] T.M. Raybold and M.C. Huff, *Proc. 16th Meeting of the North American Catalysis Society*, Boston, MA, May 30 - June 4, 1999, paper C-028.

[2.390] T. Ioannides and X. Verykios, *Catal. Lett.* **2000**, 36, 165.

[2.391] A.K. Prabhu and S.T. Oyama, *Chem. Lett.* **1999**, 3, 213.

[2.392] W.J. Onstot, R.G.Minet, and T.T. Tsotsis, *Ind. Eng. Chem. Res.* **2001**, 40, 242.

[2.393] K. Hou, M. Fowles, and R. Hughes, *Chem. Engng. Sci.* **1999**, 54, 3783.

[2.394] K. Hou, M. Fowles, and R. Hughes, *Catal. Today* **2000**, 56, 13.

[2.395] K. Aasberg-Petersen, C.S. Nielsen, and S.L. Jorgensen, *Catal. Today* **1998**, 46, 193.

[2.396] P.E. Hojlund Nielsen, K. Aasberg-Petersen, and S. Laegsgaard Jorgensen, *Proc. III International Conference on Catalysis in Membrane Reactors*, Copenhagen, Sept. 8-10, 1998, paper O26.

[2.397] Y.M. Lin, G.L. Lee, and M.H. Rei, *Catal. Today* **1998**, 44, 343.

[2.398] S. Hara, W.C. Xu, K. Sakaki, and N. Itoh, *Ind. Engng. Chem. Res.* **1999**, 38, 488.

[2.399] Y.M. Lin and M.H. Rei, *Catal. Today* **2001**, 67, 77.

[2.400] Y.M. Lin and M.H. Rei, *Int. J. Hydrogen Energy*, in press.

[2.401] N. Itoh, S. Hara, and K. Sazaki, *Proc. International Congress on Catalysis with Membrane Reactors-2000*, Zaragoza, Spain, July 3-5, 2000, 91.

[2.402] F. Frustreri, A. Mezzapica, F. Arena, A. Parmaliana, A. Tavolaro, A. Basile, A. Regina, and E. Drioli, *Proc. International Congress on Catalysis with Membrane Reactors-2000*, Zaragoza, Spain, July 3-5, 2000, 151.

[2.403] B. Obradovic and J.H. Meldon, *Proc. 10th Annual Meeting, NAMS*, Cleveland, p. 52, May 16-20, 1998.

[2.404] B. Cales and J. F. Baumard, *High Temp. - High Press* **1982**, 14, 681.

[2.405] J. Lede, F. Lapicque, J. Villermaux, B. Cales, A. Ounali, J. F. Baumard, and A.M. Antony, *Int. J. Hydrogen Energy* **1982**, 7, 939.

[2.406] H. Naito and H. Arashi, *Solid State Ion* **1995**, 79, 366.

[2.407] Y. Nigara, K. Watanabe, K. Kawamura, J. Mizusaki, and M. Ishigame, *J. Electrochem. Soc.* **1997**, 144, 1050.

[2.408] A.K. Demin and L.A. Dunyushkina, *Solid State Ion* **2000**, 135, 749.

[2.409] J. Guan, S.E. Dorris, U. Balachandran, and M. Liu, *Solid State Ion* **1997**, 100, 45.

[2.410] H. Iwahara, *Solid State Ion* **1988**, 28-30, 573.

[2.411] M. Nagata and H. Iwahara, *J. Appl. Electrochem.* **1993**, 23, 275.

[2.412] H. Iwahara, T. Hibino, and T. Sunano, *J. Appl. Electrochem.* **1996**, 26, 829.

[2.413] A. Kogan, *Int. J. Hydrog. Energy* **1998**, 23, 89.

[2.414] A. Kogan, *Int. J. Hydrog. Energy* **2000**, 25, 1043.

[2.415] A. Kogan, E. Spiegler, and M. Wolfshtein, *Int. J. Hydrog. Energy* **2000**, 25, 739.

[2.416] V.M. Gryaznov, S.G. Gulyanova, Y.M. Serov, and V.D. Yagodovskii, *Zh. Fiz.* **1981**, 55, 1306.

[2.417] Y.M. Serov, V.M. Zhernosek, S.G. Gulyanova, and V.M. Gryaznov, *Kinet. and Catal.* **1983**, 24, 303.

[2.418] H. Ohya, J. Fun, H. Kawamura, K. Itoh, H. Ohashi, M. Aihara, S. Tanisho, and Y. Nagishi, *J. Membr. Sci.* **1997**, 131, 237.

[2.419] K.R. Sridhar and B.T. Vaniman, *Solid State Ion* **1997**, 93, 321.

[2.420] Y. Nigara and B. Cales, *Bull. Chem. Soc. Jpn.* **1986**, 59, 1997.

[2.421] N. Itoh, C.M.A. Sanchez, W.C. Xu, K. Haraya, and M. Hongo, *J. Membr. Sci.* **1993**, 77, 245.

[2.422] H. Nishiguchi, A. Fukunaga, Y. Miyashita, T. Ishihara, and Y. Takita, *Adv. Chem. Conv. Mit. Carb. Diox.* **1998**, 114, 147.

[2.423] Y. Uemura, Y. Ohzuno, and Y. Hatate, *Proc. 5th International Congress on Inorganic Membranes,* Nagoya, June 22-26, 1998, 620.

[2.424] R.P.W.J. Struis, S. Stucki, and M. Wiedorn, *J. Membr. Sci.* **1996**, 113, 93.

[2.425] J. Vital, A.M. Ramos, I.F. Silva, H. Valenic, and J.E. Castanheiro, *Catal. Today* **2000**, 56 167.

[2.426] J. Vital, A.M. Ramos, I.F. Silva, and J.E. Castanheiro *Catal. Today* **2001**, 67, 217.

[2.427] J.N. van de Graaf, M. Zwiep, F. Kapteijn, and J.A. Moulijn, *Appl. Catal. A- Gener.* **1999**, 178, 225.

[2.428] J.N. van de Graaf, M. Zwiep, F. Kapteijn, and J.A. Moulijn, *Chem. Engng. Sci.* **1999**, 54,1441.

[2.429] R. Pruschek, G. Oleljeklaus, V. Brand, G. Haupt, G. Zimmermann, and J.S. Ribberink, *Energy Conversion and Management* **1995**, 36, 397.

[2.430] M. Bracht, P.T. Alderliesten, R. Kloster, R. Pruschkek, G. Haupt, E. Xue, J.R. Ross, M.K. Koukou, and N. Papayannakos, *Energy Conv. Mangmnt.* **1997**, 38, S159.

[2.431] E. Xue, M.O. Keefe, and J.R.H. Ross, *Catal. Today* **1996**, 30, 107.

[2.432] E. Kikuchi, S. Uemiya, N. Sato, H. Inoue, H. Ando, and T. Matsuda, *Chem. Lett.* **1989**, 489.

[2.433] S. Uemiya, N. Sato, H. Ando, and E. Kikuchi, *Ind. Engng. Chem. Res.* **1991**, 30, 585.

[2.434] T. Esaka, T. Fujii, K. Suwa, and H. Iwahara, *Solid State Ion.* **1990**, 40, 544.

[2.435] V. Violante, A. Basile, and E. Drioli, *Fus. Engng. Des.* **1993**, 22, 257.

[2.436] A. Basile, A. Criscuoli, F. Santella, and E. Drioli, *Gas Sep. Pur.* **1996**, 10, 243.

[2.437] S. Tosti and V. Violante, *Fus. Engng. Des.* **1998**, 43, 93.

[2.438] S. Tosti, V. Violante, A. Basile, G. Chiappetta, S. Castelli, M. De Francesco, S. Scaglione, and F. Sarto, *Fusion Eng. Des.* **2000**, 49, 953.

[2.439] S.A. Birdsell and R.S. Willms, *Fus. Engng. Des.* **1998**, 39-4, 1041.

[2.440] R.L. Espinoza, E. du Toit, J. Santamaria, M. Menendez, J. Coronas, and S. Irusta, *Proc. International Congress on Catalysis with Membrane Reactors-2000*, Zaragoza, Spain, July 3-5, 2000, 5.

[2.441] T.M. Gür and R.A. Huggins, *Solid State Ion* **1981**, 5, 563.

[2.442] T.M. Gür and R.A. Huggins, *Science* **1983**, 219, 967.

[2.443] T.M. Gür, H. Wise, and R.A. Huggins, *J. Catal.* **1991**, 129, 216.

[2.444] V. Sinha and K. Li, *Desalination* **2000**, 127, 155.

[2.445] X.Y. Tan and K. Li, *Chem. Engng. Sci.* **2000**, 55, 1213.

[2.446] K. Li and X.Y. Tan, 5073 *Chem. Eng. Sci.*, **2001**, 56, 5073.

[2.447] V. Flores and C. Cabassud, *J. Membr. Sci.* **1999**, 162, 257.

[2.448] V. Flores and C. Cabassud, *Desalination* **1999**, 126, 101.

[2.449] P. Drogui, S. Elmaleh, M. Rumeau, C. Bernard, and A. Rambaud, *J. Membr. Sci.* **2001**, 186, 123.

[2.450] J.S. Choi, I.K. Song, and W.Y. Lee, *Korean J. Chem. Engng.* **2000**, 17, 280.

[2.451] J.S. Choi, I.K. Song, and W.Y. Lee, *Catal. Today* **2000**, 56, 275.

[2.452] J.S. Choi, I.K. Song, and W.Y. Lee, *J. Membr. Sci.* **2000**, 166, 159.

[2.453] M.A. Salomon, J. Coronas, M. Menendez, and J. Santamaria, *Appl. Catal. A-Gen.* **2000**, 200, 201.

[2.454] M.T. Hicks and P.S. Fedkiw, *J. Electrochem. Soc.* **1998**, 145, 3728.

[2.455] J. Feldman, I.W. Shin, and M. Orchin, *J.Appl. Polym. Sci.* **1987**, 34, 969.

[2.456] J. Feldman and M. Orchin, *J. Mol. Catal.* **1990**, 63, 213.

[2.457] M. Naughton and R. Drago, *J. Catal.* **1995**, 155, 383.

[2.458] J.S. Kim and R. Datta, *AIChE J.* **1991**, 37, 1657.

3 Pervaporation Membrane Reactors

Pervaporation membrane reactors (PVMR) are an emerging area of membrane-based reactive separations. An excellent review paper of the broader area of pervaporation-based, hybrid processes has been published recently [3.1]. The brief discussion here is an extract of the more comprehensive discussions presented in that paper, as well as in an earlier paper by Zhu *et al.* [3.2]. Mostly non-biological applications are discussed in this chapter. Some pervaporation membrane bioreactor (PVMBR) applications are also discussed; additional information on the topic can be found in a recent publication [3.3], and a number of other examples are also discussed in Chapter 4.

Pervaporation goes back to the beginning of the membrane field. The term seems to have been coined first by Kober [3.4], who studied the phenomenon of selective evaporation of water out of a closed cellulose nitrate bag. It was not until the late seventies, however, that significant interest in pervaporation got started [3.5]. Pervaporation (PV) has established, itself, in recent years, as a promising membrane technology that is potentially useful in such applications as the dehydration of organic compounds, the removal/recovery of organic compounds from aqueous solutions, and the separation of close-boiling organic mixtures (e.g., structural isomers) and azeotropic mixtures. Polymeric membranes have been traditionally utilized in pervaporation (see Table 3.1 for a list of various polymeric membranes and their most common applications), though more recently zeolite membranes are also finding applications (see further discussion to follow).

Table 3.1. Polymeric pervaporation membranes and their respective applications. Adapted from [3.5].

Polymer	Application
Cellulose and derivatives	Extraction of water from of ethanol, separation of benzene/cyclohexane mixtures
Chitosan	Extraction of water from an aqueous solution of ethanol
Collagen	Extraction of water from solutions of alcohols and acetone
Cuprophane	Extraction of water from an aqueous solution of ethanol
Ion-exchange resins	Extraction of water from solutions of ethanol, pyridine
LDPE	Separation of C_8 isomers
NBR	Separation of benzene/n-heptane
Nylon-4	Extraction of water from an aqueous solution of ethanol
PA	Extraction of water from solutions of ethanol, acetic acid
PA-co-PE	Separation of dichloroethane/trichloroethylene mixtures
PAA	Extraction of water from solutions of ethanol, acetic acid
PAN	Extraction of water from an aqueous solution of ethanol
PAN-co-AA	Extraction of water from an aqueous solution of ethanol
PB	Extraction of 1-propanol, ethanol from an aqueous solution
PC	Extraction of water from solutions of ethanol and acetic acid

Table 3.1. Polymeric pervaporation membranes and their respective applications. Adapted from [3.5].

Polymer	Application
PDMS filled with silicalite, molecular sieves, etc	Extraction of alcohols from an aqueous solution, separation of butanol from butanol/oleyl alcohol mixture
PEBA	Extraction of alcohols and phenol from an aqueous solution, recovery of natural aromas
PI	Extraction of water from solutions of ethanol and acetic acid, separation of benzene/cyclohexane and acetone/ cyclohexane mixtures
Plasma polymerized fluorine-containing polymers	Extraction of ethanol from an aqueous solution
Plasma polymerized PMA	Separation of organic/organic mixtures
Polyion complexes	Extraction of water from an aqueous solution of ethanol
Polysulfones	Extraction of water from solutions of ethanol, acetic acid
POUA	Separation of benzene/n-hexane mixtures
PP	Separation of xylene isomers
PPO	Extraction of water from an aqueous solution of alcohols, separation of benzene/cyclohexane mixtures
PTMSP/PDMS composite	Extraction of ethanol from an aqueous solution
PTMSP and derivatives	Extraction of ethanol from an aqueous solution
PUR	Extraction of ethanol from an aqueous solution
PVA	Extraction of water from an aqueous solution of alcohols, acetic acid, ethers, pyridine, etc.
PVC	Extraction of water from an aqueous solution of ethanol
Silicone rubber (PDMS, etc.)	Extraction of alcohols, ketones, hydrocarbons, halogenated hydrocarbons, amines, acetic acid, natural aromas, etc., from an aqueous solution

AA: acrylic acid; LDPE: low density polyethylene; NBR: poly(butadiene-acrylonitrile); PA: polyamide; PAA: poly(acrylic acid); PAN: polyacrylonitrile; PB: polybutadiene; PC: polycarbonate; PDMS: polydimetylsiloxane; PE: polyester; PEBA: polyetheramide-block-polymer; PI: polyimide; PMA: poly(methyl acrylate); POUA: poly(oxyethylene urethane acrylate); PP: polypropylene; PPO: poly(phenylene oxide); PTMSP: poly(trimethylsilylpropyne); PUR: polyurethane; PVA: poly(vinyl alcohol); PVC: poly(vinyl chloride).

PV is often applied in combination with another technology as a hybrid process. Of these, PV-distillation and PV-reaction hybrid processes are already finding industrial applications [3.1]. For PV-membrane-based reactive separations, as shown in Figure 3.1, the membrane either removes the desired product (mostly in biotechnological/wastewater applications – see further discussion to follow) or the undesired product (e.g., water for esterification reactions). In either case, as is true throughout the book, to qualify for further discussion the membrane and reactor must be mutually affecting each other's operation; that is systems for which the membrane simply separates the reaction products without any further close coupling to the reactor are not discussed. Existing or proposed industrial processes for application of PVMR include the production of ethyl and butyl acetate, ethyl and *n*-butyl oleate, diethyltartrate, dimethyl urea, ethyl valerate, isopropyl and propyl propionate, and methylisobutylketone, just to name a few.

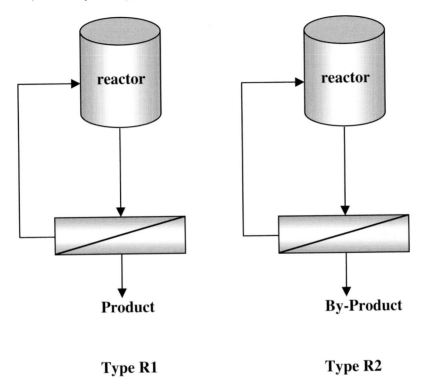

Figure 3.1. Two different types of PVMR. Adapted from [3.1].

Pervaporation membrane reactors are not a recent discovery. The use of a PVMR was proposed in a U.S. patent dating back to 1960 [3.6]. Though the technical details on membrane preparation and experimental apparatus were rather sketchy, the basic idea was described there, namely, the use of a water permeable polymeric membrane to drive an esterification reaction to completion. A more detailed description of a PVMR can be found in a later European patent [3.7], which described the use of a flat membrane (commercial PVA or Nafion®) placed in the middle of a reactor consisting of two half-cells. The reaction studied was the acetic acid esterification reaction with ethanol. For an ethanol to acetic acid ratio of 2, liquid hourly space velocities (LHSV) in the range of 2–5, and a temperature of 90 °C complete conversion of the acetic acid was reported. The use of PVMR for this reaction shows promise for process simplification, as indicated schematically in Figure 3.2, which shows a side-by-side comparison of a conventional and a proposed PVMR plant for ethyl acetate production.

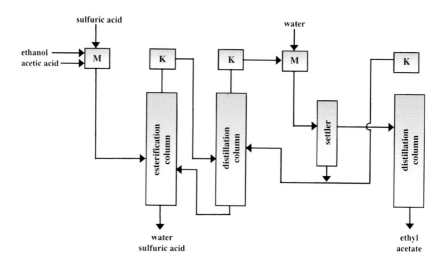

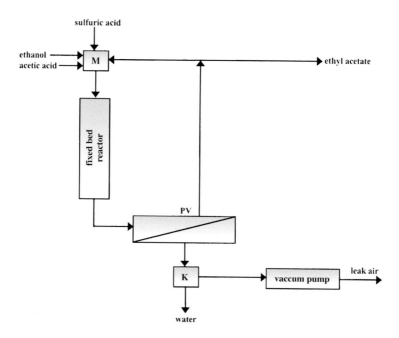

Figure 3.2. Process flow charts for ethyl acetate production. Top conventional plant, bottom proposed PVMR plant. M: mixer, K: condenser. Adapted from [3.8].

As with the more conventional membrane reactor applications, PVMR and PVMBR systems range, as shown schematically in Figure 3.3, from the simpler designs, where the reactor and membrane are housed in separate units, to the more elaborate configurations, for which the membrane and reaction functions are incorporated into the same unit. The advantages and disadvantages of the two different configurations relevant to PVMBR applications can be found in Table 3.2. Some of the same advantages and disadvantages also apply equally to PVMR systems.

Table 3.2. Advantages and disadvantages of the two PVMBR configurations of Figure 3.3. Adapted from [3.3].

Layout	Advantages	Disadvantages
External PV Unit	High efficiency (membrane area per reactor volume) Simple membrane module replacement during operation High cell density combined with slow addition of nutrition stimulate the productivity of some microorganisms	Risk of oxygen limitation in external loop Problems of sterilization of the external loop may occur Membrane flux may decline due to fouling Fermentation broth may be very viscous Physical stress on cells may increase during recirculation caused by shear forces Large reactors may exhibit inhomogeneity in recirculation loop and reactor
Internal PV Unit	No extra circulation in reactor is required Simple operation High cell density combined with slow addition of nutrients stimulate productivity of certain microorganisms	Inflexible system Membrane flux susceptible to fouling Limited membrane area per volume Fermentation medium may be viscous and contribute to concentration polarization

The principal classes of reactions that have been studied by PVMR are esterification reactions. Kita *et al.* [3.9], for example, combined a batch reactor, containing the reacting mixture of ethanol with oleic acid (2:1 or 3:1 molar ratio at 85 °C), together with *p*-toluenesulfonic acid as a catalyst, with a tubular polymeric/ceramic membrane placed on its top (see Figure 3.4). The membrane was prepared by dip-coating a commercial polyetherimide (Ultem®, GE) on the top of a mesoporous alumina membrane. The alcohol and water vapors escaping from the reactor would pass through the membrane, which removed water selectively, and then condensed and recycled them back to the reactor. The PVMR resulted in complete conversion, with the more permselective membranes resulting in superior performance. In a later study the same group [3.10] used a polyimide hollow-fiber laboratory module. The combined process provided almost complete conversion in a short reaction time. The membrane module worked stably for a long period. Bagnell *et al.* [3.11] used Nafion® membranes to study the esterification reactions of acetic acid with methanol and butanol respectively.

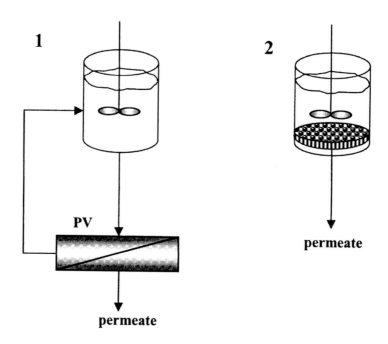

Figure 3.3. Two different PVMR designs. Adapted from [3.3].

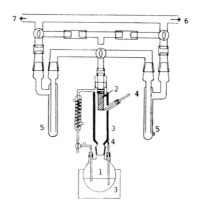

Figure 3.4. The PVMR of Kita and coworkers [3.9]. With permission from The Chemical Society of Japan.

Their laboratory PVMR consisted of a reservoir in which the reactants were placed together with Nafion® pellets, which acted as the catalyst. The liquid in the reservoir was continuously recirculated through the membrane tube, which was placed externally to the reactor. The membrane, itself, was also shown to be catalytic. A flow of inert gas (rather than vacuum) was used to remove the vapors and water from the membrane permeate. For the methanol esterification reaction the improvement in yield was modest (final conversion 77 % vs. 73 % corresponding to equilibrium), because the membrane was not very permselective towards the reaction products. Significant improvements, on the other hand, were observed with the butanol reaction (final conversion 95 % vs. 70 % corresponding to equilibrium), as the membrane is more permselective towards the products of this reaction. Exchanging the acidic protons in the Nafion® membranes with cesium ions significantly improved the permselectivity, but also reduced membrane permeance.

Dams and Krug [3.12] analyzed three different PVMR configurations, shown in Figure 3.5, in which the reactor and the hydrophilic membranes are housed in separate units. The first process design combines a MR with a PV unit and a distillation column. In the MR ethanol is reacted with acetic acid to produce ethyl acetate and water. The vapor phase of the reactor is fed into a distillation column. The top stream from the distillation column containing 87 % ethanol and 13 % water is fed for further treatment into the PV unit. The bottom stream is mostly water.

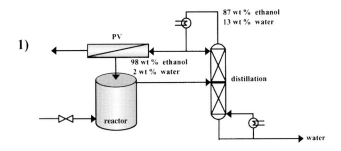

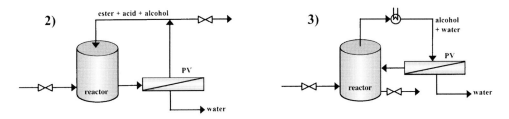

Figure 3.5. Three different PVMR process designs analyzed by Dams and Krug. Adapted from [3.3].

The retentate steam from the PV unit is recycled back into the reactor. In the second design the PV unit treats directly the liquid phase from the esterification reactor. This design attained a high conversion. It required, however, an acid resistant membrane and the PV unit permeate, which contained the acid and ester, needed appropriate disposal. In the third process design, shown in Figure 3.5, the PV unit treated the MR vapor phase. The advantage of such process layout is that it does not require an acid-resistant membrane and that the PV unit permeate, which is free from ester and acid, does not require further processing. The second design above was shown to be most efficient requiring only 7 % of the energy required for the conventional process. The energy savings for the first process design were 58 %, and for the third design 78 %. The first design is more likely, though, to be adapted, as a retrofit to an existing conventional process.

Bitterlich and coworkers [3.13] studied a PVMR process for the production of butyl acetate. The conventional process uses mineral acid (H_2SO_4) as a catalyst. In the process, the acid must be neutralized with NaOH, before removing the water by distillation. In the process studied by Bitterlich *et al.* [3.13], the catalytic action is provided by the acidic functionality of an anion exchange resin, thus, eliminating the need for neutralization. The distillation column is replaced by a PV unit, resulting in energy savings. Domingues *et al.* [3.14] studied the reaction of benzyl alcohol with acetic acid using *p*-toluene-sulfonic acid as the catalyst. A PVA membrane external to the batch reactor was to remove the produced water and to recycle the rest of the reacting mixture to the reactor vessel. The PVMR attained a conversion, which was much higher than equilibrium. Bruschke *et al.* [3.15] described the process layout of a medium-sized esterification plant employing a PVMR system that was used for the production of various esters. As shown in Figure 3.6 it consists of a cascade of two batch reactors, each individually coupled to two PV units. The plant attained a 97 % conversion with a final water concentration in the product of less than 0.1 %. The construction of a larger esterification plant utilizing PVMR technology was also mentioned in this report.

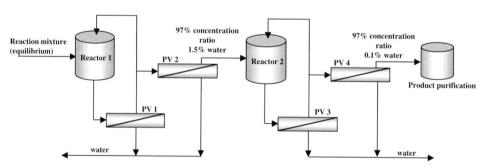

Figure 3.6. A PVMR pilot-plant for esterification. Adapted from Lipnizki *et al.* [3.1].

Some of the earlier reported applications, incorporating the membrane and reactor in the same unit were by Kita and coworkers [3.16, 3.17, 3.18, 3.19]. Kita *et al.* [3.16] used a batch reactor system containing flat, asymmetric hydrophilic membranes (polyetherimide, polyimide, Nafion®, chitosan) to study the esterification of oleic acid with ethanol, and acetic acid with ethanol. The presence of the membrane improved the reactor conversion beyond equilibrium, and, thus, reduced the reaction time and the amount of reactants needed to attain a target production rate. For the oleic acid and ethanol esterification reaction, for example, for ethanol to oleic acid feed ratios of 2 to 3 at 75 °C using the polyetherimide membranes and for long residence times of over 6 h complete conversion of the acid was reported. They also presented a simple model to account for the membrane reactor behavior. Interestingly, they noted that the esterification reaction did not observe mass action kinetics (based on concentrations). The ester and the acid were also found to plasticize the membrane especially at the higher temperatures. Okamoto *et al.* [3.17] studied in the same PVMR, using an asymmetric polyimide membrane, the condensation reaction of acetone with phenol catalyzed by a cation-exchange resin to produce bisphenol-A. The use of the PVMR significantly shortened the reaction time, and made it possible to repeatedly use the catalyst without requiring a drying treatment. The use of PVMR is quite promising for this reaction, since water acts as an inhibitor, and its *in situ* removal accelerates the reaction rate. Ni *et al.* [3.20] prepared modified aromatic polyimide membranes, which exhibited a moderate water/ethanol selectivity. The membranes were utilized in a PVMR to study the esterification reaction of $CH_3(CH_2)_3COOH$ (valeric acid) with ethanol to produce the ester, which finds use as a food additive. Removing the water through the membrane allowed the conversion of valeric acid to reach 95.2 %.

A batch PVMR containing a flat poly-vinyl-alcohol (PVA) membrane was utilized by David *et al.* [3.21, 3.22]. They studied the esterification reaction between propanol and propionic acid using *p*-toluenesulfonic acid as a catalyst. The presence of the membrane was found to accelerate the esterification reaction with the conversion of the PVMR exceeding the calculated equilibrium conversion. David *et al.* [3.21, 3.22] also presented a simple model to simulate the effect of the various operating parameters on reactor performance (a mathematical analysis for this type reactor was also presented by Feng and Huang [3.23]). The PVA membrane was shown, however, to be affected by the reaction environment. This was due to the fact that the hydroxyl groups on the PVA chains were, themselves, esterified by the acid in the presence of the homogeneous *p*-toluenesulfonic acid catalyst. David *et al.* [3.24] have suggested, instead, the use of a blend membrane consisting of a cross-linked PVA matrix together with poly-styrene-sulfonic acid (PSSA). The PSSA component embedded in the PVA matrix provides the catalytic function, while the PVA matrix, itself, provides the separation. To enhance separation David *et al.* [3.24] recommend casting the PVA-PSSA blend membrane as a thin layer on the top of a PVA pervaporation membrane. A PVMR using flat PVA membranes supported on porous steel plates was used by Chen and coworkers [3.25, 3.26] for the esterification of *n*-butanol with acetic acid. The membranes are rendered catalytic by impregnating them with $Zr(SO_4).4H_2O$ an inorganic salt. Significant improvements in yield were observed when using these membranes.

Keurentjes *at al.* [3.27] modelled the production of diethyltartrate in a PVMR by the reaction between tartaric acid and ethanol, which is described as follows:

Keurentjes *at al.* [3.27] carried out experiments to study the kinetics of this reaction, and fitted their data in terms of an activities dependent mass-action kinetic scheme. They utilized a simple model of a PVMR to study the effect of the ratio of membrane area (A) to reactor volume (V) on PVMR performance based on the measured rate constants and literature values for the water and ethanol permeances through a PVA membrane (PVMR models are further discussed in Chapter 5). Keurentjes *et al.* [3.27] report that with the aid of the PV membrane the equilibrium can be significantly shifted towards the diethyltartrate product. There is a certain optimum (A/V) ratio. This is because when the (A/V) ratio is too low, the water removal rate is too slow to have any influence, while for high values of this ratio too much ethanol is removed.

Tubular continuous PVMRs have been studied by Waldburger *et al.* [3.28], Waldburger and Widmar [3.8], and by Zhu *et al.* [3.2] for the production of ethyl acetate. In the reactor of Waldburger *et al.* [3.28] acetic acid and ethanol were mixed, and then recycled through the membrane reactor. The membrane reactor, shown schematically in Figure 3.7, consisted of a PVA supported membrane and a bed of heterogeneous catalysts placed in the annular volume between the membrane and the reactor. For a single reactor at 80 °C the ethyl acetate yield achieved was 92.1 % and the water concentration in the product was 0.5 wt. %. For three such reactors in series the ethyl acetate yield was 98.7 % and the water content in the product was 0.1 wt. %. Waldburger *et al.* [3.28] also discussed the economics of a number of process layouts involving the PVMR. Two such designs are shown in Figure 3.8. The first design consists of a cascade of two interconnected PVMR, while the second design consists of three PVMR connected serially. These two designs were compared with a conventional ethyl acetate plant shown at the bottom of the same Figure. The design consisting of the three PVMR in series had the lower capital and investment costs, followed by the conventional plant, and the cascade of the two PVMR. Zhu *et al.* [3.2] presented a mathematical model for this type of reactor, which took into account the thermodynamic non-idealities of the reactive liquid phase. A more comprehensive version of this model has, since, been developed by Park [3.29]. These models will be discussed further in Chapter 5.

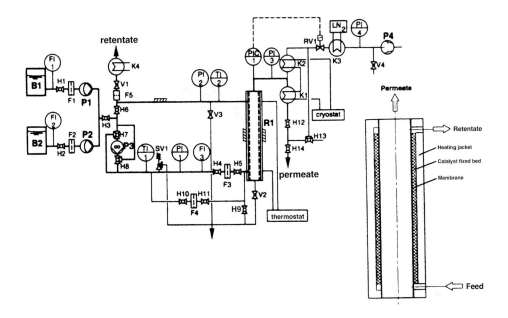

Figure 3.7. The PVMR system of Waldburger *et al.* [3.8]. Left, flow chart of the PVMR system. B: container, F: filter, H: stopcock, K: condenser, P: pump, R1: membrane reactor, V: valve. Right, schematic of the membrane reactor. With permission from John Wiley and Sons.

There are also a limited number of studies reporting the application of PVMR's to re-actions other than esterification. Herion *et al.* [3.30], for example, reported an experi-mental pilot plant scale investigation, in which a PV unit employing a PVA-based hydro-philic membrane was coupled to a reactor producing dimethyl urea (DMU). DMU production according to the BASF process is based on the reaction of methylamine with CO_2 to form DMU and water. It results in an aqueous solution of DMU, which also con-tains unreacted CO_2 and methylamine. In the conventional process this solution is sepa-rated by a distillation column. Prior to entering the distillation column the CO_2 in the so-lution is neutralized by NaOH in order to avoid the formation of solid carbamate deposits in the column. The use of the PV unit utilizing a PVA membrane allowed the removal of most of the water, and direct recycle of the concentrated CO_2 and amine into the reactor, thus, resulting in reduced energy costs, and in a reduction of the amount of by-products that must be disposed. For example, in the PVMR process there is an 86 % reduction in the amine to be separated, and a 91 % reduction in the CO_2 by-product, which implies similar reductions in the amount of NaOH required for neutralization and in the amount of salt content of the wastewater that is being produced. With over 3000 h of operation the membrane showed no decrease in performance.

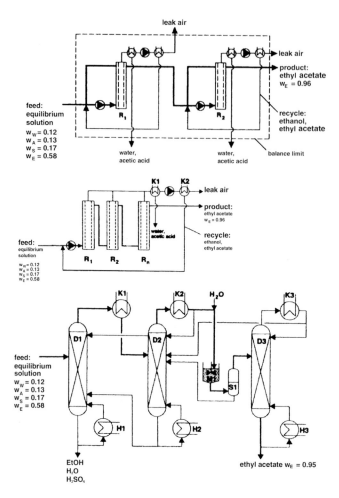

Figure 3.8. Various ethyl acetate production plants. Top, cascade of two PVMR's. Middle, three PVMR's in series. Bottom, a conventional plant. D: distillation column, H: heating, K: condenser, M: mixer, R: reactor, S: separator. From [3.8], with permission from John Wiley and Sons.

Matouq *et al.* [3.31] tested two types of catalysts an ion exchange-resin (the H⁺ form of Amberlyst® 15) and a heteropolyacid (HPA) in the production of MTBE from methanol and *t*-butyl alcohol (TBA). Both were shown, active, but the ion-exchange resin showed poor selectivity, producing substantial amounts of by-product isobutylene (IB). Matouq *et al.* [3.31] tested the production of MTBE using the ion-exchange resin in a reactive distillation column. It was difficult to test the HPA catalyst in the reactive distillation system, however, because its particle size was too small and was carried out by the liquid phase. Matouq *et al.* [3.31] proposed, instead, the use of a PVMR incorporating a PVA membrane. As shown in Figure 3.9, in the proposed system the PVMR is coupled with a conventional distillation unit. Both types of catalysts were tested in this PVMR/distillation hybrid system. Irrespectively of the choice of catalyst, the product from the top of the column was an azeotropic mixture of MTBE with ethanol. Matouq *et al.* [3.31] report that the removal of water through the PVA membrane favorably impacted on system performance. The same group [3.32] utilized a similar PVMR system in order to study the production of ETBE from the reaction of EtOH and TBA. In this study they utilized a hollow fiber membrane module external to the batch reactor through which the liquid mixture was circulated for water removal. The membrane was also more permselective towards water. Combining the PVMR and the distillation unit provided a top product with a higher fraction of ETBE. In addition, in the absence of the membrane a phase separation occurred in the top product because of the presence of larger fraction of water.

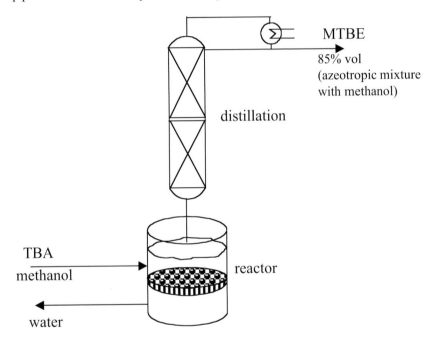

Figure 3.9. A PVMR system coupled to a distillation column. Adapted from Matouq *et al.* [3.31].

The use of a PVMR system has been suggested by Staudt-Bickel and Lichtenthaler [3.33, 3.34] for the production of methylisobutylketone (MIBK), an important solvent for paints and protective coatings. MIBK is produced according to the following three-step synthesis scheme:

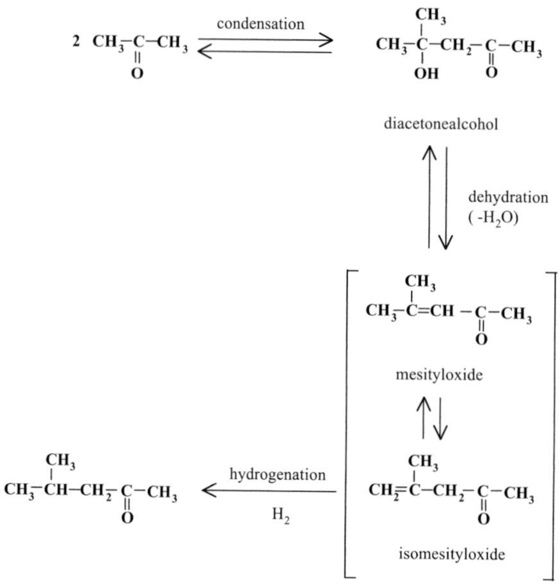

$$2\ CH_3\text{-}C\text{-}CH_3 \underset{\longleftarrow}{\overset{\text{condensation}}{\longrightarrow}} CH_3\text{-}\underset{OH}{\overset{CH_3}{C}}\text{-}CH_2\text{-}\underset{O}{C}\text{-}CH_3$$

diacetonealcohol

dehydration
(-H$_2$O)

$$\left[\quad CH_3\text{-}C\text{=}CH\text{-}\underset{O}{\overset{CH_3}{C}}\text{-}CH_3 \right.$$

mesityloxide

$$CH_3\text{-}\underset{O}{\overset{CH_3}{C}}\text{-}CH_2\text{-}C\text{-}CH_3 \left. \quad \right]$$

isomesityloxide

$$CH_3\text{-}CH\text{-}CH_2\text{-}\underset{O}{\overset{CH_3}{C}}\text{-}CH_3 \underset{H_2}{\overset{\text{hydrogenation}}{\longleftarrow}}$$

The first step, in the above scheme, is the aldol condensation of acetone, the second is a dehydration step, and the third step is a hydrogenation reaction. Today a multifunctional Pd/sulfonated organic acid ion-exchange catalyst is utilized for this reaction. Water, however, acts as an inhibitor for the reaction and must, therefore, be removed, as it accumulates in the pores of the polar ion-exchange catalyst, and prevents the less polar organic intermediate mesityloxide from reaching the Pd surface. In the conventional process shown on the top part of Figure 3.10, the product mixture after leaving the reactor (which operates at 30 bar and 150 °C) is cooled down to 100 °C, and H$_2$ is first separated in a gas-liquid separator. By-products (isopropanol or 2-methyl-pentane) and unreacted acetone are separated in a separation column, and acetone is returned to the batch reactor. The bottom mixture containing the MIBK is separated in a liquid-liquid separator into an aqueous phase containing small amounts of MIBK that must be distilled, and an organic phase that is treated in the product recovery column. The proposed PVMR system is shown at the bottom of Figure 3.10. After hydrogen separation the product mixture enters the PV unit without cooling. The permeate contains water and some acetone. The retentate containing very little water (~0.1 wt %) is fed for further processing into a second reactor. The advantage of the PVMR system is that acetone conversion is almost double that of the conventional system, significantly simplifying further downstream processing.

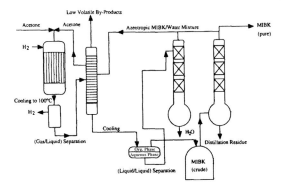

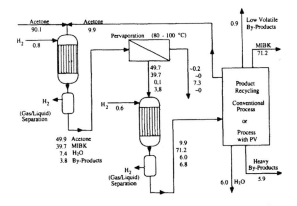

Figure 3.10. Conventional and PVMR systems for MIBK production. From Staudt-Bickel and Lichtenthaler [3.33], with permission from Elsevier Science.

In recent years the use of silica or zeolite microporous membranes has been attracting growing interest in PVMR applications. For example, the use of silica membranes for the production of coating resins by a polycondensation reaction in a PVMR system was reported by Bakker *et al.* [3.35]. The membrane showed good permselectivity to water coupled with a high permeance. Van der Gulik *et al.* [3.36] have studied the hydrodynamics of this type of PVMR, since they are important when attempting to reduce concentration and temperature polarization to obtain high water fluxes during operation. The influence of secondary flow, induced by small density differences, on polarization, was studied using CFD in a model system operated in three parallel flow situations, horizontal, vertical opposed, and vertical adding flow. Density-induced convection is found to be most effective in the horizontal situation increasing water fluxes by up to 50 %.

Zeolite membranes, in particular, show potential for widespread pervaporation-type applications. They are, typically, prepared either by the hydrothermal route (direct or secondary growth, in which a layer of seed crystals are first applied on the support and during hydrothermal synthesis only crystal growth occurs), or by the dry-gel conversion technique [3.37, 3.38]. In the latter technique the gel of the parent zeolite is first applied on the support, dried and then crystallized in the presence of vapors of ethylenediamine, triethylamine and steam. Zeolite membranes have shown good performance in the pervaporation separation of water/ethanol mixtures. Depending on the type of zeolite either ethanol or water permeate selectively through the membrane. For example, ethanol has been reported to permeate selectively through a silicalite membrane with a permeance as high as 0.2 kg/m^2·h and a separation factor of 60 at 30 °C [3.39]. Zeolite A membranes, on the other hand, are hydrophilic. For the water/ethanol mixture separation factors as high as 10,000 have been reported with a water permeance of 1.1 kg/m^2·h [3.40]. Mitsui Engineering and Shipbuilding Co, Ltd. has reported the installation of a prototype commercial PV unit using zeolite A membranes. An 87 wt. % isopropanol mixture was dehydrated to 99.95 wt. % at 95 °C. A stable (over 3000 h) performance of the unit was reported. Most recently the same company reported [3.41] the delivery of the first large-scale pervaporation plant using NaA zeolite membranes delivered to Sanwa Oil Chemical company in Japan, which produces 600 l/h of solvents (EtOH, IPA, acetone, etc.) containing less than 0.2 % wt. of water from a feed solution containing 90 % wt. solvents at 120 °C.

A membrane reactor using a H-ZSM5 membrane was used by Bernal *et al.* [3.42] to carry out the esterification reaction of acetic acid with ethanol. An equimolar ethanol/acetic acid liquid mixture was fed in the membrane interior, while He gas was used as an inert sweep on the shell-side. In this particular application the membrane, itself, provides the catalysis for the reaction. NaA and T-type zeolite membranes have been utilized for esterification reactions in a PVMR and in a vapor permeation membrane reactor (VPMR) by Tanaka *et al.* [3.43, 3.44]. Both membranes are hydrophilic and show good separation characteristics towards a number of alcohols. The NaA membrane was used to study the oleic/acid esterification in a vapor permeation membrane reactor (VPMR) at 383 K. The reaction is carried out in a batch reactor in the presence of *p*-toluenesulfonic acid as a catalyst, and the water and ethanol vapors circulate through the NaA membrane, which removes the water vapor. The VPMR was shown to reach complete conversion. The NaA, which is unstable in the presence of acids, performed stably under such conditions. The T-zeolite membrane is more stable under acidic conditions, and was used for the study of the esterification of acetic acid with ethanol at 343 K in a PVMR with the membrane being in direct contact with the reaction medium.

An interesting application of a PVMR has recently been reported by Theis *et al.* [3.45]. They have prepared PEBA membranes loaded with various catalytic metals (Pd, Rh, and combinations of Pd/Rh). They studied in their PVMR the catalytic hydrogenation of 4-chlorophenol described by the following set of reactions.

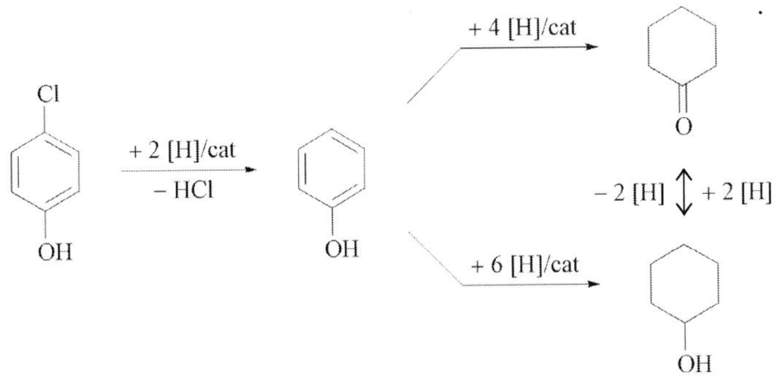

Cl

+ 4 [H]/cat

+ 2 [H]/cat
– HCl

O

– 2 [H] ↕ + 2 [H]

+ 6 [H]/cat

OH   OH   OH

A 4-chlorophenol (0.45 % in water) solution was placed on one side of the membrane and hydrogen was bubbled through it. The permeate side was kept at 1 mbar. Theis *et al.* [3.45] report that increasing the PVMR temperature increases the overall flux but it decreases the concentration of the organics in the permeate side. Surprisingly the production of phenol and cyclohexanone decreased with increasing temperature, which the authors attributed to the incomplete supply of hydrogen (which has a very low solubility in water). Comparing the action of the various catalysts, Pd alone was shown to produce pure phenol, while increasing amounts of added Rh enhanced the yield of cyclohexanone and cyclohexanol.

A number of pervaporation membrane bioreactor (PVMBR) applications have also been reported [3.3]. These represent a special class of membrane bioreactor applications, which are discussed more extensively in Chapter 4. A number of studies utilizing PVMBR involve esterification reactions. Van der Padt *et al.* [3.46], for example, studied in a PVMBR the synthesis of triglycerides from glycerol and fatty acids. The reaction is equilibrium limited, and is described as follows:

Fatty acid + glycerol ⇌ monoglyceride + water

Fatty acid + monoglyceride ⇌ diglyceride + water

Fatty acid + diglyceride ⇌ triglyceride + water

It can be chemically or enzymatically catalyzed (enzymatic synthesis proceeds at moderate temperatures, thus, avoiding the polymerization of unsaturated fatty acids, with an additional advantage that the products can usually be classified as natural ingredients). During equilibrium a mixture of mono-, di-, and triesters are produced.

One way to produce an excess of triglycerides is to *in situ* remove the water during synthesis. Van der Padt *et al.* [3.46] proposed doing that using an enzymatic PVMBR. The enzyme (lipase from *Candida rugosa*) is immobilized in the lumen side of hollow-fiber cellulose acetate membranes, in which the organic phase (which consists of a mixture of mono-, di-, and triglycerides and decanoic acid in hexadecane as the solvent) was circulated. Air circulating in the shellside of the membrane was used to control the water activity (α_w) with the use of a condenser. Van der Padt *et al.* [3.46] report an excess of triglyc-erides for sufficiently low water activities in the shellside. At the start of each experiment the enzymatic activity was independent of water activity. Enzymatic stability was influ-enced by the time on stream. After 600 h of operation the remaining activity is 26 % of the initial activity for α_w=0.1, and 71 % for α_w=0.45. A similar PVMBR system, shown schematically in Figure 3.11, utilizing the same enzyme was studied by Ujang and co-workers [3.47, 3.48] for the esterification of decanoic acid with dodecanol. They observed that there is a certain water activity of the sweep stream, which optimizes PVMBR per-formance. Under the optimum conditions the PVMBR was shown to give ester yields as high as 97 %.

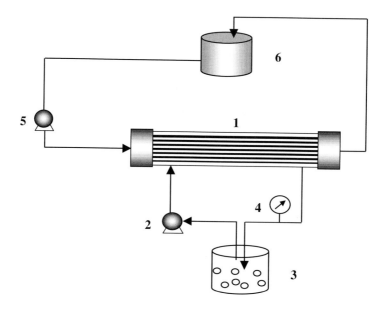

Figure 3.11. A hollow fiber PVMBR. Adapted from Ujang and Hazri [3.48].

The enzymatic esterification reaction between erucic acid and hexadecyl alcohol to form hexadecyl eruciate was studied by Nijhuis *et. al.* [3.49] in a PVMBR system utiliz-ing an external PV unit employing hollow-fiber membranes. The reaction is catalyzed by the lipase Lipozyme (produced by the fungus *Mucor Miehel*), which was immobilized on a synthetic anionic resin. The PVMBR attained over 90 % conversion, while the esterifi-cation reactor alone was limited by equilibrium (~53 % conversion). Nijhuis *et al.* [3.49] note that in the optimal design of the PVMBR one must be cautious, as there is a certain concentration of water that is needed for optimal enzyme performance. Rhee and cowork-ers [3.50, 3.51] studied, in a similar PVMBR system, the production of butyl oleate by li-pase-catalyzed esterification in *n*-hexane, using a hydrophilic cellulose acetate membrane. Lipase from *Candida rugosa* showed optimal activity in the range of water activities (α_w) of 0.52 to 0.65 at 30 °C. During the experiments the rate of ester formation from the enzymatic esterification was allowed to be the same as the rate of water removal in order to maintain an optimal (α_w) of the reaction system. The reaction rate with (α_w) control was increased two-fold from the respective value without PV, and the PVMBR conversion was 92 % as opposed to 61.1 % without pervaporation.

In other PVMBR applications that have been studied, so far, the PV membrane is used to either remove secondary metabolites (e.g., benzaldehyde, 6-phenyl-a-pyrone) or main fermentation reaction products, and by-products. Some of the key applications are shown in Table 3.3. One such application is ethanol production. Ethanol is used increasingly as an oxygenate gasoline additive. Due to high feedstock prices and competition from other products for its gasoline uses, the economics of the production of this renewable fuel have been marginal for many manufacturing facilities. Improvements of the fuel ethanol production process resulting in cost-savings of even 2–5 cents per gallon could significantly increase its demand [3.52]. One approach to accomplish this would be to convert the conventional batch or sequential-batch fermentation process employed in the industry to a continuous fermentation, by integrating the ethanol recovery operation with the fermentation step; this would have the added advantage of maintaining the ethanol concentration in the fermentor at levels, which are not inhibitory to the fermenting organism. A number of reactive separation processes for continuous ethanol recovery have been proposed; among these membrane pervaporation [3.53, 3.54, 3.55, 3.56, 3.57, 3.58, 3.59, 3.60, 3.61, 3.62, 3.63] shows good promise. The primary benefit of the PVMBR-based systems (and of the other reactive separation-based processes) is an increased fermentor volumetric productivity, which means needing a smaller fermentor capacity. Advantages of the PVMBR-based processes are their simplicity, lesser toxicity to fermenting organisms, and the recovery of a concentrated ethanol stream requiring less distillation capacity and energy consumption. Disadvantages include the need for low temperature condensation of permeate vapors and, of course, membrane cost.

Table 3.3. Some key PVMBR applications (the reactor type refers to Figure 3.3). Adapted from [3.3].

Microorganism	Reactor Type	Membranes	Module	Productivity and Conversion Rate	Reference
Separation of Acetone and Butadienol					
Bacillus subtilisK AJI992	1	PDMS PEBAX®	Plate	Increased productivity	Detterwiler *et al.* [3.64, 3.65]
Separation of Acetone-Butanol-Ethanol (ABE)					
C. acetobutylicum	1	Polyorgano-siloxane	Hollow Fiber	Production costs within an economically interesting range	Qureshi and co-workers [3.66, 3.67]
C. acetobutylicum	1	Polyorgano-siloxane	Tube	3 to 4 times higher separation in comparison to stripping, hence increased productivity	Sodeck *et al.* [3.68]

Table 3.3. (cont.)

Microorganism	Reactor Type	Membranes	Module	Productivity and Conversion Rate	Reference
C. saccharoper-butyl-acetonicum DSM 2152	1	PDMS/POMS	Pocket	Increased productivity and glucose conversion rate	Bengtson *et al.* [3.69]
C. acetobutylicum P262 supported on packed bed of bonechar	1	Polypropylene	Hollow Fiber	20 times higher productivity than the batch process	Friedl *et al.* [3.70]
Separation of Benzaldehyde					
Bjerkandera adusta	1	PDMS	Plate	Increased productivity by shifting the Benzaldehyde <=> Benzylalcohol equilibrium	Lamer *et al.* [3.71]
Separation of Butanol					
C. saccharoper-butyl-acetonicum DSM 2152	1	Polyorgano-siloxane	Tube	+ 10 % sugar conversion	Groot and Luyben [3.72]
C. saccharoper-butyl-acetonicum DSM 2152	1	Polyorgano-siloxane	Tube	Increased glucose conversion rate, 130 % in comparison to batch process	Groot *et al.* [3.73]
C. acetobutylicum DSM 792^T, ATCC 824	2	A silicone coil immersed in the reactor	Tube	Increased Glucose conversion and higher solvent production than batch process	Larrayoz and Puigjaner [3.74]
C. acetobutylicum B18	2	Polyorgano-siloxane	Tube	Increased glucose conversion rate.	Geng and Park [3.75]
Separation of Isopropanol- Butanol-Ethanol (IBE)					
C. beyerinckii LMD 27.6, Clostridium species. DSM 2152	1	Silicone tube, or silicone loaded with 60 wt. % Silicalite	Tube, Plate	Increased productivity and glucose conversion 4 times that of batch process	Groot *et al.* [3.76, 3.77, 3.78, 3.79]
Separation of Ethanol					
S. cerevisiae. ATCC 26603	1	PDMS	Hollow Fiber	Stable 1 month operation	Gudernatsch *et al.* [3.80]

Table 3.3. (cont.)

Microorganism	Reactor Type	Membranes	Module	Productivity and Conversion Rate	Reference
Baker's yeast	1	Polysiloxane, PP, PTFE	Plate	+ 306 % productivity in comparison to batch process; also increased glucose conversion and ethanol production	Nakao *et al.* [3.81]
S.cerevisiae CBS 8066	1	PV membranes silicone rubber; MF membrane	Hollow Fiber	More than 3 times gain in substrate conversion and ethanol productivity	Groot *et al.* [3.57, 3.58]
C. thermohydro-sulfuricum.YM3	1	PTFE impregnated with silicone	Plate	+225 % Productivity +346 % Ethanol production +314 % Starch conversion in comparison to batch process	Mori and Inaba [3.56]
k.A.	1	PDMS	Plate	Higher Ethanol production costs	Bemquerer-Costa [3.82]
S. cerevisiae supported on polymeric beads	1	Polysiloxane on asymmetric polysulfone (Udel® 3600) support	Plate, Tube	+ 500 % in comparison to batch process. 50 days of stable continuous operation.	Shabtai *et al.* [3.53]
Candida pseudotropicalis	1	Polysiloxane	Tube	+ 500 % in comparison to batch process, + 100 % in comparison to continuous process	Shabtai and Mandel [3.55]
Separation of 6-pentyl-a-pyrone (Lactone 6-PP)					
Trichoderma viride IMB TV 4.89	1	PDMS. PDMS/ POMS, PEBA	Plate	+ 700 % in comparison to batch process	Bengtson *et al.* [3.83, 3.84]

In one of the early studies Cho and Hwang [3.54] studied the integration of continuous fermentation and membrane separation using ethanol selective silicone rubber hollow-fiber membranes. Relative to conventional continuous fermentation, the performance of PVMBR resulted in high yeast cell densities, reduction of ethanol inhibition, longer substrate residence time of, more glucose consumption, and recovery of clean and concentrated ethanol. A 10–20 % increase in ethanol productivity was achieved. Kaseno *et al.* [3.62] have carried out three types of fermentation experiments, namely batch fermentation as the standard process, fed-batch fermentation without PV, and fed-batch fermentation with PV in order to evaluate the effect of ethanol removal on ethanol fermentation. Glucose and immobilized baker's yeast were used in the fermentations. A hydrophobic porous membrane made of polypropylene (PP) was used for the PVMBR process (coupled to a MF membrane used to remove suspended impurities and cell debris in the broth before entering the PV module – see further discussion below on the coupling of PV with MF or UF). The rate of ethanol production in the PVMBR process was 2 times higher than in the fermentation without PV. The amount of wastewater was 38.5 % of that discharged from the conventional batch process. O' Brien *et al.* [3.61] in a laboratory study coupled a continuous yeast fermentation system with a flat-plate PV membrane system to continuously recover an ethanol enriched stream from the fermentation broth. The feed was a concentrated dextrose solution, whose flow rate was automatically adjusted to match the permeate flow through the membrane in order to maintain the liquid level in the fermentor constant. The PV module contained commercially available PDMS membranes (Membrane Products, Kiryat Weizmann Ltd, Rehovot, Israel), which consistently produced an ethanol enriched (~20-23 wt.%) permeate while maintaining the ethanol content in the fermentor at a level of 4-6 wt.%. The PVMBR system exhibited very good operational stability, ethanol productivities, and substrate utilization. The PV system required periodic washing during operation in order to limit fouling due to the presence of the biomass.

O'Brien *et al.* [3.52] have recently published a study, whose purpose was to evaluate the economic feasibility of a PVMBR process for ethanol production. Total annual costs for the fermentation, distillation, and dehydration sections were compared for a standard corn dry-milling process and a process for which the batch fermentors have been replaced by a PVMBR system. The process flow diagram for a conventional 50 million gallon/year dry milling ethanol plant is shown in Figure 3.12. The process flow diagram for the PVMBR-based process is also shown in the same figure. A process simulator with cost estimating and economic evaluation capabilities (Aspen Plus, Aspen Technology, Inc., Cambridge, MA) was used for the design and costing of the plants. The costs for both cases are referenced back to July 1992. The inputs to the model for the PVMBR process were from laboratory experiments utilizing commercial polydimetylsiloxane membranes [3.63]. The cost comparison between the base case and the fermentation pervaporation case is presented in Table 3.4. Major differences exist in required equipment sizes and utility requirements for the two processes. For the conventional process sixteen 250,000-gallon fermentors were required. Due to its much higher volumetric productivity the PVMBR process uses only one 330,000-gallon fermentor. In the distillation section, for both cases, the overhead stream is 95 % vol. ethanol, but column designs differed. The column feed for the base-case contained soluble and suspended solids, which required a column of 87 disc and donut trays in the stripping section, and 28 sieve trays in the rectifying section. For the PVMBR the feed was a clean 42 wt % ethanol/water mixture, and a column with 30 sieve trays was required. The PVMBR required a refrigeration plant. There were differences in the costs between the two processes. Fermentor capital costs for the PVMBR were only 25 % of that for the base-case. For distillation and dehydration, total annual costs for the PVMBR case were 61.5 % lower, with the savings coming from reduced capital costs (42.4 %) and utility costs (57.6 %). However, the PVMBR process entails additional costs for the membrane modules and the refrigeration system and, as a result, the costs for the PVMBR case are slightly higher. The costs of the PVMBR process are sensitive to the membrane variables and, of course, to the cost of the membranes (for the base-case the membrane cost was assumed $200/m^2). The effect of membrane charac-teristics (flux and selectivity) and costs are shown in Figure 3.13. The analysis indicates that only modest improvements in either flux, or selectivity, or cost from the values used in the base case could make the PVMBR cost competitive.

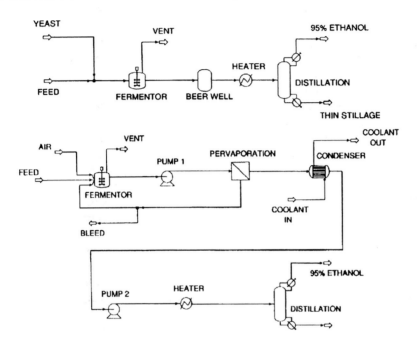

Figure 3.12. A conventional (top), and a PVMBR-based ethanol plant (bottom). From [3.52], with permission from Elsevier Science.

Table 3.4. Ethanol production costs in US$: fermentation-pervaporation case vs. batch fermentation case. Adapted from [3.52].

	Fermenta-tion	Pervapo-ration	Distillation and dehy-dration	Refrigera-tion	Total an-nual	Unit costs (US$/Gal)
Fermenta-tion/pervaporation						
Capital cost total	4,797,900	19,847,600	7,410,600	8,579,900	4,636,000	0.813
Capital cost annualized (based on 9 year life)	533,000	2,205,000	823,000	953,000	4,515,000	0.090

Table 3.4. (cont.)

	Fermentation	Pervaporation	Distillation and dehydration	Refrigeration	Total annual	Unit costs (US$/Gal)
Utilities						
Steam	0	0	1,582,200	505,700	2,087,900	0.042
Cooling water	159,400	0	156,400	140,500	456,300	0.009
Electric power	18,600	35,800	18,200	37,600	110,200	0.002
Membrane replacement		898,400			898,400	0.018
Annual utilities	178,000	934,000	1,757,000	684,000	3,553,000	0.071
Plant labor	126,000	523,000	195,000	226,000	1,070,000	0.021
Supplies	119,900	496,000	185,300	214,500	1,016,000	0.020
General works	48,000	198,000	74,300	85,700	406,000	0.008
Total annual cost	1,004,900	4,356,600	3,304,300	2,163,200	10,560,000	0.211
Batch fermentation costs						
Capital cost total	19,175,314	0	3,565,679		32,740,993	0.655
Capital cost annualised (based on 9 year life)	2,131,000	0	1,507,000		3,638,000	0.073

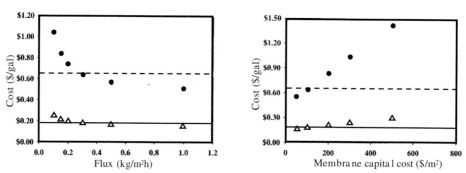

Figure 3.13. The effect of membrane flux and costs on process economics. ●: fermentation/pervaporation capital cost; Δ: fermentation/pervaporation anual cost, --- : base case capital cost; —— : base case annual cost. Adapted from [3.52].

The production of butanol using a PVMBR has also been studied by a number of investigators [3.85]. Interest in the production of butanol (and as noted above of alternative fuels in general) from renewable resources has been growing. Butanol offers a number of advantages, when compared to other fuels (e.g., ethanol and 2,3-butanodiol), which include a higher calorific value, low freezing point, and immiscibility with water. Production of butanol by fermentation, however, is hampered by serious product inhibition, with a maximum allowable total solvent acetone+butanol+ethanol (ABE) concentration of 20 g/l when using either *Clostridium acetobutylicum* or *C. beijerinckii* strains. This, of course, means that in the final product one must remove 980 g of water to recover 20 g of solvents, out of which butanol is only 8–13 g. The amount of energy required to recover the solvents by distillation, under such conditions, is unfortunately higher than their total energetic content [3.70]. Two approaches to improve the situation have been followed. The first involves the development of more efficient strains, which are tolerant to higher solvent concentrations. The other involves the use of reactive separation techniques for *in situ* removal of the solvents, most notable among them PVMBR systems.

Groot *et al.* [3.86] investigated the technical feasibility of five reactive separation technologies (fermentation coupled to stripping, adsorption, liquid-liquid extraction, pervaporation, and membrane solvent extraction). They concluded that liquid-liquid extraction and pervaporation reactive separation processes show the greatest potential, with PVMBR systems particularly attractive due to their operational simplicity. Membranes utilized include silicone [3.76, 3.77, 3.74, 3.87, 3.75, 3.85, 3.88], supported liquid membrane systems [3.87, 3.89], polypropylene [3.70], and silicalite filled PDMS membranes [3.90, 3.91]. The results with PVMBR systems have been very promising.

Friedl *et al.* [3.70] studied ABE production in an integrated process using *C. acetobutylicum* immobilized onto a packed bed of bonechar coupled with continuous product removal by pervaporation. Using a concentrated feed solution containing lactose at 130 g/l, a lactose utilization value of 98.7 % was observed. The PVMBR system showed low acid loss, a high solvent yield of 0.39 g solvents/g lactose utilized, and a solvent productivity of 3.5 g/l.

Qureshi and Blaschek [3.85, 3.88] used a silicone membrane and a hyper-butanol producing mutant strain *C. beijerinckii BA101*. The integrated PVMBR process provided up to 200 % improvement in productivity. The PVMBR system allowed the use of a much more concentrated glucose solution and produced no acids. The PV membrane did not negatively affect the performance of *C. beijerinckii BA101*. Though in the presence of the active broth butanol selectivity decreased by a factor of 2–3, membrane flux remained unchanged. Most recently Qureshi and Blaschek [3.92] used a PVMBR to compare the performance of *Clostridium beijerinckii* BA101 (mutant strain) and *C. beijerinckii* 8052 (wild type) in terms of their substrate and butanol inhibition. The wild-type strain is more strongly inhibited by butanol than is the mutant strain. In the batch reactor *C. beijerinckii* BA101 produced 25.3 g/l of total ABE solvents, whereas the PVMBR experiment produced 165.1 g/l of total solvents. Solvent productivity increased from 0.35 (batch reactor) to 0.98 g/l·h (fed-batch reactor). The current membranes in addition to butanol allow also acetone and ethanol to permeate through. The development of more permselective membranes or the use of a hybrid distillation/PVMBR system may offer advantages here.

ABE production was studied by Qureshi *et al.* [3.91] in a laboratory PVMBR system, which coupled a fed-batch fermentor (using *C. acetobutylicum*) with an UF membrane and a PV product recovery system using a silicalite-silicone composite membrane. Cells of *C. acetobutylicum* were removed from the culture using the 500,000 molecular weight cut-off UF membrane (A/G Technology Corporation, Needham, MA) and returned to the fed-batch fermentor. The ABE was removed from the ultrafiltration permeate using the silicalite-silicone composite pervaporation membrane (306 µm thick) prepared in their laboratory. The use of the UF membrane was dictated by the fact that the fermentor operates at 35 °C while the PV unit operates at 78 °C. The fed-batch reactor was operated for 870 h (ten cycles). Totally 155 g/l of solvents were produced at solvent yield of 0.31–0.35, both significantly higher than the corresponding values in the batch reactor. During the operation the silicalite-silicone composite membrane was exposed to fermentation broth for 120 h. Membrane flux remained constant and acetone/butanol selectivity was also not affected, indicating that the membrane was not fouled by the ABE fermentation broth. Acetic acid and ethanol did not diffuse through the silicalite-silicone composite membrane at low concentrations.

Qureshi and Maddox [3.67] studied the economics of producing ABE from whey permeate or industrial molasses in a PVMBR system consisting of a continuous fluidized-bed reactor using *C. acetobutylicum* cells immobilized on bonechar and coupled to a silicone membrane system for product removal. Interestingly in their study Qureshi and Maddox conclude that the single most important factor influencing ABE price is the cost of the whey permeate (or the molasses). A membrane selectivity of 10 was considered in their calculations. Raising the selectivity to 180 seemed to have little effect on the ABE price (<10%). Similarly, increasing the flux through the membrane seemed to also have little of an effect (~$0.06/l price reduction for a fivefold improvement in flux).

Groot *et al.* [3.78, 3.79] studied glucose fermentation using either a CSTR or a fluidized bed bioreactor coupled to a pervaporation unit for isopropanol-butanol-ethanol (IBE) production. They utilized *C. beijerinckii* LMD 27.6 or *Clostridium* DSM 2152, which they immobilized on alginate beads. They used two types of PV modules, one built in their laboratory containing silicone membrane tubes, and a commercially available plate-and-frame type PV module with a flat silicone membrane containing silicalite (60 % vol.). The experiments with the laboratory PV module were promising. The substrate consumption rate, for example, could be increased by a factor of 4 when compared with the case without any product recovery. Experiments with the commercial PV module were less successful, however, encountering problems with sterilization and plugging by microbial cells. Groot *et al.* [3.79] also presented a model, which fitted reasonably well the experimental data. The model was used to provide further insight into the changes of process parameters that are needed for further optimization. The model indicates that high productivity and high substrate consumption is possible by matching the feed flow rate and the alcohol recovery rate. The accumulation of by-products in the medium, due to the simultaneous removal of water, had a significant impact on performance, and this effect was dependent on membrane selectivity.

As noted above (and as is true more generally with membrane bioreactors) PVMBR show significant improvements in productivity over the more conventional non-integrated bioreactor systems. Some of the challenges facing the PVMBR systems are the same facing the more conventional membrane bioreactor systems. The continuous removal of fermentation products by the PV membrane leads, for example, to an enrichment in the fermentor of the various impermeable components. Hydrophobic PV membranes are impermeable, for example, to salts and various organic acids, and, in particular, acetic and lactic acid, which are key by-products of many fermentation processes. Mori and Inaba [3.56], for example, studied the production of ethanol from the direct fermentation of uncooked starch using the anaerobic, thermophilic bacterium *Clostridium thermohydrosulfuricum* YM3, and a silicone rubber membrane impregnated in the pores of a microporous polytetrafluoroethylene (PTFE) membrane. Thermophilic microorganisms may be better suited for PVMBR systems as they allow operation at higher temperatures more favorable to membrane operation. They report significant gains in starch consumption (3.1 times), ethanol production (3.4 times), and ethanol productivity (2.2 times) over the case in the absence of alcohol recovery. They also report, however, that the PVMBR operation had to be stopped after 168 h because of the accumulation of acetic and lactic acids in the fermentor leading to the death of the cells. To avoid this problem one of the simplest approaches is using a continuous side cell (or medium) bleed from the fermentor and replacing it with fresh feed (see Figure 3.14). Such an approach was utilized, for example, by Nakao *et al.* [3.81] during ethanol production from glucose fermentation.

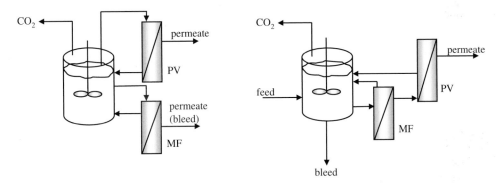

Figure 3.14. PVMBR incorporating a MF membrane. Adapted from Lipnizki *et al.* [3.3].

Another key challenge in PVMBR applications is membrane fouling as a result of the adsorption of macromolecular or colloidal components, when the membrane comes in contact with the fermentation broth. Selecting the appropriate membrane materials and conditions of operation in the PV unit or using immobilized cells and enzymes tends to minimize fouling problems. An idea is to combine PV with another membrane operation like microfiltration (MF) or ultrafiltration [3.91] to remove the offensive colloidal/macromolecular components before they cause damage to the PV membrane. Two different PVMBR configurations, which incorporate a MF unit are shown in Figure 3.14. If membrane fouling becomes unavoidable, then membrane cleaning becomes part of the operation (with a negative impact on economics – see further discussion in Chapter 4). Membrane cleaning can take place by either periodically shutting down the whole operation [3.71] for membrane clean-up, or continuously by employing two separate parallel PV units, one of which stays in use while the other is being cleaned. Shabtai *et al.* [3.53] followed the latter approach. By combining membrane periodic cleaning with reactor medium bleeding and replacement they were able to maintain a stable, continuous operation for over 50 days.

Combining MF and PVMBR systems has been systematically studied by Groot *et al.* [3.57, 3.58] for ethanol production. Groot *et al.* [3.57] investigated experimentally three PVMBR systems, shown schematically as III, IV, and V in Figure 3.15, for continuous EtOH production from glucose fermentation using a baker's yeast strain *Saccharomyces cerevisiae* CBS8066 and commercial PV and MF membranes. In the conventional PVMBR system the fermentation broth is directly recycled through the PV module, where ethanol and water permeate through the membrane. EtOH removal leads into a decrease of its concentration in the broth, thus, in principle, decreasing product inhibition and increasing productivity and substrate conversion. A cell-bleed, as previously noted, helps prevent the build-up of toxic by-products.

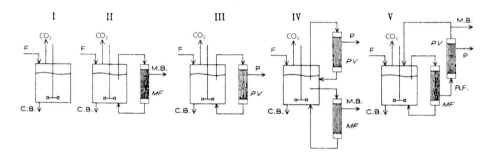

Figure 3.15. Different processes for EtOH production. F: feed; C.B.: cell bleed; M.B.: medium bleed; P: product of pervaporation; R.F.: recirculation flow. From Groot *et al.* [3.58], with permission from Springer-Verlag.

Since the membrane also removes substantial amounts of water (only a very modest EtOH/H_2O selectivity < 5.7 was reported for the commercial silicone rubber membranes utilized) this leads to a higher biomass concentration. In system IV the broth is, in addition, recycled over a MF module for cell retention, thus, providing an additional means for controlling biomass concentration. In system V pervaporation is only applied to the cell-free broth. In comparison to systems III and IV, the PV membrane in system V can be operated at higher temperatures for optimal performance, which results in an increase in flux, and a decrease in the required PV surface area. On the other hand, a larger membrane area for the MF membrane is required, as it must produce the large flow required for product recovery in the PV step. The authors report that over 900 h of continuous operation they encountered no problems with either the MF or the PV membrane modules. The substrate conversion was increased from 118 kg/m^3 in a system without EtOH recovery to 360 kg/m^3 in a system with PV. The productivity was 14 kg/m^3·h in a system with PV and 42 kg/m^3·h in a system, which combines MF with PV. For the PVMBR systems they noted a decrease in the growth and specific substrate production rates when compared to systems without *in situ* EtOH removal, which is in contrast to the basic premise of such systems. They attributed this observation to the inhibitory effects of compounds other than ethanol, which would tend to accumulate in the broth; this observation points out the significance of optimizing the cell and medium bleeds when operating PVMBR systems.

In a subsequent study Groot *et al.* [3.58] studied the economics of the PVMBR systems III, IV and V for EtOH production from industrial molasses, and compared them with the conventional process (I) as well as with a more conventional MBR utilizing a MF membrane (II). The schematic process layout for PVMBR system IV is shown in Figure 3.16. The feed (molasses, nutrients, and process water) is prepared in a mixing tank and is fed to the fermentor. The PV and MF membrane sections are divided in separate loops, each provided with its own circulation pump. These loops can be optimized independently with respect to hydrodynamics, mass and heat transfer, in order to achieve maximum flux. For the PV membrane an EtOH/H_2O selectivity of 20 was assumed, which is higher than the selectivity of the membranes used in the experiments of Groot *et al.* [3.57]. The cell and medium bleeds and the pervaporate are fed directly into a distillation column to produce a 95 wt. % EtOH product. Further details about the other process layouts and the assumptions incorporated in the economic evaluation of the various process designs can be found in the original paper [3.58]. Table 3.5 provides a detailed comparison between the various process designs. The conventional MBR (design II) shows the lower production costs. The authors showed that the PVMBR system IV becomes competitive to the MBR, when the cost of the PV membrane decreases by a substantial factor of 4.

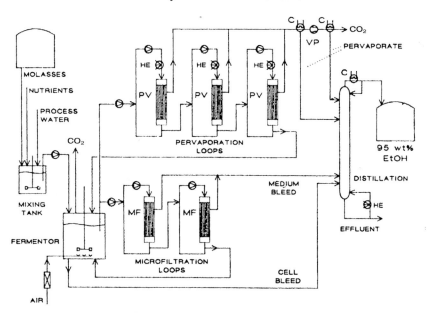

Figure 3.16. Process layout for PVMBR system IV. HE: heat exchanger; C: condenser; VP: vacuum pump. From Groot *et al.* [3.58], with permission from Springer-Verlag.

Table 3.5. Base case cost analysis of the five fermentation membrane configurations. Adapted from [3.58].

Scheme		I	II	III	IV	V
Capacity of the plant	(106 l,/a)	46.0	29.4	28.4	29.4	29.4
Fermentation volume	(m^3)	1,000	50	50	50	50
Productivity	(kg/m^{3}h)	4.69	60	57.8	60	60
Biomass concentration	(kg/m^3)	5.63	150	91.2	150	150
Ethanol concentration	(kg/m^3)	38.0	68.2	28.5	67.4	67.3
Acetic acid concentration	(kg/m^3)	0.23	0.48	4.27	1.13	1.19
Substrate consumption	(kg/m^3)	89.3	172.9	551	350	350
Ethanol in pervaporate	(kg/kg)	0	0	0.371	0.593	0.506
Inlet dilution rate	(h^{-1})	0.125	0.826	0.25	0.408	0.408
Product dilution rate	(h^{-1})	0	0	0.15	0.0726	0.1066
Medium bleed rate	(h^{-1})	0	0.718	0	0.289	0.255
Cell bleed rate	(h^{-1})	0.123	0.0271	0.0767	0.0271	0.0271
Recirculation rate	(h^{-1})	0	0	0	0	1.47
Pervaporation area	(m^2)	0	0	2.500	1.211	1.066
Microfiltration area	(m^2)	0	1,177	0	482	2.449
Investment fermentor	(M$)	7.9	0.79	0.79	0.79	0.79
Investment distillation	(M$)	0.099	0.083	0.082	0.083	0.083
Investment pervaporation	(M$)	0	0	1.97	0.95	0.84
Investment microfiltration	(M$)	0	0.93	0	0.38	1.93
Total capital cost	(M$)	39.08	12.61	22.78	16.60	23.78
Depreciation membranes	($/1)	0	0.024	0.052	0.034	0.070
Molasses and nutrients	($/1)	0.436	0.436	0.436	0.436	0.436
Steam	($/1)	0.017	0.012	0.021	0.014	0.017
Electricity	($/1)	0.001	0.001	0.012	0.005	0.003
Process water	($/1)	0.009	0.005	0.002	0.003	0.004
Cooling water	($/1)	0.003	0.002	0.002	0.002	0.004
Total production costs	($/1)	0.665	0.566	0.674	0.602	0.670

Often the permeate stream resulting from the PV unit requires further processing. This can be accomplished either by a second pervaporation unit or by the use of a distillation column. The use of two PV units, one containing a hydrophobic and the other containing a hydrophilic membrane has been suggested by Strathmann and Gudernatsch [3.93] (see Figure 3.17). In this PVMBR the permeate of the hydrophobic PV unit is further dehydrated in the hydrophilic PV unit. The combination of the two PV units with different membrane characteristics has resulted in ethanol concentrations of up to 99.7 %. In the design of Bemquerer-Costa [3.82], shown in Figure 3.18, the hydrophobic PV membrane is used to produce an ethanol-rich permeate stream from a fermentation broth, followed by a hydrophilic PV membrane to obtain an ethanol concentration of 95 %. The retentate stream of the hydrophobic PV unit is further concentrated using a centrifuge and drum dryer. The water-rich ethanol solution obtained from these operations is further processed by a distillation column to also produce a final ethanol concentration of 95 %. The complexity of the processes shown in Figures 3.17 and 3.18 make them uneconomical, when, for example, compared with the simpler MBR process involving a MF unit and culture re-circulation [3.3]. As Lipnizki *et al.* [3.3] note PVMBR can be an economic alternative to conventional, non-integrated processes.

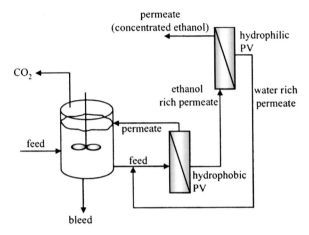

Figure 3.17. PVMBR incorporating multiple PV membranes, adapted from Lipnizki *et al.* [3.3].

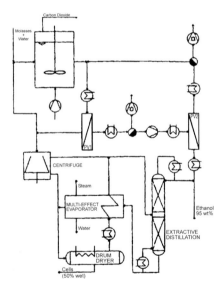

Figure 3.18. A novel PVMBR for ethanol production. From Lipnizki *et al.* [3.3], with permission from John Wiley and Sons.

They do not fare as well, however, when compared to the more mature, and simpler MBR processes as those involving, for example, a bioreactor and a coupled MF unit. What is required to make the PVMBR processes more competitive (as noted in the study of Groot *et al.* [3.58]) is the development of new cheaper, and more permselective PV systems.

Conceptually related to the PVMR concept is a membrane reactor system proposed by Chemseddine and Audinos [3.94] for the esterification reaction between methanol and oleic acid. With their reactor Chemseddine and Audinos [3.94] aim to overcome one of the key problems with esterification reactions, which is the separation of the conventional homogeneous mineral acid catalysts from the final products. The membrane reactor is shown schematically in Figure 3.19. It utilizes an ion exchange membrane. On one side of this membrane one circulates the mineral acid (H_2SO_4) dissolved in MeOH, while on the other side one circulates the mixture of MeOH and oleic acid. The protons from the sulfuric acid side penetrate through the membrane and act as catalysts for the esterification reaction. The compensating cations, however, cannot penetrate the membrane and remain on the MeOH/H_2SO_4 side. Water that is produced on the MeOH/Oleic acid side also permeates to the MeOH/H_2SO_4 side, thus, shifting the equilibrium to produce more ester. The concept's technical feasibility was successfully tested experimentally utilizing a number of ion-exchange membranes.

The above examples have shown that pervaporation membrane-based reactive separation processes are attracting significant attention and that the technology has found some industrial applications. This is an area, where significant activity is already under way, and many more advances are expected in the future.

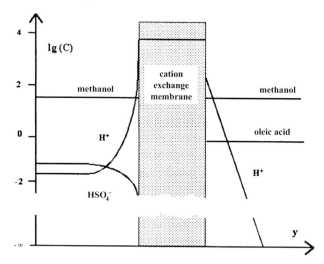

Figure 3.19. The membrane reactor concept of Chemseddine and Audinos. Adapted from Chemseddine and Audinos [3.94].

References

[3.1]   F. Lipnizki, R.W. Field, and P.K. Ten, *J. Membr. Sci.* **1999**, 153, 183.

[3.2]   Y. Zhu, R.G. Minet, and T.T. Tsotsis, *Chem. Engng. Sci.* **1996**, 51, 4103.

[3.3]   F. Lipnizki, S. Hausmanns, G. Laufenberg, R. Field, and B. Kunz, *Chem. Engng. Technol.* **2000**, 23, 7.

[3.4]   P.A. Kober, *J. Am. Chem. Soc.* **1917**, 39, 1944.

[3.5]   S. Zhang and E. Drioli, *Sepn. Sci. Technol.* **1995**, 30, 1.

[3.6]   J.F. Jennings and R.C. Binning, Organic Chemical Reactions Involving Liberation of Water, *U.S. Patent*, 2, 956, 070, **1960**.

[3.7]   G.K. Pearse, Esterification Process, *European Patent*, 0,210,055, **1987**.

[3.8]   R. Waldburger and F. Widmer, *Chem. Engng. Technol.* **1996**, 19, 117.

[3.9]   H. Kita, K. Tanaka, K. Okamoto, and M. Yamamoto, *Chem. Lett.* **1987**, 2053.

[3.10] K. Okamoto, M. Yamamoto, S. Noda, T. Semoto, Y. Otoshi, K. Tanaka, and H. Kita, *Ind. Engng. Chem. Res.* **1994**, 33, 849.

[3.11] L. Bagnell, K. Cavell, A.M. Hodges, A.W.H. Mau, and A.J. Sheen, *J. Membr. Sci.* **1993**, 85, 291.

[3.12] A. Dams and J. Krug, *Proc. 5th Int. Conf. on Pervap. Proc. in the Chem. Ind.*, R. Brakish, Ed., Brakish Material Corporation, Englewood, NJ, USA, 1991, 338.

[3.13] S. Bitterlich, M. Meissner, and W. Hefner, *Proc. 5th Int. Conf. on Pervap. Proc. in the Chem. Ind.*, R. Brakish, Ed., Brakish Material Corporation, Englewood, NJ, USA, 1991, 273.

[3.14] L. Domingues, F. Recasens, and M.A. Larrayoz, *Chem. Engng. Sci.* **1999**, 54, 1461.

[3.15] H.E.A. Bruschke, G. Erlington, and W.H. Schneider, *Proc. 7th Int. Conf. on Pervap. Proc. in the Chem. Ind.*, R. Brakish, Ed., Brakish Material Corporation, Englewood, NJ, USA, 1995, 310.

[3.16] H. Kita, S. Sasaki, K. Tanaka, K.I. Okamoto, and M. Yamamoto, *Chem. Lett.* **1988**, 2025.

[3.17] K. Okamoto, T. Semoto, K. Tanaka, and H. Kita, *Chem. Lett.* **1991**, 167.

[3.18] K. Okamoto, N. Tanihara, H. Watanabe, K. Tanaka, H. Kita, A. Nakamura, Y. Kusuki, and K. Nakagawa, *J. Membr. Sci.* **1992**, 68, 53.

[3.19] K. Okamoto, M. Yamamoto, Y. Otoshi, T. Semoto, M. Yano, K. Tanaka, and H. Kita, *J. Chem. Engng Japan.* **1993**, 26, 475.

[3.20] X. Ni, Z. Xu, Y. Shi, and Y. Hu, *Water Treat.* **1995**, 10, 115.

[3.21] M.O. David, T.Q. Nguyen, and J. Neel, *Trans. Inst. Chem. Engng.* **1991**, 69, 335.

[3.22] M.O. David, T.Q. Nguyen, and J. Neel, *Trans. Inst. Chem. Engng.* **1991**, 69, 341.

[3.23] X. Feng and R.Y.M. Huang, *Chem. Engng. Sci.* **1996**, 51, 4673.

[3.24] M.O. David, T.Q. Nguyen, and J. Neel, *J. Membr. Sci.* **1992**, 73, 129.

[3.25] Y. Zhu and H.F. Chen, *J. Membr. Sci.* **1998**, 138, 123.

[3.26] Q.L. Liu, Z.B. Zhang, and H.F. Chen, *J. Membr. Sci.* **2001**, 182, 173.

[3.27] J.T.F. Keurentjes, G.H.R. Janssen, and J.J. Gorissen, *Chem. Engng. Sci.* **1994**, 49, 4681.

[3.28] R. Waldburger, F. Widmer, and W. Heinzelmann, *Chem. Ing. Tech.* **1994**, 66, 850.

[3.29] B. Park, Models and Experiments with Pervaporation Membrane Reactors Integrated with a Water Adsorbent System, Ph.D Thesis, University of Southern California, Los Angeles; USA, 2001.

[3.30] C. Herion, W. Spiske, and W. Hefner, in: R. Brakish, Ed., *Proc. 5th Int. Conf. on Pervap. Process. in the Chem. Ind.*, Brakish Materials Corp., Englewood, NJ, USA, 1991.

[3.31]   M. Matouq, T. Tagawa, and S. Goto, *J. Chem. Engng. Jpn.* **1994**, 27, 302.
[3.32]   B.L. Yang and S. Goto, *Sepn. Sci. Technol.* **1997**, 32, 971.
[3.33]   C.R. Staudt-Bickel and R.N. Lichtenthaler, *J. Membr. Sci.* **1996**, 11, 135.
[3.34]   C.R. Staudt-Bickel and R.N. Lichtenthaler, *Proc. International Congress on Organic Membranes, 96*, 394, 1996.
[3.35]   W.J.W. Bakker, I.A.A.C.M. Bos, W.L.P. Rutten, J.T.F. Keurentjes, and M. Wessling, *Proc. 5th International Congress on Inorganic Membranes, Nagoya, Japan,* p. 448, 1998.
[3.36]   G.J.S. van der Gulik, R.E.G. Janssen, J.G. Wijers, and J.T.F. Keurentjes, *Chem. Engng. Sci.* **2001**, 56, 371.
[3.37]   W. Xu, J. Dong, J. Li, and F. Wu, *J. Chem Soc. Chem. Commun.* **1995**, 1967.
[3.38]   M. Matsukata, M. Ogura, T. Osaki, P.R.H. Prasad Rao, N. Nomura, and E. Kikuchi, *Topics in Catalysis,* **1999**, 9, 77.
[3.39]   T. Sano, H. Yamagashita, Y. Kiyozumi, D. Kitamoto, and F. Mizukami, *Chem. Lett.* **1992**, 2413.
[3.40]   H. Kita, K. Horii, Y. Ohtoshi, K. Tanaka, and K. Okamoto, *J. Mater. Sci. Lett.* **1995**, 14, 206.
[3.41]   Y. Morigami, M. Kondo, J. Abe, H. Kita, and K. Okamoto, *6th International Congress on Inorganic Membranes*, Program and Book of Abstracts, p 55, Montpellier, France, June 26-30, 2000.
[3.42]   M.P. Bernal, J. Coronas, M. Menendez, and J. Santamaria, *4th International Conference on Catalysis in Membrane Reactors, ICCMR 2000*, 71, 2000.
[3.43]   K. Tanaka, H. Kita, and K. Okamoto, 4th International Conference on Catalysis in Membrane Reactors, ICCMR 2000, 73, 2000.
[3.44]   K. Tanaka, R. Yoshikawa, C. Ying, H. Kita, and K. Okamoto, *Catal. Today* **2001**, 67, 121.
[3.45]   J. Theis, G. Bengtson, H. Scheel, and D. Fritsch, 4th *International Conference on Catalysis in Membrane Reactors*, ICCMR 2000, 57, 2000.
[3.46]   A. van der Padt, J. J. Sewalt, and K. Van't Riet, *J. Membr. Sci.* **1993**, 80, 199.
[3.47]   Z. Ujang, N. Al-Sharbati, and A.M. Vaidya, *Biotechnol. Prog.* **1997**, 13, 39.
[3.48]   Z. Ujang and A. Hazri, *J. Membr. Sci.* **2000**, 175, 139.
[3.49]   H.H. Nijhuis, A. Kempermann, J.T.P. Derksen, and F.P. Cuperus, in: R. Brakish, Ed., *Proc. 6th Int. Conf. On Pervap. Proc. In the Chem. Ind.,* Brakish Material Corporation, Englewood, NJ, USA, 368, 1992.
[3.50]   S.J. Kwon, K.M. Song, W.H. Hong, and J.S. Rhee, *Biotechnol. BioEngng.* **1995**, 46, 393.
[3.51]   S.J. Kwon and J.S. Rhee, *J. Microbiol. Biotechnol.* **1998**, 8, 165.
[3.52]   D.J. O'Brien, L.H. Roth, and A.J. McAloon, *J. Membrane Sci.* **2000**, 166, 105.
[3.53]   Y. Shabtai, S. Chaimovitz, A. Freeman, E. Katchalski-Katzir, C. Linder, M. Nemas, M. Perry, and O. Kedem, *Biotechnol. BioEngng.* **1991**, 38, 869.
[3.54]   C.W. Cho and S.T. Hwang, *J. Membr. Sci.* **1991**, 57, 21.
[3.55]   Y. Shabtai and C. Mandel, *Appl. Microbiol. Biotechnol.* **1993**, 40, 470.
[3.56]   Y. Mori and T. Inaba, *Biotechnol. BioEngng.* **1990**, 36, 849.
[3.57]   W.J. Groot, R.H. Kraayenbrink, R.G.J.M. van der Lans, and K.Ch.A.M. Luyben, *Bioprocess Engng.* **1992**, 8, 99.
[3.58]   W.J. Groot, R.H. Kraayenbrink, R.H. Waldram, R.G.J.M. van der Lans, and K.Ch.A.M. Luyben, *Bioprocess Engng.* **1993**, 8, 189.
[3.59]   W. Zhang, X.J. Yu, and Q.A. Yuan, *Biotechnol. Tech.* **1995**, 9, 299.

[3.60]   W. Gudernatsch, K. Kimmerle, N. Stroh, and H. Chmiel, *J. Membr. Sci.* **1988**, 36, 331.

[3.61]   D.J. O'Brien and J. C. Craig, *Appl. Microbiol. Biotechnol.* **1996**, 44, 699.

[3.62]   A. Kaseno, I. Miyazawa, and V. Kokugan, *J. Ferment. BioEngng.* **1998**, 86, 488.

[3.63]   D.J. O'Brien, L.H. Roth, G.E. Senske, and J.C. Craig, Jr., *Appl. Microbiol. Biotechnol.*, in press.

[3.64]   B. Detterwiler, I.J. Dunn, E. Heinzel, and J.E. Prenosil, *Biotechnol. Engng.* **1991**, 41, 791.

[3.65]   B. Detterwiler, I. J. Dunn, E. Heinzel, and J.E. Prenosil, *Biotechnol. Engng.* **1991**, 41, 791.

[3.66]   N. Qureshi, I.S. Maddox, and A. Friedl, *Biotechnol. Progress* **1992**, 8, 382.

[3.67]   N. Qureshi and I.S. Maddox, *App. Biochem. Biotechnol.* **1992**, 34/35, 441.

[3.68]   G. Sodeck, H. Effenberger, E. Steiner, and W. Salzbrum, in, R. Brakish, ed., *Proc. 2nd Int. Conf. on Pervap. Proc. in the Chem. Ind.*, Brakish Material Corporation, Englewood, NJ, USA, 157, 1987.

[3.69]   G. Bengston, H. Pingel, K.W. Böddeker, J.P.S.G. Crespo, and K.H. Kroner, *Proc. 5th Int. Conf. on Pervap. Proc. in the Chem. Ind.*, R. Brakish, Ed., Brakish Material Corporation, Englewood, NJ, USA, 508, 1991.

[3.70]   A. Friedl, N. Qureshi, and I. S. Maddox, *Biotechnol BioEngng.* **1991**, 38, 518.

[3.71]   T. Lamer, H.E. Spinnler, I. Souchon, and A. Voilley, *Process Biochem.* **1996**, 31, 533.

[3.72]   W.J. Groot and K.C.A.M. Luyben, *Biotechnol Lett.* **1987**, 9, 867.

[3.73]   W.J. Groot, R.G.J.M. van der Lans, and K.C.A.M.Luyben, *Proc. 3rd Int. Conf. on Pervap. Process. in the Chem. Ind.*, Bakish Materials Corp., Englewood, NJ, USA, p. 439, 1988.

[3.74]   M.A. Larrayoz and L. Puigjaner, *Biotechnol. BioEngng.* **1987**, 30, 692.

[3.75]   Q. Geng and C.H. Park, *Biotechnol. BioEngng.* **1994**, 43, 978.

[3.76]   W.J. Groot, C.E. van den Oever, and N.W.F. Kossen, *Biotechnol. Lett.* **1984**, 6(11), 709.

[3.77]   W.J. Groot, G.H. Schoutens, P.N. Van Beelan, C.E. van den Oever, and N.W.F. Kossen, *Biotechnol. Lett.* **1984**, 6, 789.

[3.78]   W.J. Groot, M.C.H. den Reyer, T. Baart de la Faille, R.G.J.M. van der Lans, and K.Ch.A.M. Luyben, *Chem. Engng. J.* **1991**, 46, B1.

[3.79]   W.J. Groot, M.C.H. den Reyer, R.G.J.M. van der Lans, and K.C.A.M.Luyben, *Chem. Engng. J.* **1991**, 46, B11.

[3.80]   W. Gudernatsch, H. Mucha, T. Hoffman, H. Strathmann, and H. Chmiel, *Ber. Bunsenges. Phys. Chem.* **1989**, 93, 1032.

[3.81]   S.I. Nakao, F. Saitoh, T. Asakura, K. Toda, and S. Kimura, *J. Membr. Sci.* **1987**, 30, 273.

[3.82]   A. Bemquerer-Costa, Ph.D. Dissertation, RWTH, Aachen, 1989.

[3.83]   G. Bengston, K.W. Böddeker, H.P. Hanssen, and I. Urbasch, *Biotechnol. Techn.* **1992**, 6, 23.

[3.84]   G. Bengston, K.W. Böddeker, V. Brockmann, and H.P. Hanssen, *Proc. 5th Int. Conf. on Pervap. Proc. in the Chem. Ind.*, R. Brakish, Ed., Brakish Material Corporation, Englewood, NJ, USA, 430, 1992.

[3.85]   N. Qureshi and H.P. Blaschek, *Biotechnol. Prog.* **1999**, 15, 594.

[3.86]   W.J.Groot, R.G.J.M. van der Lllans, and K.C.A.M. Leuben, *Process Biochem.* **1992**, 27, 61.

[3.87]   M. Matsumura and H. Kataoka, *Biotechnol. BioEngng.* **1987**, 30, 887.

[3.88]   N. Qureshi and H.P. Blaschek, *Biomass Bioeneng.* **1999**, 17, 175.

[3.89]   M. Matsumura, S. Takehara, and H. Kataoka, *Biotechnol. BioEngng.* **1992**, 39, 148.

[3.90]   N. Qureshi, *Recovery of Alcohol fuels using selective membranes by pervaporation*, Ph.D. Thesis, University of Nebraska, Lincoln, NE, 1997.

[3.91]   N. Qureshi, M.M. Meagher, J. Huang, and R.W. Hutkins, *J. Membr. Sci.*, **187**, 93, 2001.

[3.92]   N. Qureshi and H.P. Blaschek, *Appl. Biochem. Biotechnol.* **2000**, 84-6, 225.

[3.93]   H. Strathmann and W. Gudernatsch, *Bioprocess. Technol.* **1991**, 11, 67.

[3.94]   B. Chemseddine and R. Audinos, *J. Membr. Sci.* **1996**, 115, 77.

4 Membrane Bioreactors

4.1 Membrane Reactors for the Production of Biochemicals

The concept of coupling reaction with membrane separation has been applied to biological processes since the seventies. Membrane bioreactors (MBR) have been extensively studied, and today many are in industrial use worldwide. MBR development was a natural outcome of the extensive utilization membranes had found in the food and pharmaceutical industries. The dairy industry, in particular, has been a pioneer in the use of microfiltration (MF), ultrafiltration (UF), nanofiltration (NF), and reverse osmosis (RO) membranes. Applications include the processing of various natural fluids (milk, blood, fruit juices, etc.), the concentration of proteins from milk, and the separation of whey fractions, including lactose, proteins, minerals, and fats. These processes are typically performed at low temperature and pressure conditions making use of commercial membranes.

MBR are finding fertile ground for application in biochemical synthesis [4.1, 4.2] for the production of a broad spectrum of products. These range from food, liquid fuels (e.g., ethanol), and plant metabolites, to fine chemicals, including medical products, flavoring agents, food colors, fragrances, etc. Biochemical synthetic processes are important in the pharmaceutical industry, because they allow the production of complex molecules, like hormones, which cannot be produced safely and efficiently with the more conventional techniques [4.3].

In the biochemical synthesis area coupling the biological reaction with membrane separation provides potential advantages. Biological reactions, for example, typically generate a complex product mixture. Only a limited number of such products are of value, however; in fact several, often, turn out to inhibit enzymatic or biological function, or even to be toxic. Removing these components *in situ* through the membrane prolongs biocatalyst life, and increases the product turnover rate. Biocatalysts, used for biochemical synthesis, including enzymes, mammalian and plant cells are costly [4.4]. One must be able, therefore, to separate them from the product effluent, and to recycle them back into the reactor. Membranes are well-suited for performing this function. An additional benefit they offer is to increase cellular concentration, thus, increasing productivity and reactor throughput. This is a key advantage of the MBR systems, as can be seen in Figure 4.1. The data in this figure are from a study by Liew *et al.* [4.5]. They studied the production of (*R*)-(-)-phenylacetylcarbinol (PAC) using *Candida utilis*. The PAC is an essential intermediate for the synthesis of (*1R*, 2*S*)-(-)-ephedrine and (*1S*, 2*S*)-(+)-ephedrine, which are major ingredients for several pharmaceutical products used as decongestants and anti-asthmatic drugs. They utilized both a batch bioreactor, and a MBR with a Ceraflo® MF membrane unit. Their studies showed that cell concentration in the MBR was higher than that in the conventional reactor at all levels of dilution.

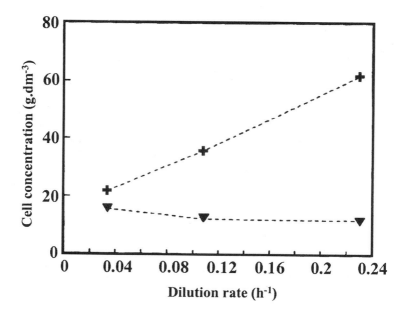

Figure 4.1. Comparison of a MBR process and a classical continuous fermentation process. (▼) no membrane; (+) MBR. Adapted from Liew *et al.* [4.5], with permission from SCI.

Often the membrane's primary role in a MBR system is to separate *in situ* the valuable products from the unreacted raw materials and biocatalysts. This has many potential advantages. If the membrane is successful in separating the produced chemicals, one then avoids additional downstream separation steps. For the synthesis of high purity pharmaceutical and food products membranes may offer advantages over the more conventional separation techniques (distillation, evaporation, crystallization, etc.). This is because they are simpler, less energy intensive, and can operate under the mild conditions required to maintain biological, or enzymatic activity for the synthesis of biochemicals, which are sensitive to heat. Membranes can be tailor-made (in terms of their molecular sieving properties and surface specificity through the use of immobilized, ion-exchange or affinity ligands [4.4]) in order to be able to separate the products from the biocatalysts. This is important, particularly, when pathogenic microorganisms or enzymes are used for the bioconversion [4.6]. Finally, the membrane provides the ability to independently adjust the residence times in the reactor of products, reactants, and biocatalysts in order to improve operational flexibility and provide effective process control.

The way membranes (in various forms, i.e., cylindrical, coaxial, flat-sheet, spiral-wound, and hollow fiber, etc.) couple with the bioreactor depends on the role the membrane performs. As with catalytic and pervaporation membrane reactors, the simplest configuration consists of two separate but coupled units, one being the bioreactor the other the membrane module. The biocatalyst (e.g., enzymes, bacteria, yeasts, mammalian cells) could, in this case, be suspended in a solution and continuously circulated through the membrane unit. The alternate configuration involves coupling the membrane and biore-actor into the same unit. In this configuration the biocatalyst may be immobilized on the membrane's surface or within its pore structure. The first configuration still remains the most popular one. Concerns remain with the second configuration coupling the membrane and bioreactor in the same unit, relating to membrane biofouling, especially for fermenta-tion and other whole-cell conversion applications [4.2]. Furthermore, when the biocatalyst is placed within the membrane structure, the substrate and/or the products have to be transported through the membrane. This, generally, creates additional mass-transport re-sistances, which may influence the performance of these bioreaction systems.

In the simplest MBR application, as previously noted, the membrane separates the products or metabolites, while retaining the biocatalysts, which are then recycled back to the reactor (Figure 4.2).

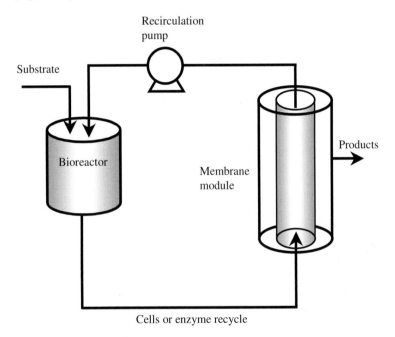

Figure 4.2. A schematic of a simple MBR.

For whole-cell processes MF or UF membranes are generally used to separate the biomass (average size 0.1 to several μm) from the product molecules. In enzymatic transformations UF membranes with weight cut-off from 5,000 to 10,0000 Da are utilized, in order to re-tain the enzymes, whose molecular weight ranges from 10,000 to 100,000 Da. This is the configuration that was utilized, for example, in the early MBR systems by Michaels [4.7] and Flaschel *et al.* [4.8]. These simple MBR processes have been widely studied in the in-dustry in the production of many chemicals, with efficiency improvements reported over the conventional bioreactors. The production of lactic acid from lactose is a good example

[4.9, 4.10, 4.11]. Lactic acid is an important additive and preservative agent in the chemical, cosmetics, pharmaceutical, and food industries. It is also used as the base for the production of biodegradable polymers like polylactates [4.12]. Its current worldwide production is estimated to be 40,000 tons per year. The results reported by Olmos-Dichara and coworkers [4.13] are typical of the results reported in many of the prior studies of this reaction system. They carried out a study comparing the performance of a batch reactor and a MBR for the production of lactic acid using *L. cassei sp. rhamnosus* as a biocatalyst. The MBR consists of the batch bioreactor coupled with a cross-flow mineral membrane filtration unit. MBR productivity was eight times that of the batch reactor, while the biomass concentration ($77g \cdot l^{-1}$) in the MBR was 19 times that found in the batch culture.

A different application involves using the membrane for the delivery of one of the reactants (e.g., bubble-free aeration [4.14]). One recent example of such an application is that reported by Onken and Berger [4.15]. They used a microporous polypropylene hollow-fiber membrane for the controlled addition of oxygen in the biotransformation of citronellol into 3,7-dimethyl-1,6,7-octanetriol by *Cystoderma carcharias*.

As was described above in a number of MBR processes the membrane, in addition to performing the separation functions previously discussed, also acts as a host for the biocatalysts (whole cells or enzymes) which are immobilized in the membrane's pore structure. Concerns with such MBR configurations include membrane biofouling, mass transport limitations and biocatalyst activity loss and denaturation. In the two sections that follow we discuss further some of the key aspects of MBR for biochemical synthesis. We classify these reactors into two types, namely whole-cell and enzymatic MBR.

4.1.1 Whole-cell Membrane Bioreactors

Coupling whole-cell (plant or mammalian cells, bacteria or yeast) biocatalysis (fermentation, oxidation, etc.) with membranes is a common application of MBR. As previously noted, one interesting example is lactose transformation into lactic acid. In recent years inorganic membranes are becoming popular for this application [4.16, 4.17, 4.18]. Moueddeb *et al.* [4.18], for example, have used a novel MBR, which contained two coaxial asymmetric porous alumina tubes (Figure 4.3) both incorporating an α-alumina MF thin layer. The membranes are coupled in such way so that the two MF layers confine in an annular space the support macrostructure, where bacteria are fixed. This reactor system is well-suited for mass transfer, reaction kinetics, and biomass growth studies. A similar idea was used by Flinders Technology [4.19]. In this case the bacteria are gel-immobilized in a composite membrane, and total substrate biotransformation takes place when the feed stream crosses the composite membrane where the bacteria are fixed. Crespo *et al.* [4.20] studied the production of lactic (and propionic) acid in a MBR using two different types of bacteria. The MBR was able to sustain a high cell concentration, and metabolite production. They observed, however, a problem, frequently reported with MBR, namely a decrease in membrane permeation rate with time on stream. This could either be due to biofouling (see further discussion below) or simply due to changes in viscosity and the other fluid properties with time, as biomass accumulates in the reactor. Whatever the reason, the decrease in permeation negatively impacts MBR efficiency. The problem can be minimized somewhat by maintaining a cellular concentration which is below a critical level [4.16].

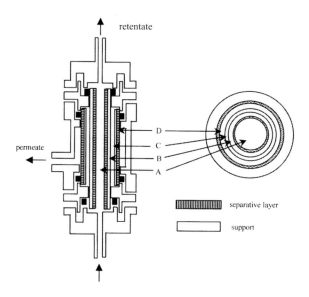

Figure 4.3. A schematic of the MBR; A, inner compartment, B and D membranes, C annular space. Adapted from Moueddeb *et al.* [4.18], with permission from Elsevier Science.

Ethanol production is another reaction that has been studied in a MBR. Cheyran and Mehaia [4.21] using *Saccharomyces cerevisiae*, obtained MBR conversions, which were 30 times that of a batch reactor. In a number of studies the bioreactor was coupled with pervaporation (PV) or membrane distillation (MD). PV, for example, was utilized by Cho and Hwang [4.22] for the *in situ* removal of ethanol. They report a 10–20 % increase in productivity. The use of pervaporation membrane bioreactors (PVMBR) for the production of ethanol was discussed more extensively in Chapter 3. Gryta *et al.* [4.23] coupled the reaction with a MD process using hydrophobic, polypropylene hollow-fiber membranes. The MBR process exhibits a better productivity than the classical fermentation (Figure 4.4). For this reaction, but also for many other applications involving whole-cell MBR, use has been made of carbon or ceramic membranes. This is because membranes in such MBR require frequent cleaning and sterilization, for which inorganic membranes, due to their better thermal and chemical resistance, are well-suited for. Inorganic membrane modules, on the other hand, have typically lower filtration areas than their polymeric counterparts. The latter, though less resistant to sterilization and harsh chemical treatments, can be manufactured in configurations with high surface/volume (s/v) ratios. Hollow-fiber and spiral-wound membranes, for example, have (s/v) between 10^3–10^4 $m^2 \cdot m^{-3}$ (Rautenbach and Albrecht, [4.24]). Large filtration areas are, often, necessary in whole-cell MBR, because the mass transfer coefficients are typically low.

Whole-cell MBR have been utilized in a number of biochemical synthesis reactions. An example used industrially, is growth hormone biosynthesis by the bacteria *E. coli* (Legoux *et al.* [4.25]). Using the MBR allows the synthesis of this hormone free from pathogens, like those causing the Creutzfeld-Jacob disease, for example. Other industrial examples include the synthesis of homochiral cyanohydrins (Bauer *et al.* [4.26]), the production of L-aspartic acid [4.16, 4.27], and the biotransformation of acrylonitrile to acrylamide [4.28]. Bosetti *et al.* [4.29] have reported the oxidation of naphthalene by *Pseudomonas fluoroscens* to optically pure *cis*-1,2-dihydroxy-1,2-dihydronaphtalene. They reported that the MBR, after 25 h of reaction, had a production rate three times that of a batch reactor. Recently Miyano *et al.* [4.30] reported improvement in the rate of vitamin B12 production by *Propionibacterium freudenreichi* using a hollow-fiber MBR. The strain used produced also propionic and acetic acid, which inhibit its activity. The authors compared the MBR with a batch reactor using a co-culture. For the batch reactor application, a second strain (*Ralstodia. Eutropha*), which was able to metabolize partially the toxic propionic acid, was introduced into the fermentor. The experimental results of Miyano *et al.* [4.30] showed that the MBR was much more efficient than the co-culture batch operation.

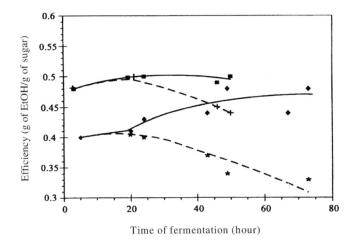

Figure 4.4. Variation of the efficiency of the fermentation process carried out in the system with MD (solid line); without MD (broken line); for the broths with different initial concentrations of sugar at 309 K (sugar concentration (g/dm^3): ■ , ✛ : 50; ◆✱: 150). Adapted from Gryta *et al.* [4.23], with permission from Elsevier Science.

Whole-cell MBR have also been applied for two-phase biotransformations encountered, when one or more reactants and/or products have low aqueous solubility. For such processes, the product's better solubility into an organic solvent is, sometimes, exploited for continuous separation out of the aqueous phase in order to reduce the potential for product inhibition [4.31]. However, for conventional, two-phase bioreactors, mass transfer limitations between the two phases and emulsification are key problems, the latter making it difficult to separate the desired products. MBR, using either porous or dense membranes provide advantages here, since they provide a fixed interface to separate the aqueous and organic phases within the bioreactor, thus, preventing phase mixing. Doig *et al.* [4.32] used this type of reactor for the reduction of geraniol to citronellol (both poorly water-soluble) by *Saccharomyces cerevisiae*. The biocatalyst and the water-soluble substrates were maintained under agitation in a reactor vessel, which also contained a silicon rubber tubular membrane, inside which circulated a hexadecane solution containing the geraniol. The authors report MBR biotransformation rates, which were comparable to those of the direct contact, two-phase system, but in the MBR, the emulsification and bulk-phase break-through problems were avoided. A schematic of the MBR used is shown in Figure 4.5. More recently Doig *et al.* [4.33] reported the epoxidation of 1,7-octadiene by *P. oleovorans* using the same MBR type. They reported that the MBR was able to produce epoxide during 1250 h of continuous and stable operation, during which time 84 % of the product was accumulated in the uncontaminated, pure solvent phase. Recently Krieg *et al.* [4.34, 4.35] published a study on the enantioselective catalytic hydrolysis in a MBR of 1,2-epoxyoctane to (*S*)-1,2-epoxyoctane plus (*R*)-1,2-octanediol by the yeast *Rhodospridium tiruloides*. This reaction was carried out in a MBR in emulsion form because of the low solubility of the substrate. The results have demonstrated that effective hydrolysis of the substrate with good stereo- and regioselectivity is possible. It was also shown that this reaction was suitable for scaling-up for larger scale production.

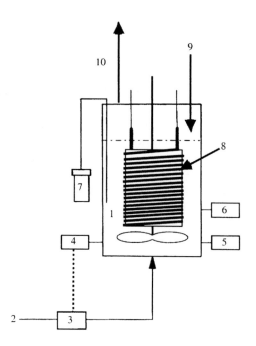

Figure 4.5. Schematic of MBR for the reduction of geraniol to citronellol. (1) biomedium, (2) oxygen supply, (3) control valve, (4) oxygen probe, (5) pH probe, (6) temperature monitor, (7) sampling port, (8) silicon rubber membrane, (9) solvent inlet, (10) solvent exit. Adapted from Doig *et al.* [4.32], with permission from John Wiley & Sons Inc.

As previously noted, hollow-fiber MBR are compact systems with high (s/v) ratios. This is an important feature for biological processes, which are often diffusionally controlled, like, for example, when mammalian cells are used for the synthesis of monoclonal antibodies. In an interesting MBR (called perfusion MBR) utilized for this reaction the cells grow in the extra-capillary space (i.e., in between the fibers). Nutrients circulating within the fibers, themselves, diffuse through the thin fiber skin to nourish the cells, while the metabolites are also continuously extracted through the hollow fibers [4.36]. Perfusion MBR mimic the function of the organs of living organisms, which also make use of a system of small capillaries. For mammalian cells (also called hybridomas) this type of MBR has been shown to support cell concentrations as high as 10^8 cells/ml [4.37]. These high cell concentrations result in very concentrated antibody and protein harvests, with antibody harvests as high as 17 mg/ml [4.38, 4.39] being reported. Such high yields, and the relatively simple and inexpensive operation of hollow-fiber MBR has made them an attractive choice in a number of commercial applications for production of monoclonal antibodies [4.40, 4.36].

MBR and perfusion MBR are also looked upon as an attractive concept in the development of bioartificial organs (liver, pancreas). There is an urgent need today for the development of such bioartificial organs, as there is critical shortage of organ donors. One of the proposed solutions of this problem is the use of an artificial organ able to maintain the toxic metabolites concentration in the body under critical levels, while a donor is being found. An interesting review article on bioartificial livers (BAL), which discusses the use of MBR using hepatocytes (human or porcine) has been recently written by Legallais *et al.* [4.41]. For the design of MBR, which mimic the function of human organs, two approaches have been used. In the first approach, shown schematically in Figure 4.6a, the membrane separates the blood or plasma flow from the cells, which are also continuously fed with oxygen and other nutrients. A number of technical issues have to be considered for the effective operation of these MBR. They include the anchorage of cells onto the membrane surface or matrix, and the effective exchange rate of oxygen, nutrients and metabolites between blood and cells. Protection against immunological response is also of critical importance [4.41]. The second approach, which involves the use of a perfusion MBR, is shown schematically in 4.6b. In this case the cells are perfused directly by plasma or blood, whereas the hollow fibers provide the means for oxygen supply and for carbon dioxide removal.

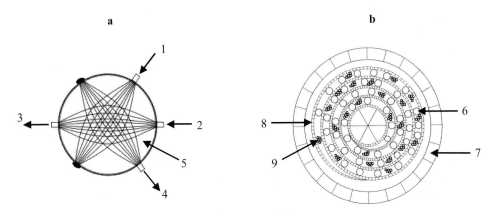

Figure 4.6. The two MBR types for bioartificial organs. **a**: separate plasma and cells configuration. 1: plasma or culture medium feed, 2: oxygen feed, 3: oxygen exit, 4: plasma or culture medium exit, 5: living cells and medium. **b**: perfusion MBR, 6: hollow fibers for oxygen feed, 7: external shell, 8: spirally wound polyester film, 9: anchored cells in a 3D matrix. Adapted from Legallais *et al.* [4.41].

Hollow-fiber MBR have also been used for the production of a number of other fine chemicals. Molinari *et al.* [4.42] used such a MBR for the production of isovaleraldehyde from isoamyl alcohol using *Gluconobacter oxidans*. In their work hydrophobic hollow-fiber membranes were used in order to continuously extract the aldehyde, thus, avoiding its oxidation to the corresponding acid. Hollow-fiber MBR have also been used by Koyama *et al.* [4.27] in the synthesis of L-aspartic acid by *E. coli,* and by Cantarella *et al.* [4.28] for the production of acrylonitrile from acrylamide by *Brevibacterium imperialis* CBS 489-74. Hollow-fiber MBR have also been used with reactions, which involve an active cell extract instead of whole cells (called cell-free translation reaction). Yamamoto *et al.* [4.43], for example, used such a MBR for enhancing the protein production rate during cell-free translation reaction with a wheat germ extract.

Whole-cell, hollow-fiber MBR are still under development. Despite their significant potential they have, so far, found only limited application for biochemicals production. One of the reasons is that cleaning of the hollow-fiber membranes is difficult, especially when whole-cell biocatalysts are immobilized in the small fibers. The mass transfer between the nutrients and cells has also to be taken into consideration and enhanced. Immobilizing the biocatalysts in porous beads, instead of directly on the membrane, may tend to avoid some of these problems, and to simplify membrane cleaning. The concept of using MBR as bioartificial organs is technically very attractive; the various MBR under development, however, must still be validated with clinical results. One can expect, however, that their development will follow the success of artificial kidneys, which are currently employed worldwide.

4.1.2 Enzymatic Membrane Bioreactors

MBR are finding application for many important enzymatic reactions. In MBR enzymes are, often, immobilized onto the membrane surface or in its porous structure. Free enzymes and also enzymes grafted onto a soluble polymer (to increase their molecular size) are utilized. In the latter case the membrane may simply act as a barrier to retain the enzyme, while allowing the removal of products and/or the addition of reactants. A key advantage of immobilization is that it permits easier enzyme recovery; in addition, grafting the biocatalyst does, on occasion, increase its thermal stability and its resistance towards organic solvents.

MBR find widespread application in the enzymatic hydrolysis of macromolecules (for detailed reviews see Prazeres and Cabral [4.44] and Cheyran and Mehaia [4.45]). They have been applied, for example, to protein hydrolysis, and to enzymatic reactions, requiring co-factor recycling [4.2, 4.46]. For the latter class of reactions, one of the earliest successes was L-phenylalanine synthesis from acetamidocinnamic acid utilizing NADH (β-nicotinamide adenine dinucleotide) as a co-factor [4.47]. Coenzymes like NADH act as co-substrates during the enzymatic reaction and, in so doing, they are converted into an inactive form. Since they are expensive ($1000/mol) they must be recovered for regeneration. In the study of Schmidt *et al.* [4.47], the co-factor was grafted onto a soluble polymer, and was separated by an UF membrane. Protein hydrolysis is the largest area of MBR application. One of the first and more important applications is whey protein hydrolysis by pancreactine [4.48]. Casein hydrolysis by alcalase has been studied in a MBR by Mannheim and Cheyran [4.49]. The membrane allowed them to independently control the enzyme and substrate concentrations, residence time and permeate flow; this, in turn, resulted in MBR productivity which was twelve times that of a batch reactor. Pouliot *et al.* [4.50] studied casein hydrolysis using trypsin or chymotrypsin in a hollow-fiber MBR. The membrane allowed them to separate *in situ* the various amino acids produced. The same system has been studied more recently by Curcio *et al.* [4.51] for the synthesis of *para-k*-casein and glyco-macro-peptides by chymosin. Enzymatic transformation of hemoglobin, an important animal slaughterhouse by-product, has been studied in a MBR by Cempel *et al.* [4.52] and Belhocine *et al.* [4.53], who used papain enzyme as a biocatalyst. These authors reported that MF or UF membranes could be successfully utilized, without enzyme loss, because a dynamic protein layer formed, which acted as the real membrane during the process. Recent studies have also reported the application of MBR in the hydrolysis of proteins derived from vegetables, like soy or alfalfa. D'Alvise *et al.* [4.54] reported, for example, a pilot-plant scale study for alfalfa protein hydrolysis using endopeptidase delvolase. They obtained good peptidic fraction productivities (340 g of nitrogen in product·h^{-1}·m^{-2}), and a continuous and reproducible permeate rate. The synthesis of many other important antibiotic precursors like, for example, 6-acylaminopenicillanic acid by free penicillin acylase using different MBR configurations has been also reported [4.55].

Enzymatic hydrolysis of polysaccharides (cellulose, starch) or oligosaccharides (maltose, saccharose, lactose) for the synthesis of food products is another class of processes MBR have been applied to. Paolucci-Jeanjean *et al.* [4.56] have recently reported, for example, the production of low molecular weight hydrolysates from the reaction of cassava starch over α-amylase. In this case the UF membrane separates the enzyme and substrate from the reaction products for recycle. Good productivity without noticeable enzyme losses was obtained. Houng *et al.* [4.57] had similar good success with maltose hydrolysis using the same type of MBR.

Fruit juice clarification by pectinases and cellulases is another interesting application. In the conventional process after the enzymatic reaction in the pulp treatment step takes place, filtration over diatomaceous earth follows. This filtration-type process produces a lot of solid waste, and results in costly enzyme loss. MBR are appropriate for such application either for enzyme recovery and recycle or in the form of a more compact CMR type system, with the biocatalyst immobilized on the membrane itself [4.58].

Hydrolysis or transesterification reactions of lipids catalyzed by lipases and phospholipases are other reactions, which MBR have been applied to. With the use of MBR one aims to continuously separate the products, while retaining the enzymes. For these systems one deals with complex micellar solutions of reactants and products [4.59]. The lipase-catalyzed hydrolysis of oils takes place in two-phase reactors, with fatty acids and glycerin as products. Fatty acids are obtained in the hydrophobic phase, and glycerin in the hydrophilic phase. Hoq *et al.* [4.60] studied the enzymatic hydrolysis of olive oil by lipase in a MBR using a hydrophobic membrane, see Figure 4.7. The enzyme was adsorbed on the membrane surface in contact with the aqueous phase, while the olive oil was recirculated on the other membrane side. In this case the membrane separates the two phases, thus, avoiding emulsion formation; it also provides a means for immobilizing the lipase. Lipase catalyzed hydrolysis of olive and sunflower oil was also studied by Giorno *et al.* [4.61], and Gan *et al.* [4.62]. Both studies concluded that changes in the hydrodynamic conditions have a strong influence on MBR productivity and enzyme activity. More recently this reaction, being carried out in a MBR, was modeled by Ceynowa and Adamczak [4.63]. Their studies indicate that improved productivity in the MBR can only be obtained for the case in which the reactor is kinetically limited, and that such technology could be used to produce di- and/or mono-glycerides, simply by maintaining the system working at steady state at the desired conversion.

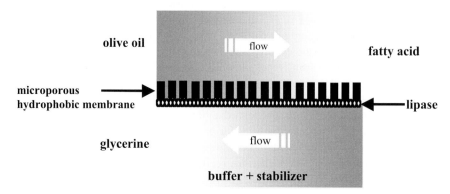

Figure 4.7. Schematic of the enzymatic olive oil conversion in a hollow-fiber bioreactor. Adapted from Hoq *et al.* [4.60].

MBR have also found use in the production of polysaccharides. MBR provide a simplified method for producing these materials, which are, typically, obtained in small quantities from natural sources by elaborate procedures. Hicke *et al.* [4.64] reported the synthesis of inulin (poly-β-(2,1)-frutan) in a MBR with the aid of fructosyltransferase immobilized in the membrane's porous structure. Inulin, with a high molecular weight and low polydispersity, was continuously synthesized from sucrose, as the substrate passed through the MF membrane.

For enzymatic conversions, in particular, hollow-fiber MBR find common use, with enzymes typically immobilized in the porous structure of the hollow fibers. One advantage these MBR offer over the conventional bioreactors, is the longer contact times of reactants with the enzymes, due to their high (s/v) ratios. Low residence times are a problem with conventional enzymatic bioreactors, and good efficiency is obtained only with rapid reactions [4.65]. One example of the use of an enzymatic hollow-fiber MBR was reported by Furui *et al.* [4.66] and Lopez and Matson [4.67], who used lipase for the biosynthesis of optically active (2*R*,3*S*)-3-(4-methoxyphenyl)-glycidic acid methyl ester (–MPGM) (a diltiazem chiral intermediate, which is an important calcium channel blocker used for hypertension and angina treatment) by asymmetric hydrolysis of the racemic compound (+/– MPGM). For this purpose they used a process, which combines two different reservoirs (one containing an aqueous and the other a toluene organic solution) separated by a commercial hollow-fiber membrane, together with a crystallizer, which was used for further product purification (Figure 4.8). The lipase was immobilized onto the hydrophilic UF fibers. A by-product, the methoxyphenylacetaldehyde, which is an enzyme inhibitor, is also produced, however, it is eliminated by an adduct formation with bisulfite in the aqueous phase. The organic solution recirculates between the MBR, its reservoir, and the crystallizer. The bioreaction takes place at the membrane interface. The crystallizer helps to enrich the slurry with (–MPGM). The productivity of this system was enhanced by 160 % over the MBR system in the absence of the crystallizer. The industrial production of the (–MPGM), was developed by Sepracor, and was realized at an industrial scale by Tanabe in Japan.

The synthesis of other important pharmaceutical products including the PCA (noted in Section 4.1.1) has been successfully realized by using enzymes or active cells extract instead of whole cells [4.68]. The racemic resolution or synthesis of the (*S*)-(+)-2-(6-methoxy-2-naphtyl) propionic acid (naproxen), which is an important member of the family of 2-aryl-propionic acid derivatives used as non-steroidal anti-inflammatory drugs, has been also realized with a MBR.

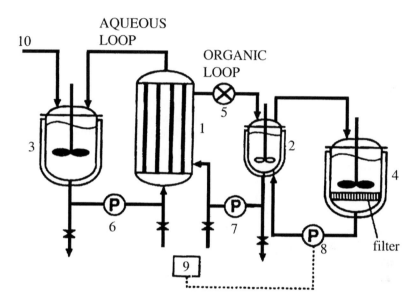

Figure 4.8. Schematic diagram of a membrane bioreactor with a crystallizer. 1: hollow-fiber MBR, 2: organic solution tank, 3: aqueous solution tank, 4: crystallizer, 5: pressure control valve, 6, 7, 8: pumps, 9: timer, 10: introduction of basic or acid solution for pH control, 11:filter. Adapted from Furui *et al.* [4.66].

The racemic resolution of this molecule is very important because the *S*-enantiomer is 28-fold more active than the *R*-enantiomer. Sakaki and co-workers [4.69] realized the production of (*S*)-naproxen from the racemic naproxen methyl ester using lipase immobilized in hollow fibers. Their results showed that the MBR had good enzyme stability and enantiomeric excess of up to 0.92. The stereoselective hydrolysis of racemic 2-substituted propionates catalyzed by carboxyl esterase has been performed by Cretich and coworkers [4.70]. The enzymatic reaction was realized in a MBR, which coupled the principle of electrophoresis with the enzymatic reaction. The optically pure (*S*)-naproxen was continuously recovered in an acidic chamber purified from unreacted substrate and enzyme by using an electric field as driving force.

A hollow-fiber MBR has also been used in the enzymatic synthesis of (^{11}C) formaldehyde by Hughes and Jay [4.71]. This molecule is used for the production of radio-pharmaceuticals like (^{11}C) thymidine [4.72]. (^{11}C) formaldehyde can be produced by the oxidation of (^{11}C) methanol over a metal catalyst or by the enzymatic oxidation by alcohol oxidase and catalase enzymes immobilized onto glass beads. In both cases the radio-chemical yield is small due to the total oxidation of formaldehyde as a result of the high temperatures utilized in the catalytic reaction or because of the long bioreactor contact times. Hollow-fiber MBR help avoid long contact times, and provide for the controlled addition of the substrate. Hughes and Jay [4.71] obtained encouraging results. The MBR gave stable high yields during a 4-week period, twice that of the conventional synthesis process.

As one may conclude from the above discussion, many factors must be taken into account in order to optimize MBR performance, including enzyme and substrate concentrations, residence time, and permeate flow. Productivity and long-term enzyme activity depend on these parameters. In the case of immobilized enzymes it is important to control the quantity of the enzyme attached to the membrane, since enzyme loading has an effect on activity. A number of methods based on physical entrapment, and ionic, covalent, and van der Waals interactions between the enzymes and the membrane surface have been used for enzyme immobilization. Covalent binding between the amino or carboxyl groups of the enzymes and the membrane surface using reagents such as glutaraldehyde, for example, has been extensively used [4.3]. For polymeric membranes immobilization can be easily accomplished, if the polymer has chemical groups, which are able to interact directly with the enzyme or with the immobilization reagent. In the case of inorganic membranes the immobilization task is more difficult. A number of efforts have been undertaken to increase the surface-enzyme interaction in order to increase the enzyme stability and to render the bonds between the biocatalyst and the solid surface stronger and more stable. Reagents like γ-aminopropyltriethoxysilane (APTES) have been used to activate the inorganic surface in order to improve the stability of the fixed enzyme [4.73, 4.74]. This molecule has been recently used by Ida *et al.* [4.75] for the activation of a cylindrical porous membrane utilizing a surface corona discharge induced plasma. Using this method the authors were able to immobilize glucoamylase onto the surface of the membrane pores. They claim an enhancement of the thermal and pH stability of the immobilized enzyme in comparison to the same enzyme in the free state.

The importance of controlling the quantity and the activity of the enzyme attached to the membrane, was discussed in a recent study by Ganapathi *et al.* [4.76, 4.77] for the amidase activity of papain. These authors studied the influence of the immobilization method, including the effect of product adsorption, and the choice of membrane material, on the enzyme's activity. Their results, interestingly enough, indicate that when the enzyme loading increases the activity decreases (Figure 4.9). This result may be explained by the crowding of the enzyme on the surface, resulting in the blocking of the active site or in protein denaturation. An alternate explanation is that it may be due to multiple point attachment of the enzyme, which would result in a decrease of the conformational flexibility at the active site, thereby, inhibiting the ability of the enzyme to bind to the substrate. A review paper on these effects was published recently by Butterfield *et al.* [4.78]. These authors reported that, in spite of the enzyme stability increase, the random immobilization on a membrane could induce a dramatic decrease on the enzyme activity. This activity decrease could be related to many factors including blocking of the active site, thus, making them inaccessible to the substrate, to multiple point binding or to enzyme denaturation. Membrane hydrophobicity is an important factor. In fact, when the membrane is hydrophobic multiple point attachment could occur, thus, blocking the active sites. This could be avoided by activating the membrane material with aldehydic or other groups, which are able to react with the lysine amino groups of enzymes. Other techniques like oriented site-specific immobilization, using molecular biological methods have also been used [4.78].

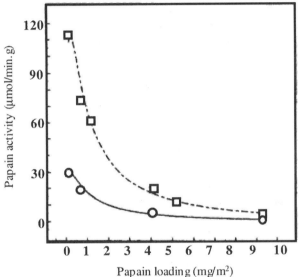

Figure 4.9. Effect of papain loading on modified polysulfone membrane on the activity. □ : activity corrected by the adsorption of the reaction product, *p*-nitroaniline. O : activity. Adapted from Ganapathi *et al.* [4.77], with permission from the American Chemical Society and the American Institute of Chemical Engineers.

4.1.3 Technical Challenges for Membrane Bioreactors

Though MBR offer advantages over the more conventional bioreactors, they, themselves, are not completely free of problems. One such key problem, as previously noted, relates to changes in biocatalyst activity. This is a serious concern for whole-cell MBR, when the cells are immobilized in the membrane's pore structure. Diffusional limitations for nutrients may cause reduction in cell division and metabolite production [4.79]. Mass transfer limitations for metabolites may also be detrimental to biocatalyst activity, since they are often toxic for the microorganisms. These phenomena are more pronounced at high cell densities, often encountered in MBR, and can produce noticeable metabolic changes. Borch *et al.* [4.80] observed, for example, that the homolactic fermentation by *Lactobacillus* became heterolactic at high cell concentrations. Another common cause of biocatalyst deactivation in whole-cell MBR is the high mechanical stress microorganisms find themselves exposed to under the high circulation rates required to maintain a good transmembrane flux. Depending on the cell's natural resistance, this may, on occasion, result in severe biocatalyst deactivation [4.20]. One must, therefore, always strive to strike a fine balance between maintaining both good cell activity and adequate transmembrane permeate flux.

For enzymatic MBR an important problem is the intrinsic decrease in activity as a result of the enzyme's immobilization on the membrane support or the grafting onto various macromolecules in order to increase its molecular size and improve its retention by the membrane. One could compensate for this effect by providing an additional quantity of grafted enzyme; however, there is a limit to the quantity of enzyme one may add without causing important physicochemical changes in the retentate solution. When enzymes are immobilized into the membrane structure, the loading must also be limited, because high enzyme loadings may result in activity losses due to sterical overcrowding.

One of the most serious problems encountered with MBR is biofouling, which typically manifests itself by a dramatic decrease over time in permeate flow [4.81]. This problem has many origins. For example, it may be caused by adsorption on the membrane surface and in its internal porous structure of the various metabolites cells produce, and of the coagulated proteins from lysed cells. These accumulate in the reactor over time, and also tend to increase the solution viscosity. Another cause of biofouling, when cells are fixed in the membrane, is the normal cell growth process, which may result in pore plugging, thus, considerably decreasing mass transfer. Whatever the cause of biofouling, the net result is a decrease in membrane flux over time as the MBR conversion and biomass density increase (Figure 4.10). Biofouling is a significant technical challenge. The conventional technique for dealing with it is to stop the reaction, after the MBR has run for a predetermined period of time, in order to carry out membrane cleaning and reactor sterilization operations. On occasion, if biofouling is slow, this may not present much of a problem, because cleaning could be coupled with the sterilization procedure, which is necessary in most bioreactors. When biofouling is severe or when membrane cleaning time is too long, however, one may have to overdesign the MBR, using, for example, two membrane modules so that one of them is in use while the other is being cleaned and sterilized. A technique for reducing biofouling, and for maintaining low viscosity and high mass transfer coefficients involves trying to keep the MBR cell density below a critical value through, for example, the addition of a fresh feed solution [4.16]. Warren *et al.* [4.82] followed such an approach. They studied ethanol fermentation with the biomass concentration stabilized between 50 and 100 $kg \cdot m^{-3}$, and were able to avoid the presence of foam caused by proteins coming from lysed cells.

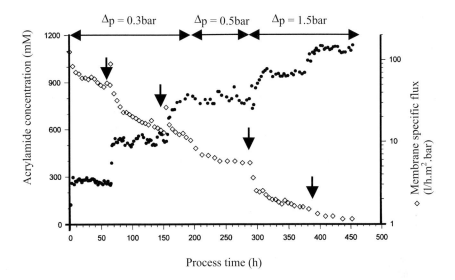

Figure 4.10. Acrylonitrile conversion to acrylamide and membrane specific flux evolution in a ultrafiltration MBR using the strain *Brevibacterium imperialis* CBS 489-74. Each arrow indicates the restoration of the initial acrylonitrile concentration, The Δp indicates the transmembrane pressure applied. Adapted from Cantarella *et al.* [4.28], with permission from Elsevier Science.

A key factor determining MBR success is the choice of membranes and appropriate cleaning techniques, in order to avoid irreversible plugging. Because of the increased downtimes, due to the need to clean the membranes, it is important to maintain as high a productivity as it is possible during the time the MBR is running. In recent years a number of smart ideas have been proposed for *in situ* cleaning of the membranes, like back- and pulsed-flushing, aeration, etc. Significant improvements in performance have been reported in a number of cases [4.83]. Arroyo and Fonade [4.84] report, for example, the use of an intermittent jet, which generates a whirling flow next to the membrane surface. This approach results in doubling the permeate flux during the cross-flow filtration of *S. cerevisiae* [4.85]. A MBR using a rotating membrane (~4500 rpm) has been patented and commercialized by Chemap [4.86]. More recently the PallSep VMF® membrane system, working through the aid of a torsion spring system, has been proposed [4.87]. The inventors report, that the vibrational energy creates high shear rates (~10^5 s^{-1}) near the membrane surface, resulting in high permeate flow rates. Generally speaking, mechanical solutions have demonstrated their efficiency only when membrane fouling results from particle deposition. When biofouling is caused by the adsorption of biochemicals, only efficient chemical cleaning techniques are able to restore the membrane.

In contrast to whole-cell MBR, enzymatic MBR seem to have fewer problems of fouling and mass transfer limitations. In fact, if the enzymes have a good stability, the system is free of adsorption or decomposition products. Also, mass transport limitations are generally not a significant concern. For example, a recent study by Ujang and Hazri [4.88] indicates, that the overall mass transfer coefficient for the removal of water from an organic phase through a hollow-fiber membrane with immobilized lipase is not influenced by the enzyme loading.

4.2   Environmental Applications of Membrane Bioreactors

A growing application of MBR is in wastewater treatment. Conventionally, wastewater treatment is carried out either by physicochemical techniques or by biological processes. The physicochemical techniques often simply transfer the contaminants in the wastewater streams into a different medium that must, itself, be disposed of, or, when they destroy them, they often also generate toxic by-products, which are difficult to eliminate. The biological processes have an advantage, in that they transform the complex organic contaminants into simple, harmless gaseous or water-soluble compounds, together with residual sludge. The conventional biological treatments, on the other hand, have the disadvantage that one must, at some point, physically separate the biocatalyst from the treated water. Immobilization of the biomass on porous supports [4.89] alleviates the problem somewhat. For heavily polluted wastewaters, however, fast growing biomass clogs the beds, and results in bed shutdowns, and the need for frequent regeneration. MBR processes provide a solution to the problem of biomass/wastewater separation, since the membrane provides an effective barrier for microbes and other particles. Its use, furthermore, allows for more effective process control, since one can independently adjust the wastewater residence time in the bioreactor, and the product (i.e., purified water) withdrawal rate through the membrane. MBR, in general, require a smaller volume and footprint than the more conventional (e.g., sedimentation) systems, an important advantage in many European countries, and in Japan. The use of MBR often reduces the number of treatment steps required for wastewater or groundwater treatment. Coté *et al.* [4.102], for example, compared a conventional activated sludge (AS) biological process with two processes involving the use of MBR (Figure 4.11) for water reclamation and reuse. The MBR process with an immersed membrane module is the best in terms of reduced sludge production, water quality, safety, and compactness.

The first environmental applications of MBR date back to the late sixties and early seventies. These early studies focused on the biological treatment of sewage water using UF plate-type membranes [4.90, 4.91, 4.92]. Since then MBR have been utilized for a variety of other applications including the purification of underground and surface water [4.93; 4.94, 4.95, 4.96, 4.97] and the treatment of a variety of domestic and industrial wastewaters [4.98, 4.99, 4.100]. MBR for wastewater treatment have been operated under both aerobic and anaerobic conditions. The two processes differ in terms of the rate of sludge production, their hydraulic residence time, biomass concentration, and the type of pollutants, which are preferentially treatable by each process. Lower organic pollutant removal rates, the potential for odors, and a higher start-up time are among the main characteristics of the anaerobic processes. They require, on the other hand, less energy to operate than aerobic processes. Figure 4.12 shows schematically some of the key technical differences between the two different types of processes [4.101].

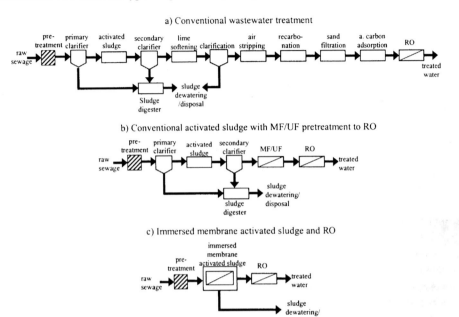

Figure 4.11. Comparison between different wastewater treatment processes. MF: microfiltration unit, UF: ultrafiltration unit, RO : reverse osmosis unit. From Coté *et al.* [4.102], with permission from Elsevier Science.

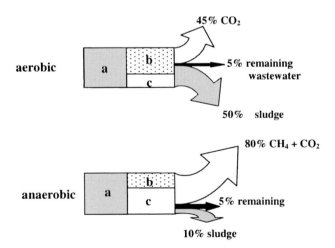

Figure 4.12. Comparison between aerobic and anaerobic wastewater biological treatment. **a**: wastewater, **b**: biomass content, **c**: hydraulic retention time.

The majority of industrial wastewater treatment applications are aerobic. However, the type of microorganisms used and the MBR process parameters vary depending on the wastewater's origin (domestic or industrial). Air bubbling is utilized to supply oxygen for biomass growth and to create turbulent flow conditions next to the membrane surface, in order to decrease membrane fouling, which is the most difficult problem hindering the broader application of MBR for wastewater treatment. Among the first to develop a commercial MBR process for industrial wastewater treatment, called the Membio process, was the Memtec Company [4.103]. However, it turned out that the membrane flux was not sufficiently large to make the process economic. Membrane flux is of key importance in this area, because the water price is, generally, low and the volumes to be treated enormous. The same problem has attracted the attention of a number of other industrial companies [4.99]. A process developed by the Kubota Corporation in Japan [4.104, 4.105] uses flat-sheet, organic membranes suspended in the bioreactor. An air-diffuser provides the oxygen (from air) necessary for the aerobic treatment to take place, and to scour the surface of the membranes in order to prevent fouling. Yamagiwa *et al.* [4.106] have reported the development of a MBR process, in which the recycle line from the membrane is immersed into the reactor, entraining air bubbles. In this configuration only one pump was used for providing the air and increasing the transmembrane pressure. This type of configuration, however, provides for a limited amount of oxygen transfer, and can only be used in small-scale applications. The Biosep® process developed by CGE [4.102, 4.107] in France consists of a bioreactor, in which a membrane in the form of a hollow-fiber bundle is immersed in the activated sludge (see Figure 4.13). Air is blown into the reactor through its bottom to create agitation in the reactor vessel, important in terms of improving reactor efficiency, and diminishing biofouling. Recently Bouhabila and co-workers [4.108] have reported on the efficiency of air blowing to limit particle deposition and polarization phenomena for this type of MBR. They reported that increasing the air flow-rate increased filtrate flux by a factor of 3. In addition, periodic backwashing gave an additional improvement in efficiency by decreasing the membrane resistance by 3-5 fold. Membrane plugging is also avoided by using hollow-fiber membranes, whose dense, permselective skin on the outside part of the fiber is in contact with the biological suspension. The clean water is extracted by applying vacuum on the inner membrane side. The continuous operation of the system has proven that regular back-washing is no longer required. The process has a lower cost of operation than its more conventional counterpart, while attaining a remarkable reduction in floor space requirements.

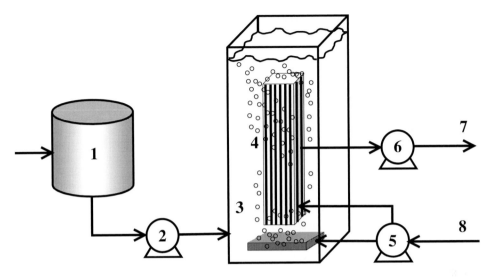

Figure 4.13. Aerated MBR for wastewater treatment. 1: feed tank; 2: feed pump; 3: reactor vessel and diffuser; 4: hollow-fiber or flat-sheet membranes; 5: air pump for diffuser or back-washing; 6: permeate vacuum pump; 7: permeate; 8: air feed.

Several authors have proposed improvements of the aforementioned immersed membrane, aerated MBR system. Ueda and Hata [4.109] used a gravitational instead of a suction filtration type unit, in which the transmembrane pressure is due to the pressure-head of the mixed liquor over the flat MF membrane module. The transmembrane pressure varied from 2 to 17 kpa, the maximum value corresponding to a liquor level of 1.69 m. In spite of the small pressure gradient present, the average permeate flow obtained was $2 \cdot 10^{-2}$ m$^3 \cdot$m$^{-2} \cdot$h^{-1}, which compares favorably with other MBR processes. The low transmembrane pressure used causes less plugging, and makes the process more economical and less energy demanding. A number of companies have developed MBR, which couple a MF, UF, or NF membrane unit with the bioreactor through a side stream. The Rhodia/Orelis process uses this principle, in which the classical settler unit is replaced by a cross-flow, flat-sheet membrane filtration unit. No air-diffuser is used. Table 4.1 lists a number of MBR processes currently in operation for wastewater treatment. As indicated in the table, MBR systems have been applied to many types of effluents. Dufresne *et al.* [4.110], for example, have employed an aerated MBR with hollow-fiber membranes for the treatment of pulping process wastewaters. The MBR operated at a much higher activated sludge concentration than the conventional system involving an AS bioreactor unit and a secondary clarifier. The lignin removal was improved, and the produced effluent was free of suspended solids.

Table 4.1. Commercial aerobic MBR plants for treating wastewater effluents. Adapted from Gander and Judd [4.111] and Gauthier and Praderie [4.112].

Company	MBR process	Membrane configuration	N° of plants treating domestic waste-water	N° of plants treating indus-trial waste-water	Capacity (m³day⁻¹)	Type of wastewater
OTV/ CGE* France	Biosep	Submerged, hollow fibers	3	15	300–13000	Domestic and municipal grey, industrial: - paper, cosmet-ics, chemicals, food and auto-mobile
Rhodia/ Orelis France	Pleiade	Submerged and side stream, flat, UF	56	14	7–500	Urban, grey and black
Kubota Japan	Kubota	Submerged, flat MF	92	50	10–2000	Domestic and municipal grey
Zenon* Canada	Zeeweed and ZenoGem	Submerged, hollow fibers MF, HF	150 (total)		340–5500	Municipal and industrial
AquaTech Korea	Biosurf	Side stream, UF	12	6	40–3000	Municipal
Degremont France	MBR	Side stream	2	3		Municipal

(* OTV and Zenon utilize the same technology)

The treatment of fermentation wastewater was studied by Lu *et al.* [4.113]. The authors used a rotary disk UF module coupled with an aerated bioreactor. The treated sludge was highly concentrated, and high organic carbon removal was obtained; nitrification of am-monia nitrogen was also successfully achieved. This reactor was able to work continu-ously during 130 days without noticeable loss of performance.

Recently Ahn *et al.* [4.114] reported the retrofitting of a municipal sewage treatment plant with a MBR using an AS bioreactor coupled to a series of four UF membrane modules. Since their sludge concentration was relatively low, a continuous operation of the MBR unit for over 40 days without chemical washing was possible. Carbon elimination and nitrogen and phosphorous removal efficiencies proved to not depend on the feed composition. Nah *et al.* [4.115] reported nitrogen removal (nitrification and denitrification) from a wastewater from a small apartment complex, using an intermittently aerated, hollow-fiber submerged MBR. The operation included alternating cycles of aeration and anoxic mixing, conditions which allowed for the simultaneous removal of carbon and nitrogen, while the permeate flux was maintained almost constant during 160 days of operation. The above examples show that MBR systems are, indeed, capable of working successfully under real conditions.

The Lyonnaise des Eaux in France [4.116] has developed a process for the denitrification of underground waters in order to produce drinking water. This process combines a bioreactor with adsorption by powdered activated carbon, together with a hollow-fiber UF unit. This process allows the elimination of nitrates, nitrites, pesticides, and herbicides (atrazine, diethylatrazine, simazine, metabenzthiazuron, and urea derivatives, etc.) as well as taste and odor compounds. These molecules are frequently present in underground waters in Europe, as a result of past intensive agricultural practices. The UF membrane unit also disinfects the water by removing protozoa, bacteria, and viruses.

In recent years there have been numerous reports of MBR for environmental applications making use of ceramic membranes. As already noted, these membranes present advantages over their polymeric counterparts, in that they have a higher transmembrane flux and temperature resistance, and are more robust to various chemicals in cleaning and disinfection operations. Scott *et al.* [4.117] have reported the use of a MBR with a MF ceramic membrane, operating at low transmembrane pressures of <0.3 bar, in order to reduce energy costs and plugging, coupled to the bioreactor through a side stream (Figure 4.14). Though aeration was partially supplied by a sparger immersed in the bioreactor, the membrane (in addition to separating the purified water) also provided additional aeration during regular back-flushing. Use of this MBR system has led to markedly improved removal rates of organic contaminants. Çiçek *et al.* [4.118] compared the performance of a conventional AS process and a MBR process using monolith type, ceramic membranes for the treatment of a wastewater containing high-molecular weight compounds. The MBR gave better performance, both in terms of solids retention and in wastewater treatment. A study of the microbial populations in both systems showed that the MBR sludge contained higher amounts of free-swimming bacteria (single cells), while the AS sludge contained only few such bacteria. It was shown that this is because the recirculation loop in the MBR decreases the sludge flock size, simultaneously resulting in a higher number of free-swimming bacteria, and improved mass transfer characteristics. The overall enzymatic activity was also shown to be superior in the MBR. This result differs from the behavior of the fine chemicals synthesis MBR, where one is concerned about the effect the high fluid velocities in the recirculating loop (used to improve mass transfer characteristics) will have on cell activity and endurance.

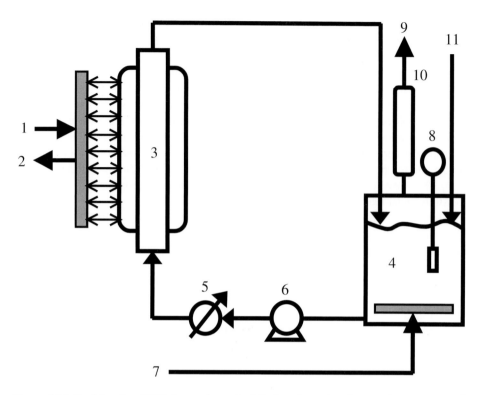

Figure 4.14. Dual function MBR. 1: membrane backflush and aeration; 2: treated waste permeation; 3: ceramic membrane module; 4: bioreactor liquor; 5: heat exchanger; 6: recycle pump; 7: sparger aeration; 8: dissolved oxygen probe; 9: gas exit; 10: volatile reflux; 11: untreated waste feed. Adapted from Scott *et al.* [4.117].

Though there has been significant interest in the application of ceramic membrane-based MBR for wastewater treatment, the industrial application of such systems is still lagging. These membranes are, generally, more expensive (per m^2) than polymeric membranes, which, in addition, can be made with much higher (s/v) ratios in the form of hollow fibers. Wastewater treatment applications, typically, involve large volumes of fluids to be treated, requiring large membrane areas, and are producing a low-value product; all these characteristics are more favorable towards the use of polymeric rather than ceramic membrane modules. Furthermore, membrane fouling appears to be more significant, for these types of applications, for ceramic membranes [4.119]. This is not surprising, since inorganic surfaces are known to have strong interactions with biopolymers. Flock break-up, which has a beneficial effect on mass transfer characteristics, could be contributing to membrane plugging, because it creates smaller size flocks, which could easily lodge themselves into the ceramic membrane's pore space, and because it increases the release of biopolymers and the non-settable fraction of the suspension.

Anaerobic biological processes for wastewater treatment have also been developed during the last ten years, especially for the treatment of high strength, food industry effluents. Anaerobic processes provide some advantages here. These include a lower sludge production rate, and the production of methane, which could find use as an energy source. Anaerobic biological treatment by methanogenic microorganisms is typically carried out in a single pass bioreactor with no sludge recycle. In this process the biomass concentration and the organic loading rates are generally low. The integration of a membrane separation device with an anaerobic bioreactor aims to control, independently from each other, the hydraulic and solids retention rates. These MBR processes have been reviewed recently by Stephenson *et al.* [4.120]. The authors report that the coupling of an ultrafiltration membrane with an anaerobic bioreactor allows for an increase in the biomass concentration, for higher COD (chemical oxygen demand) and TOD (total oxygen demand) loading rates, as well as high removal efficiencies (up to 90 %).

One of the most successful applications of anaerobic MBR was realized by Weir EnVig in South Africa with the Adulf® process. This process has been used for the treatment of different food industry wastewaters; in all cases studied high COD removal rates were obtained, up to 97 % [4.120]. A constant membrane permeate flux of 60 $l \cdot m^{2} \cdot h^{-1}$ was obtained during 18 months of MBR operation during the treatment of a wine distillery effluent. Beaubien *et al.* [4.121] have reported in their study of an anaerobic MBR utilizing a ceramic membrane almost complete removal of total organic matter. The high shear rates employed were not detrimental for the methanogenic bacteria, which maintained their specific activity. This is promising, in view of the commonly held view that the slow-growing methanogenic bacteria may not be appropriate for such applications. An anaerobic MBR involving the combination of a packed bed bioreactor with a microfiltration hollow fiber module [4.122] has been applied to the removal of nitrate ions from natural, underground water. The obtained results indicated high denitrification, rates up to 100 %, but only at low nitrate concentration in the raw wastewater.

Some disadvantages result from the application of anaerobic MBR for the treatment of high strength wastewaters, like kraft bleach, wine distillery or wool scouring waste. These relate to the coupling of slow growing anaerobic microorganisms with the high strength influent. The combination, generally, means that high hydraulic retention times (the inverse of the dilution rate) are needed, typically in the range of 50 to 150 h. Fouling is also a problem in the absence of air-bubbling. Increasing the tangential flow may alleviate the fouling problem somewhat, but one must always be aware that pumping can have a negative impact on bioreactor performance, because of the cell lysis that is caused by the mechanical shear caused from the pump.

Enzymatic MBR have also been used for the removal of various pollutants from wastewaters, a key class being phenolic compounds, whose toxicity, hazardous nature, and increasing presence in wastewaters are becoming well documented [4.123]. Bodzek *et al.* [4.124] have studied the phenol and cyanide enzymatic removal from coke industry wastewaters. For this purpose, they immobilized an enzymatic fraction, isolated from a bacterial strain of *Pseudomonas,* on ultrafiltration membranes. The results obtained showed that, in spite of long contact times (up to 4 h), only modest biodegradation rates of real wastewaters were obtained (up to 25 % and 50 % for phenols and cyanides respectively). Edwards *et al.* [4.125] studied the degradation of a number of phenolic compounds (phenol, methoxyphenol, cresol, and cholorophenol) by polyphenol oxidase immobilized in UF polysulfone capillary membranes. Though the immobilized enzymes exhibited significantly reduced activity, the MBR experiments still showed good oxidation conversion of the phenolic compounds into *o*-quinones. This MBR system, of course, has the additional advantage of the easy separation of the biocatalyst from the raw water and products. Jolivalt *et al.* [4.126] have reported recently the decomposition of phenyl urea pesticide in wastewater using a MBR with immobilized laccase. The authors observed that the grafted enzyme had an activity, which was comparable to that of the free enzyme and good enzyme stability. Almost complete removal of the phenyl urea was obtained after 5 min of reaction for the highest concentrations of immobilized laccase.

From the discussion, so far, it is obvious that MBR are effective for the treatment of various wastewaters. For the most part, however, all the aforementioned studies have dealt with water-soluble pollutants. Industrial wastewaters contain, in addition, other organic contaminants, which are poorly water-soluble (organochlorides, benzene, toluene, xylenes, etc.). Though present in low concentrations, they are, typically, highly toxic and generally recalcitrant to biological degradation [4.127]. Bacterial strains have been isolated which efficiently degrade these organics; they are strongly affected, however, by the extreme conditions (pH, ionic strength, etc.) that are frequently encountered in wastewaters [4.128]. Organophilic membranes have been used for the removal of such organics [4.129]. The membrane acts as contactor or physical interface between the wastewater and a solvent phase, where the organics are dissolved. In recent years innovative MBR systems have also been applied for the treatment of such compounds. Livingston [4.130] proposed the concept of extractive membrane bioreactors (EMB), in which a dense hydrophobic membrane separates the wastewater and a separate aqueous phase, where the bioreaction takes place (see Figure 4.15). The organics in the wastewater, which exhibit a strong affinity with the membrane, pass through it and are biotransformed in the compartment containing the water phase. The membrane separates effectively the two phases not allowing the ionic species found in the wastewater to interfere with the bioconversion.

The biological degradation of organochlorides has been reported by Strachan and Livingston [4.131], who used a tubular, silicon-rubber membrane, with the wastewater circulating in the membrane tube. The organochlorides have high membrane/aqueous phase partition coefficients, and pass through the membrane to the biomedium (outer side), where they are transformed. The biomedium is then recirculated to an aerated tank. Livingston *et al.* [4.132] carried out additional studies of monochlorobenzene degradation by the *Pseudomonas* strain JS150 in a pilot-plant scale unit. They concluded that the overall mass transfer resistance was strongly dependent on the biofilm characteristics on the biomedium (shell) side. Pampel and Livingston [4.133] used an EMB operating under anaerobic conditions to dechlorinate tetrachloroethylene in a synthetic wastewater. Anaerobic conditions were chosen because tetrachloroethylene, a toxic and potentially carcinogenic compound, extensively used in the past as a dry cleaning and degreasing agent, is recalcitrant under aerobic conditions. The EMB proved relatively effective in biodegrading tetrachloroethylene.

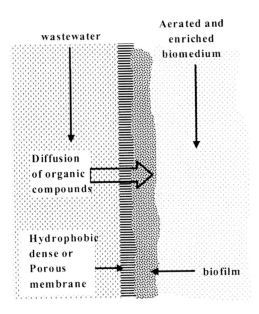

Figure 4.15. Principle of the extractive membrane bioreactor.

Other investigators have also studied EMB for various environmental applications. An EMB with a dense silicone rubber membrane was used by Freitas dos Santos and Lo Biundo [4.134] for the treatment of a pharmaceutical wastewater containing dichloromethane using *Vibrio fischeri*. More than 95 % removal efficiency was attained in a pilot-plant scale unit during 400 h of continuous operation. More recently, Xiao *et al.* [4.135] have studied the toluene biodegradation by *Pseudomonas putida* in an EMB containing a hollow-fiber, silicone rubber membrane. Tubular and spiral-wound type hollow-fiber membrane modules with the same area were studied. Both showed good removal efficiencies, with the spiral-wound modules giving a little better removal efficiency (93 % vs. 83 %).

Other authors in their EMB studies have been using hydrophobic porous membranes. Porous membranes, which are not wetted, can effectively, when their pore size is under a certain limit, separate aqueous solutions. In this case, the organic pollutant must diffuse through the pores, a more effective process than diffusion through a dense material. For this type of process the aqueous solutions placed on both sides of the membrane have to be isotonic, in order to avoid water vapor diffusion from the less towards the most concentrated solution. Aziz *et al.* [4.136] reported the use of an EMB using microporous hollow-fiber membranes to remove chlorinated solvents from water, for subsequent biodegradation by methanotrophic bacteria. Inguva *et al.* [4.137] reported the biodegradation of 1,2-dichloro- and trichloroethane by *Pseudomonas cepacia* and *Xanthobacter autotrophicus*, respectively, in an EMB using UF flat or hollow-fiber polypropylene porous membranes. They compared removal efficiencies with and without biomass present in the reactor, and concluded that the microbial reaction results in a significant enhancement. Furthermore, the flat-membrane EMB was significantly more effective than the hollow-fiber EMB. Pressman *et al.* [4.138] reported the rapid degradation of trichloroethylene (TCE) by *Methylosinus trichosporium,* a bacterium, which attaches poorly to surfaces. An EMB using an UF polypropylene hollow-fiber porous membrane was used in two different configurations (Figure 4.16). In the first configuration the biomedium fluid on the shellside flows parallel to the fibers length, and co-current contacting was used in order to provide the greatest concentration gradient at the EMB inlet, where the degradation rates are the largest. The second EMB is a radial-flow module, which has a baffle at the middle of the center-feed tube, which causes the flow to move radially outward across the fibers in the first half of the module. Because the shell flow must exit the other end of the feed tube, the flow is routed back radially inwards across the fibers and back into the feed tube. For this configuration the authors chose countercurrent contacting, since prior studies indicated this configuration gave better mass transfer. The results showed that the baffled, radial-flow EMB performs significantly better than the parallel-flow system. TCE degradation rate remained almost constant for over three weeks of steady state operation, among the best results reported in the literature, so far. Mass transfer rates decreased over time, however. Bacteria detachment may be one of the causes, but biofouling probably also results from metabolites and proteins from the lysed bacteria. EMB using porous membranes are appropriate for applications, where the selective degradation of a single pollutant is not required; otherwise a dense membrane EMB may be necessary.

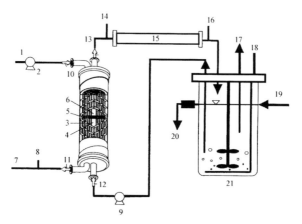

Figure 4.16. Schematic of a hollow-fiber EMB. 1: TCE polluted water, 2,9: peristaltic pumps, 3: hollow fibers membranes, 4: distribution tube, 5: baffle, 6: collection tube, 7: treated effluent, 8,14,16,17: sample ports, 10: lumen influent, 11: lumen effluent, 12: shell influent, 13: shell effluent, 15: plug flow reactor, 18: air, 19: methanol and nutrients, 20: waste, 21: growth chemostat. From Pressman *et al.* [4.138], with permission from John Wiley & Sons Inc.

When the wastewater streams contain significant quantities of inorganics, it is difficult to treat them with MBR. In fact, when the salt and acid content is sufficiently high the biological degradation is frequently inhibited. This is particularly true, when biorefractory pollutants are present in acidic waste streams. Liu *et al.* [4.139] have recently published a study on an integrated system, which treats acidic streams containing 2,4-dichlorophenol (DCP). Their system utilizes an extraction unit, which makes use of a water insoluble organic solvent in order to separate the organics from the wastewater. A stripping unit allows to the transfer of the organic pollutants to an alkaline aqueous solution. The emulsion that is formed in the stripping unit is separated by filtration with a ceramic hydrophilic membrane, the organic phase is recirculated to the stripping unit, whereas the aqueous phase is fed into the bioreactor, where a second membrane allows the separation of the biomass from the treated water, which is recirculated in order to dilute the pollutant content in the wastewater stream. This type of system was shown to be very efficient for the treatment of DCP containing streams. During 14 days of continuous operation, 98.7 % of DCP and 86.5 % of the total organic content were biodegraded (Figure 4.17).

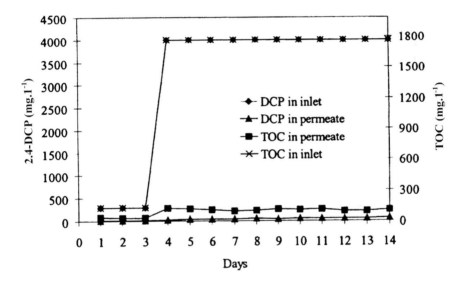

Figure 4.17. Continuous biodegradation of DCP by an extractive MBR system. From Liu *et al.* [4.139], with permission from Elsevier Science.

MBR using hydrophobic membranes have also been used in the biological treatment of waste gas. These MBR function very similarly with the EMB (a schematic of the process is shown in Figure 4.18). The gas phase supplies the carbon and oxygen necessary for the development of the biomass. The aqueous phase, where the microorganisms grow, supplies the other nutrients. In most cases a biomass film is created on (or within) the membrane. The membrane must be able to provide a large liquid-gas interface, and, as a result, large areas are required. Membrane surface hydrophobicity is an important characteristic, since it determines the interaction between the organic molecules and the membrane. As with EMB, *Pseudomonas* or *Xanthobacter* bacteria strains are commonly utilized. One of the earliest studies is by Bauerle *et al.* [4.140], who studied the biodegradation of xylenes, dichloromethane and *n*-butanol using a silicon rubber membrane, which separates the waste gas phase from an aqueous phase containing an activated sludge. Reij *et al.* [4.141] used a MBR with a mesoporous hydrophobic membrane for propene degradation by a *Xanthobacter* type bacteria strain. This is a challenging reaction system, because propene has low water solubility. The membranes had a very high porosity (up to 75 %) in order to increase the mass transfer resistance. They report a stable propene degradation rate during 20 days of continuous operation, with the mass transfer resistance mainly residing in the biofilm. One problem they encountered is excessive biomass growth, which, if not re-moved, may create serious problems with mass transfer. This was also observed by Freitas dos Santos *et al.* [4.142], who studied in a MBR the degradation of 1,2-dichloroethane by *Xanthobacter* strain. They observed a decline in reactor performance, and an increasing pressure drop across the biofilm phase. They attributed both phenomena to the excessive amounts of biofilm being produced in their spiral-wound membrane module.

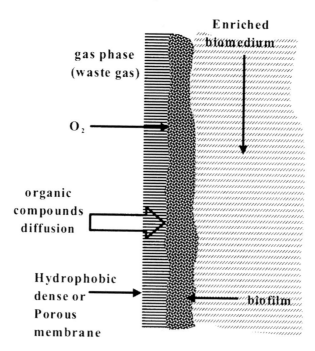

Figure 4.18. Schematic of the MBR for waste gas treatment.

MBR present a number of advantages in the biological treatment of waste gas over the conventional biofilters, in which the gas flows through a bed of microorganisms immobi-lized on solid supports like compost or soil, which consume the gaseous organic pollut-ants. One of the key advantages is that the MBR continuously eliminate the degradation products, which may potentially inactivate the biomass. Biofilters also present problems related to the starvation of the resident microorganisms during plant shut-downs, which result in a certain time-lag in performance, when plant operation is recommenced. MBR being modular systems are less sensitive to feed interruptions and upsets. For example, in a recent patent Peretti *et al.* [4.143] report the treatment of VOC containing waste gas streams using a modular MBR system, which is able to withstand fluctuations in pollutant concentrations. The VOC in the waste gas stream are perstracted with an oleic solvent phase using a bundle of hollow fiber hydrophobic porous membranes serving as a contactor between the gas and the oil phases. Once concentrated into the solvent phase, the pollutants are biodegraded in a MBR, which contains a second bundle of porous hydrophobic membranes separating the oleic and aqueous phases, and serves also as support for the biofilm. Good results have been reported in several literature publications with the biodegradation of a number of compounds (see Table 4.2), most of which are reviewed in a recent article by Reij *et al.* [4.144]. In their article they note that the MBR process can be easily scaled-up, and that the removal of poorly water-soluble pollutants from waste gas should be considered as one of the most promising future MBR applications. They conclude that the key challenge that remains is biomass growth control, and its long-term effect on reactor performance.

Table 4.2. Membrane bioreactors for waste gas treatment. Adapted from Reij *et al.* [4.144].

Pollutant	Concentration (ppm)	Type of membrane	Biofilm	Inoculant
Xylenes	30–140	Dense silicone tubes	Yes	Sludge
n-Butanol	40–180	Dense silicone tubes	Yes	Sludge
Dichloro-methane	47–220	Dense silicone tubes Mesoporous hydrophobic material sheet	Yes and no, depending on case	Sludge, strain DM21
Dichloroethane	150	Silicone spiral-wound	Yes	*Xanthobacter* GJ10
Trichloroethene	20	Polysulfone fibers	Yes	Sludge
Propene	330–2700	Mesoporous hydrophobic material fibers	Yes	*Xanthobacter* Py2
Toluene	20–56	Dense silicone tubes. Mesoporous hydrophobic material, sheet	Yes	*Pseudomonas*
n-Hexane	32	Dense silicone tubes	?	?
Mixture	Low?	Mesoporous hydrophobic material, sheet	Yes	Various strains

To summarize, MBR have been successfully utilized for both wastewater and waste gas treatment. Membranes play a variety of roles. They are used for separating and recycling the biosolids and biomass, as contactors in order to increase the mass transfer during the extraction phase of the pollutants from a wastewater or gas stream or act as a biomass support (in a biofilm form) for the organics degradation. This is certainly an area where the application of MBR shows excellent future potential.

4.3   References

[4.1]   J.M. Engasser, "Reacteurs a Enzymes et Cellules Inmobilisées", in Chap. 4.2, *Biotechnologie,* Technique et Documentation Lavoisier, Paris, 1988.

[4.2]   J.H. Hanemaaijer, J. Stahouders, and S. Visser, in, O.M. Neijssel, R.R. van der Meer and K.Ch.M. Luyben, Eds, *Proc. 4th Eur. Congr. in Biotechn.*, Elsevier, Amsterdam, 1, 119, 1987.

[4.3]   L. Giorno and E. Drioli, *Trends in Biotechnol.* **2000**, 18, 339.

[4.4]   L. Sajc, D. Grubisic, and G. Vunjac-Novakovic, *Biochem. Eng. J.* **2000**, 4, 2.

[4.5]   M.K.H. Liew, A.G. Fane, and P. Rogers, *J. Chem. Tech. Biotechnol.* **1995**, 64, 200.

[4.6]   W.H. Hanish, "Cell Harvesting," in: W.C. Mc Gregor, Ed., *Membrane Separation in Biotechnology*, Chapter 3, M. Dekker NY, 1986.

[4.7]   A.S. Michaels, *Chem. Eng. Prog.* **1968**, 64, 31.

[4.8]   E. Flaschel, C. Wandrey, and M.R. Kula, *Adv. Biochem. Eng.* **1983**, 26, 73.

[4.9]   T.B. Vick Roy, D.K. Mandel, D.K. Dea, H.W. Blanch, and C.R. Wilke, *Biotechnol. Lett.* **1983**, 5, 665.

[4.10] P. Boyabal, C. Corre, and S. Terre, *Biotechnol. Lett.* **1987**, 9, 207.

[4.11] B. Bibal, Y. Vayssier, G. Goma, and A. Pareilleux, *Appl. Microbiol. Biotechnol.* **1991**, 30, 630.

[4.12] R. Datta, S.P. Tsai, P. Bonsignore, S.H. Moon, and R.J. Frank, *FEEMS Microbiol. Rev.* **1995**, 16, 221.

[4.13] A. Olmos-Dichara, F. Ampe, J.L. Uribelarea, A. Pareilleux, and G. Goma, *Biotechnol. Lett.* **1997**, 19,8, 709.

[4.14] S. Rissom, U. Schwarz-Linek, M.Vogel, V.I. Tishkov, and U. Kragl, *Tetrahedron: Asymm.* **1997**, 8, 2523.

[4.15] J. Onken and R.G. Berger, *Appl. Microbiol. Biotechnol.* **1999**, 51, 158.

[4.16] A.M.R.B. Xavier, L.M.D. Gonçalves, J.L. Moreira, and M.J.T. Carrondo, *Biotechnol. Bioeng.* **1995**, 45, 320.

[4.17] S. Tejayadi and M. Cheryan, *Appl. Microbiol. Biotechnol.* **1995**, 43, 242.

[4.18] H. Moueddeb, J. Sanchez, C. Bardot, and M. Fick, *J. Membr. Sci.* **1996**, 114, 59.

[4.19] flinderstech@flinders.edu.au, march 2001.

[4.20] J.P.S.G. Crespo, A.M.R.B. Xavier, M.T.O. Barreta, L.M.D. Goncalves, J.S. Almeida, and M.J.T. Carrondo, *Chem. Eng. Sci.* **1992**, 47, 205.

[4.21] M. Cheyran and M.A. Mehaia, *Proc. Biochem.* **1984**, 19, 204.

[4.22] C.W. Cho and S.T. Hwang., *J. Membr. Sci.* **1982**, 10, 253.

[4.23] M. Gryta, A.W. Morawski, and M. Tomaszewska, *Catal. Today* **2000**, 56, 159.

[4.24] R. Rautenbach and R. Albrecht, *Membrane Processes*, John Wiley & Sons Inc., Chichester, 1989.

[4.25] R. Legoux, P. Lepatois, J.E. Liauzun, B. Niaudet, W.G. Roskam, and W. Roskam, French Patent, 2,597,114, 1987; U.S. Patent 4,945,047, 1990.

[4.26]   B. Bauer, H. Chmel, F. Effenberger, and H. Starthmann, *Proc. Int. Cong. on Membr. and Membr. Processes. ICOM*, Heidelberg, Germany, August 1993.

[4.27]   Y. Koyama, K. Shimazaki, K. Akashi, Y. Kawahara, K. Kubota, and H. Yoshii, in: O.M. Neijssel, R.R van der Meer, and K.Ch.A.M. Luyben, Eds., *Proc. 4th Eur. Congr. in Biotechn.*, Elsevier, Amsterdam, 1, 119, 1987.

[4.28]   M. Cantarella, A. Spera, L. Cantarella, and F. Alfani, *J. Membr. Sci.* **1998**, 147, 279.

[4.29]   A. Bosetti, D. Bianchi, N. Andriollo, D. Cidaria, P. Cesti, G. Sello, and P. Di Gennaro, *J. Chem. Tech. Biotechnol.* **1996**, 66, 375.

[4.30]   K. Miyano, K. Ye, and K. Shimizu, *Biochem. Engng. J.* **2000**, 6, 207.

[4.31]   R.D. Schwartz and C.J. McCoy, *Appl. Env. Micro.* **1977**, 34, 47.

[4.32]   S.D. Doig, A.T. Boam, D.I. Leak, A.G. Livingston, and D.C. Stuckey, *Biotechnol. Bioengng.* **1998**, 58, 6, 587.

[4.33]   S.D. Doig, A.T. Boam, A.G. Livingston, and D.C. Stuckey, *Biotechnol. Bioengng.,* **1999**, 63, 5, 601.

[4.34]   H.M. Krieg, J.C. Breytenbach, and K. Keizer, *J. Membr. Sci.* **2000**, 180, 69.

[4.35]   H.M. Krieg, A.L. Botes, M.S. Smith, J.C. Breytenbach, and K. Keizer, *J. Mol. Catal. B:Enzym.* **2001**, 13, 37.

[4.36]   Y. Shi, J. Ploof, and A. Correia, *IVD Technol. Mag.* **1999** May.

[4.37]   J. Hopkinson, *Bio/technol.* **1985**, 3, 225.

[4.38]   R. Kurkela, E. Fraune, and P. Vihko, *Biotechniques* **1993**, 15, 674.

[4.39]   J.M. Piret and C.L. Cooney, *Biotechnol. Bioengng.* **1990**, 36, 902.

[4.40]   J. Feder and R. Tolber, *Int. Biotechnol. Lab.* **1985**, June, 41.

[4.41]   C. Legallais, B. David, and E. Dore, *J. Membr. Sci.* **2001** , 181, 81.

[4.42]   F.Molinari, F. Aragozzini, J.M.S. Cabral, and D.M.F. Prazeres, *Enzym. Microb. Technol.* **1997**, 20, 604.

[4.43]   Y. Yamamoto, S. Sugimoto, X. Shen, T. Nagamune, S. Yao, and E. Suzuki, *Biochem. Engng. J.* **1999**, 3, 2, 151.

[4.44]   D.M.F. Prazeres and J.M.S. Cabral, *Enzyme Microb. Technol.* **1994**, 16, 738.

[4.45]   M. Cheyran and M.A. Mehaia, in: W.C. McGregor, Ed., "Membrane Bioreactors", *Membrane Separation in Biotechnology*, Marcel Dekker, New York, 255, 1986.

[4.46]   C. Wandrey, "Fine Chemicals", in: O.M. Neijssel, R.R. van der Meer, and K.Ch.A.M. Luyben, Eds., *Proc. 4th European Congress in Biotechnology*, Elsevier, Amsterdam, 1987, 3, 22.

[4.47]   E. Schmidt, W. Hummel, and C. Wandrey, in: O.M. Neijssel, R.R van der Meer, and K.C.A.M Luyben, Eds., *Proc. 4th Eur. Congr. in Biotechn.*, Elsevier, Amsterdam, 1987, 2, 189.

[4.48]   J.L. Maubois, L. Roger, G. Brulé, and M. Piot, French Patent N° 79 16483, 1979.

[4.49]   A. Mannheim and M. Cheyran, *J. Food Sci.* **1990**, 55, 381.

[4.50]   Y. Pouliot, S.F. Gauthier, and C. Bard, *J. Membr. Sci.* **1993** , 80, 257.

[4.51]   S. Curcio, V. Calabro, and G. Ioro, *J. Membr. Sci.* **2000**, 173, 247.

[4.52]   N. Cempel, J.M. Piot, and D. Guillochon, *J. Agric. Food. Chem.* **1994**, 42, 2059.

[4.53]   D. Belhocine, H. Mokrane, H. Grib, H. Lounici, A. Pauss, and N. Mameri, *Chem. Engng. J.* **2000**, 76, 189.

[4.54]   N. D'Alvise, C. Lesueur-Lambert, B. Fertin, P. Dhulster, and D. Guillochon, *Proc. Second European Congress of Chemical Engineering*, Montpellier, 1999.

[4.55]   J. Lopez, U.S. Patent 5,500,352, 1996.

[4.56]   D. Paolucci-Jeanjean, M.P. Belleville, G.M. Rios, and N. Zakhia, *Biochem. Engng. J.* **2000**, 5, 17.

[4.57]   J.Y. Houng, J.Y. Chiou, and K.C. Chen, *Bioprocess Eng.* **1992**, 8, 85.

[4.58]   F.P. Cuperus, S.Th. Bouwer, G. Boswinkel, R.W. van Gemert, and J.W. Veldsink, *Proc. Fourth Workshop: Optimisation of Catalytic Membrane Reactors Systems*, Oslo, Norway, May, **1997**, pp: 83-88.

[4.59]   M.A.P. Morgado, J.M.S. Cabral, and D.M.F. Prazeres, *J. Am. Oil. Chem. Soc.* **1996**, 73, 3, 337.

[4.60]   M.M. Hoq, T. Yamane, S. Shimizu, T. Funada, and S. Ishida, *J. Am. Oil Chem. Soc.* **1985**, 62, 6, 1016.

[4.61]   L. Giorno, R. Molinari, M. Natoli, and E. Drioli, *J. Membr. Sci.* **1997**, 125, 177.

[4.62]   Q. Gan, F. Baykara, H. Rahmat, and L.R. Weatherley, *Catal. Today* **2000**, 56, 179.

[4.63]   J. Ceynowa and P. Adamczak, *Sep. Pur. Technol.* **2001**, 22-23, 443.

[4.64]   H.G. Hicke, M. Ulbricht, M. Becker, S. Radosta, and A. Heyer, *J. Membr. Sci.* **1999**, 161, 239.

[4.65]   A. Marc, C. Burel, and J.M. Engasser, "Reacteurs Enzymatiques a Fibres Creuses pour l'Hydrolyse d'Amidon et Sacharose", in *Utilisation, d'Enzymes en Technologie Alimentaire*, P. Dupuy, Ed., Techiques et Documentation Lavoisier, Paris, 35, 1982.

[4.66]   M. Furui, T. Furutani, T. Shibatani, Y. Nakamoto, and T. Mori, *J. Ferm. Bioengng.* **1996**, 81, 1, 21.

[4.67]   J.L. Lopez and S.L. Matson, *J. Membr. Sci.* **1997**, 125, 189.

[4.68]   P. Iwan, G. Goetz, S. Schmitz, B. Hauer, M. Breuer, and M. Pohl, *J. Mol. Catal. B:Enzym.* **2001**, 11, 387.

[4.69]   K. Sakaki, L. Giorno, and E. Drioli, *J. Membr. Sci.* **2001**, 184, 27.

[4.70]   M. Cretich, M. Chiari, and G. Carrea, *J. Biochem. Biophys. Methods* **2001**, 48, 247.

[4.71]   J.A. Hughes and M. Jay, *Nucl. Med. Biol.* **1995a**, 22, 1, 105.

[4.72]   J.A. Hughes and M. Jay, *J. Labl.Comp. Pharm.* **1995 b**, 26, 12, 1133.

[4.73]   D.K. Tompkison, I.A. Angelo, and M.P. Arthur, *Indian J. Dairy Sci.* **1983**, 36, 328.

[4.74]   R. Puvanakrishna and S.M. Bose, *Indian J. Biochem. Biophys.* **1984**, 21, 323.

[4.75]   J. Ida, T. Matsuyama, and H. Yamamoto, *Biochem. Engng. J.* **2000**, 5,179.

[4.76]   S. Ganapathi-Desai, D.A. Butterfield, and D. Bhattacharyya, *J. Chem. Technol. Biotechnol.* **1995**, 64, 157.

[4.77]   S. Ganapathi, D.A. Butterfield, and D. Bhattacharyya,. *Biotechnol. Prog.* **1998**, 14, 865.

[4.78]   D.A. Butterfield, D. Bhattacharyya, S. Daunart, and L. Bachas, *J. Membr. Sci.* **2001**, 181, 29.

[4.79]   P. Boyabal, C. Corre, and S. Terre, *Lait.* **1988**, 68, 65.

[4.80]   E. Borch, H. Berg, and O. Holst, *J. Appl. Bacteriol.* **1991**, 71, 265.

[4.81]   J. Sanchez and T.T. Tsotsis, "Reactive Membrane Separation", in: S. Kulprathipanja, Ed., *Reactive Separation Processes*, Taylor and Francis, USA, 2001.

[4.82]   R.K. Warren, G.A. Hill, and D.G. Mac Donald, *Trans. I. Chem. E.* **1994**, 72, 149.

[4.83]   C. Fonade, and M.Y. Jaffrin, "Regimes Instationnaires: Mise en œuvre des Perturbations", Chap. 4, in : G. Dauphin, F. Rene, and P. Aimar, Eds., *Les Separations par Membrane dans les Procedes de l'Industrie Alimentaire*, Technique et Documentation Lavoisier, Paris, 1988.

[4.84]   G. Arroyo and C. Fonade, *J. Membr. Sci.* **1993**, 80, 117.

[4.85]   C. Maranges, C. Casanovas, and C. Lafforgue-Delorme, *Biotechnol. Lett.* **1995**, 9, 649.

[4.86]   A. Margaritis and C.R. Wilke, *Biotechnol. Bioengng.* **1978**, 20, 709.

[4.87]   G. Leach, I. Sellick, M. Collins, D. Caire, and C. Felisaz, *Proc. Second European Congress of Chemical Engineering*, Montpellier, 1999.

[4.88]   Z. Ujang and A. Hazri, *J. Membr. Sci.* **2000**, 175, 139.

[4.89]   J.S. Devinny, M.A. Deshusses, and T.S. Webster, *Biofiltration for Air Pollution Control*, CRC Lewis Press, 1998.

[4.90]   C.V. Smith, D.O. Gregorio, and R.M. Talcom, *Proc. 24rd Ind. Waste Conf. of Conference*, Purdue University, Ann Arbor Science, USA, p. 1300, 1969.

[4.91]   F.W. Hardt, L.S. Clesceri, N.L. Nemerov, and D.R. Washington, *J. Wat. Pollut.Control Fed.* **1970**, 42, 2135.

[4.92]   I. Bemberis, P.J. Hubbard, and F.B. Leonard, *Amer. Soc. Agric. Engng. Winter Mtg.* **1971**, 71, 1.

[4.93]   G.K. Anderson, C.B. Saw, and M.I.A.P. Fernandez, *Process. Biochem.* **1986**, 12, 174.

[4.94]   B. Langlais, P. Denis, S. Triballeau, M. Faivre, and M.M. Bourbigot, *Wat. Sci. Tech.* **1993**, 25, 219.

[4.95]   J.A. Scott, D.J. Neilson, W. Liu, and P.N. Boon, *Wat. Sci. Tech.* **1998**, 38, 4-5.

[4.96]   C.S.F. Ragona and E.R. Hall, *Wat. Sci. Tech.* **1998**, 38, 4-5.

[4.97]   C. Wisnieswski, A. Leon-Cruz, and A. Grasmick, *Biochem. Eng. J.* **1999**, 3, 61.

[4.98]   S. Chaize and A. Huyard, *Wat. Sci. Tech.* **1991**, 23, 1591.

[4.99]   S. Kimura, *Water Sci. Technol.* **1991**, 23, 1573.

[4.100]   J. Manem, E. Trouve, A. Beaubien, A. Huyard, and V. Urbain, *Proc. 66th Annu. Water Environ. Fed. Conf. Exposition*, Anaheim, Calif., USA, 1993.

[4.101]   A. Mersmann, *Chem. Engng. Process.* **1995**, 34, 279.

[4.102]   P. Coté, H. Buisson, C. Pound, and G. Araki, *Desalination* **1997**, 113, 189.

[4.103]   C. Monk, "Application of Membranes in Waste Water Treatment", *Proc. of Filtech Conference, Publ. by The Filtration Society*, p. 249, October, 1993.

[4.104]   H. Ishida, Y. Yamada, S. Matsumura, and M. Moro, *Proc. Int. Cong. on Membr. and Membr. Processes. ICOM,* Heidelberg, Germany*, Sept., 1993.

[4.105]   S. Churcose, *Proc. 1st Int. Meeting on Membrane Bioreactors for Wastewater Treatment*, Cranfield University, 5-6 March, 1997.

[4.106]   K. Yamagiwa, Y. Ohmae, M. Hatta Dahlan, and A. Ohkawa, *Bioresource Technol.* **1991**, 37, 215.

[4.107]   M. Praderie, H. Buisson, H. Paillard, and T. Vouillon, *Galvano-organo-traitements de Surface,* **1998**, 685, 390.

[4.108]   E.H. Bouhabila, R. Ben Aim, and H. Buisson, *Sep. Pur. Technol* **2001**, 22-23, 123.

[4.109]   T. Ueda and K. Hata, *Wat. Res.* **1999**, 33, 12, 2888.

[4.110]   R. Dufresne, H.C. Lavalle, R.E. Lebrun, and S.N. Lo, *Tappi Journal* **1998**, 81, 4, 131.

[4.111]   M. Gander and J.J. Judd, *Sep. Purif. Technol.* **2000**, 18, 119.

[4.112]   S. Gauthier and M. Praderie, OTV Communication Direction, *Personal Communication*, 2000.

[4.113]   S.G. Lu, M. Ukita, M. Sekine, M. Fukagawa, and H. Nakanishi, *Environ. Technol.* **1999**, 20, 431.

[4.114]   K-H. Ahn, H.Y. Cha, and K.G. Song, *Desalination* **1999**, 124, 279.

[4.115]   Y.M. Nah, K.H. Ahn, and I.T. Yeom, *Environm. Technol.* **2000**, 21, 107.

[4.116]   V. Urbain, R. Benoit, and J. Manen, *J. AWWA.* **1996**, 75.

[4.117]   J.A. Scott, J.A. Howell, T.C. Arnot, K.L. Smith, and M. Bruska, *Biotechnol. Techniq.* **1996**, 10, 4, 287.

[4.118]   N. Çiçek, J. Franco, M. Suidan, V. Urbain, and J. Manem, *Wat. Environm. Res.* **1999**, 71, 1, 64.

[4.119]   C. Wisniewski and A. Grasmick, *Coll. Surf. A.* **1998**, 138, 403.

[4.120]   T. Stephenson, S. Judd, B. Jefferson, and K. Brindle, *Membrane Bioreactors for Wastewater Treatment*, IWA Publishing, 2000.

[4.121]   A. Beaubien, M. Bâty, F. Jeannot, E. Francoeur, and J. Manem, *J. Membr. Sci.* **1996**, 109, 173.

[4.122]   E. W_sik, J. Bohdziewicz, and M. Blaszczyk, *Sep. Purif. Technol.* **2001**, 22, 383.

[4.123]   B.N. Alberti and A.M. Klibanov, *Biotechnol. Bioengng. Symp.* **1981**, 11, 373.

[4.124]   M. Bodzek, J. Bohdziewicz, and M. Malgorzata, *J. Membr. Sci.* **1996**, 113, 373.

[4.125]   W. Edwards, R. Bownes, W.D. Leukes, E.P. Jacobs, R. Sanderson, P.D. Rose, and S.G. Burton, *Enzym. Microb. Technol.* **1999**, 24, 209.

[4.126]   C. Jolivalt, S. Brenon, E. Caminade, C. Mougin, and M. Pontié, , *J. Membr. Sci.* **2000**, 180, 103.

[4.127]   T.C. Arnot and N. Zahir, *Res. Env. Technol.* **1996**, 1, 145.

[4.128]   M.L. Rochkind-Dubisky, G.S. Sayler, and J.W. Blackburn, *Microbiol. Ser.* **1987**, 18.

[4.129]   A. Mersmann, *Chem. Engng. Process* **1995**, 34, 279.

[4.130]   A.G. Livignston, *J. Chem. Technol. Biotechnol.* **1994**, 60, 117.

[4.131]   L.F. Strachan and A.G. Livingston, *J. Membr. Sci.* **1997**, 128, 231.

[4.132]   A.G. Livingston, J.P. Arcangeli, A.T. Boam, S. Zhang, M. Marangon, and L.M. Freitas dos Santos, *J. Membr. Sci.* **1998**, 151, 29.

[4.133]   L.W.H. Pampel and A.G. Livingston, *Appl. Microbiol. Biotechnol.* **1998**, 50, 303.

[4.134]   L.M. Freitas dos Santos and G. Lo Biundo, *Environm. Prog.* **1999**, 18, 1, 34.

[4.135]   Z. Xiao, M. Hatta Dahlan, X.H. Xing, Y. Yoshikawa, and K. Matsumoto, *Biochem. Engng. J.* **2000**, 5, 83.

[4.136]   C.E. Aziz, M. Fitch, L.K. Linquist, J.G. Pressman, G. Georgiou, and G.E. Jr. Speitel, *Environm. Sci. Technol.* **1995**, 29, 2574.

[4.137]   S. Inguva, M. Boensch, and G.S. Shreve, *AIChE J.* **1998**, 44, 9, 2112.

[4.138]   J.G. Pressman, G. Georgiou, and G.E. Jr. Speitel, *Biotechnol. Bioengng.* **1999**, 62, 6, 681.

[4.139]   W. Liu, J.A. Howell, T.C. Arnot, and J.A. Scott, *J. Membr. Sci.* **2001**, 181, 127.

[4.140]   U. Bauerle, K. Fischer, and D. Bardtke, *STAUB Reinhaltung der Luft* **1986**, 46, 233.

[4.141]   M.W. Reij, K.D. de Gooijer, J.A.M. de Bont, and S. Hartmans, *Biotechnol. Bioengngl.* **1995**, 45, 107.

[4.142]   L.M. Freitas dos Santos, U. Hommerich, and A.G. Livingston, *Biotechnol. Prog.* **1995**, 11, 194.

[4.143]   S.W. Peretti, S.M. Thomas, and R.D. Shepherd, US Patent 5,954,8589, 1999.

[4.144]   M.W. Reij, T.F. Keurentjes, and S. Hartmans, *J. Biotechnol.* **1998**, 59, 155.

5 Modelling of Membrane Reactors

Nomenclature

Other Notation

C_p: molar heat capacity of purge gas	$(cal \cdot mol^{-1} \cdot K^{-1})$	
D: diffusion coefficient	$(m^2 \cdot s^{-1})$	
L: total membrane reactor length	(m)	
F: molar flow rate	$(mol \cdot s^{-1})$	
J: molar flux	$(mol \cdot m^{-2} \cdot s^{-1})$	
k: reaction rate constant	$(mol \cdot m^{-3} \cdot s^{-1})$	
k': reaction rate constant	$(mol \cdot m^{-3} \cdot Pa^{-2} \cdot s^{-1})$	
p: permeability	$(mol \cdot m^{-1} \cdot Pa^{-1} \cdot s^{-1})$	
P: pressure	(Pa)	
Q: volumetric flow rate	$(m^3 \cdot s^{-1})$	
r: radial coordinate	(m)	
r_i: region I radius	(m)	
r_m : membrane radius	(m)	
r_s-r_m: support thickness	(m)	
r_p: pore radius	(m)	
R: ideal gas constant = 8.205	$(m^3 \cdot Pa \cdot mol^{-1} \cdot K^{-1})$	
t: time	(s)	
T: temperature	(K)	
U: superficial fluid velocity	$(m \cdot s^{-1})$	
$\overline{U}$: average fluid velocity	$(m \cdot s^{-1})$	
V: volume	(m^3)	
x: molar fraction	(-)	
z: axial coordinate	(m)	

Greek symbols

α: stoichiometric coefficient	(-)	
χ: catalyst loading	(-)	
ε: porosity	(-)	
ϕ: permeation modulus	(-)	
Φ: Thiele or pseudo-Thiele modulus	(-)	
η: effectiveness factor	(-)	
φ: apparent reaction rate	$(mol \cdot kg^{-1} \cdot s^{-1})$	
κ: mass transfer coefficient	$(m \cdot s^{-1})$	
λ: thermal conductivity	$(cal \cdot m^{-1} \cdot K^{-1} \cdot s^{-1})$	
Π: dimensionless pressure (P/P_0)	(-)	
Θ: dimensionless radius $(r/r_m$-r_i or $r/r_i)$	(-)	
ρ: density	$(kg \cdot m^{-3})$	
τ: tortuosity		
ζ: dimensionless radius $(r/r_m$-$r_i)$	(-)	

ξ : dimensionless axial coordinate (z/L)   (-)
ψ : dimensionless concentration ($C_i/C_{i,0}$)   (-)

Subscript

A: substrate
B: product
b: catalyst bed
C: cyclohexane
e: effective
eq: equilibrium molar concentration
f: number of differential sections (slides) which divide a tubular integral reactor (1, ...n)
H: hydrogen
i, j: component i or j
K: Knudsen diffusion
m: membrane
0: membrane reactor entrance or feed
p: pores
S: support
T: total
w: space coordinate of integration

Superscript

I: tube or lumen side
II: membrane region
III: non reactive support region
IV: shell side

5.1   Catalytic Membrane Reactors

5.1.1   Fundamentals

During the recent years, there have been a number of investigations dealing with the modelling and simulation of catalytic membrane reactors. In most of these studies the authors model and simulate the performance of particular experimental membrane reactor systems, in an effort to understand the behavior of such reactors. Often the models are used to compare the yields obtained with membrane reactors to those of the more conventional systems, in order to describe the potential improvements in conversion/selectivity that one derives (or may expect to derive) from the membrane reactors. Cyclohexane dehydrogenation is a frequently studied reaction, mostly because its kinetics are well-known, which makes it easier to model. In addition, since it takes place at relatively low temperatures, where it is easier to do experiments, because one avoids the need for very complex sealing systems, there are more data available to compare with simulations. Modelling studies have also been carried out with the dehydrogenation reaction of other hydrocarbons, like linear alkanes to alkenes or ethylbenzene to styrene, which may be of greater industrial interest.

The first relevant modeling effort we are familiar with is a simulation study by Gill *et al.* [5.1], who in 1975 developed a model for a wall-catalyzed porous tubular reactor. The papers by Itoh [5.2] and Sun and Khang [5.3], published in the eighties, are among the earliest studies explicitly dealing with the modelling of catalytic membrane reactors. Itoh [5.4] developed an one-dimensional model for a PBCMR utilizing a dense Pd membrane. Sun and Khang [5.3] simulated the performance of two types of membrane reactors, namely a PBMR and a CMR, utilizing a symmetric porous membrane. Salmon and Robertson [5.5] developed a simple mathematical solution for the model of an annular reactor containing a catalytic membrane. These authors studied the coupling of radial diffusion and reaction in the membrane, whereas only radial diffusion was considered to occur in the support. The considerable body of modeling work in this area before 1993 has been reviewed and organized by Tsotsis *et al.* [5.6]; later efforts have been reviewed by Sanchez and Tsotsis [5.7], and a more recent paper by Dixon [5.8] contains a good review of the most recent efforts up to 1999. In this chapter we will first briefly review some of the basic aspects of membrane reactor modelling in order to provide our reader with the rudimentary knowledge that is required for understanding the modeling efforts in this area. Our emphasis in the chapter will, however, be on the most recent efforts, and on modelling work on membrane bioreactors and other types of reactors, which were not covered in the aforementioned review efforts in this area.

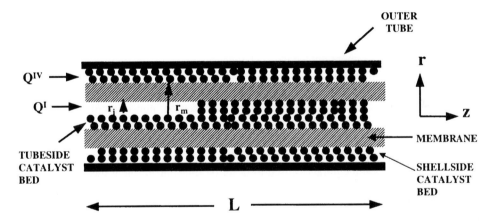

Figure 5.1. Schematic of membrane reactor for PBCMR model. Q^I and Q^{IV} are the volumetric flow rates into the tube (feed) side and on the shell (permeate) side, respectively. Adapted from Tsotsis *et al.* [5.14].

Tsotsis *et al.* [5.6] have developed a model of a generic PBCMR, shown schematically in Figure 5.1. This model takes into account mass and energy balances in the tubeside and shellside and in the membrane, itself, and accounts for the existence of pressure drops in the shellside and the tubeside. The membrane is considered to consist of a single permselective layer, either dense or mesoporous, following a Knudsen type diffusion mechanism, as it was common with most pre-1993 modelling investigations. Tsotsis *et al.* [5.6] compared the results of this model with their own experimental results and those obtained by other investigators. The same model applies to many of the different membrane reactors studies reported in Chapter 2. One of the many modelling studies utilizing this model is the study of Shu and coworkers [5.9], who modelled the data in their studies of CH_4 steam reforming in a catalytic membrane reactor using a Pd-Ag membrane deposited on porous stainless steel. The same reaction was studied in membrane reactors utilizing dense metal membranes by Barbieri *et al.* [5.10] and Oklany *et al.* [5.11]. More recently Choi *et al.* [5.12] modelled the catalytic decomposition of gaseous methyl *t*-butyl ether into isobutene and methanol in a PBMR. These authors obtained a good agreement between the experimental and simulation results. Trianto and Kokugan [5.13] simulated a PBMR for which the tubeside and shellside are filled with the same dehydrogenation catalyst. The model was validated with experiments, and was used to investigate the effect of various operating parameters on reactor performance.

There are a number of membrane reactor systems, which have been studied experimentally, that fall outside the scope of this model, however, including reactors utilizing macroporous non-permselective membranes, multi-layer asymmetric membranes, etc. Models that have been developed to describe such reactors will be discussed throughout this chapter. In the membrane bioreactor literature, in particular, but also for some of the proposed large-scale catalytic membrane reactor systems (e.g., synthesis gas production) the experimental systems utilized are often very complex, in terms of their configuration, geometry, and, of course, reaction and transport characteristics. Completely effective models to describe these reactors have yet to be published, and the development of such models still remains an important technical challenge.

The model developed by Tsotsis and coworkers [5.6, 5.14] to describe the general PBCMR reactor, shown schematically in Figure 5.1, utilizing mesoporous Knudsen type membranes, accounts for volume change due to the reaction, non-isothermal effects, and pressure drops through the packed bed regions, using the empirical Ergun equation. Furthermore, the model can be conveniently adapted for dense metal and solid oxide membranes by replacing the Knudsen model of transport, by other transport laws (e.g., Sievert's law for metal membranes) used to describe transport through these dense membranes. The model still incorporates, however, a number of key simplifying assumptions. The authors assume plug-flow behavior for both the tube (feed) and shellside regions. This assumption is valid in the case of turbulent flow, and for large reactor length/tube diameter and tube diameter/particle diameter ratios. These are conditions that are expected to prevail in large industrial reactors, but more difficult to obtain for the laboratory scale reactors discussed, for example, in Chapter 2. It does not account for external mass transport limitations (a number of groups have presented models that take into account these effect [5.15, 5.16]), which could be of concern for future industrial scale catalytic membrane reactor processes. The model, furthermore, does not account for the complex gas transport phenomena encountered with multi-layered membranes.

The model has, since, been expanded in important ways by a number of other investigators. Dixon and coworkers [5.17] and Tayakout *et al.* [5.18, 5.19], for example, have presented models, which for the case of dilute reactive mixtures account for the multilayer nature of most of the commercially available mesoporous and dense membranes. In contrast to the model of Tsotsis *et al.* [5.14], which considers the transport through the membrane to be governed by the permselective mesoporous layer, these two groups utilize detailed diffusion equations to describe mass transport in each of the other individual support membrane layer. The model of Tayakout *et al.* [5.18, 5.19], in addition, allows for the possibility of the presence of axial dispersion effects, potentially a matter of concern for some reactors [5.20, 5.21]. Both models assume, however, very dilute reactant mixtures and, therefore, neglect complications resulting from changes in the number of moles due to the reaction. Tayakout *et al.* [5.18, 5.19] have applied their model to the cyclohexane dehydrogenation reaction, while Dixon and coworkers have modeled a variety of reaction systems, including ethylbenzene dehydrogenation [5.17], methane partial oxidation using dense solid oxide membranes [5.22], and the CO_2 and NO decomposition reactions [5.23].

Sanchez and Tsotsis [5.7] have developed a more general model which incorporates the models of Tayakout *et al.* [5.18, 5.19] and Dixon and coworkers [5.17] as special cases. We revisit this model here for the special case of a membrane, which consists of a mesoporous film, which is catalytically active (region II in Figure 5.2) and of a porous support layer IV, which is inactive. As in the model of Tsotsis *et al.* [5.14] the tubeside and the shellside compartments are occupied by a packed bed of catalysts.

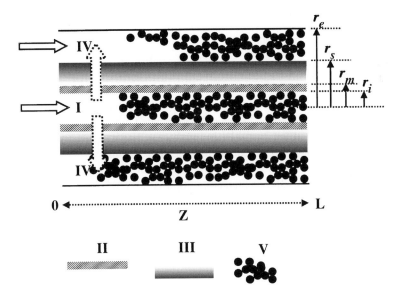

Figure 5.2. Schematic of the multilayered membrane reactor. I: feed compartment, II: active layer, III: porous support, IV: shellside, V: catalyst packed bed. Adapted from [5.7].

The gas transport through the mesoporous membrane layer is considered to be by Knudsen diffusion. No overall pressure drop is imposed across the membrane, so transport through the macroporous support structure is considered to be by molecular diffusion alone. For macroporous supports, in the presence of a pressure gradient, convective flow may turn out to be the dominant mode of transport. For such membranes several authors have utilized the Dusty Gas model, which takes into account simultaneous transport by Knudsen, molecular diffusion, and convective flow, see further discussion to follow in this chapter. Surface flow and diffusion in the mesoporous region are also considered to be negligible. In the model the reaction could potentially take place inside the catalytically active membrane and/or in the packed catalyst beds placed in the tubeside and/or shellside compartments. The models of Dixon and coworkers [5.17] and Tayakout *et al.* [5.18, 5.19] do not incorporate an energy balance equation, and are, therefore, only strictly valid under isothermal conditions. However, Dixon and coworkers [5.17] took into account the temperature gradients measured experimentally along the reactor length in the calculation of reaction rate constants and mass transfer coefficients. In their experiments the reactor and furnace lengths were of similar size, and significant temperature gradients had developed along the reactor length.

Based on the above assumptions the model equations in cylindrical coordinates, and the corresponding boundary conditions for each of the reactor regions of Figure 5.2 are described as follows:

In the tubeside (zone I)

$$U_I \frac{\partial C_i^I}{\partial z} = D_i^I \frac{1}{r} \frac{\partial}{\partial r} \left(r \frac{\partial C_i^I}{\partial r} \right) - \rho_b \alpha_i \varphi^I \tag{5.1}$$

with $0 \leq r \leq r_i$ and $0 \leq z \leq L$

where C_i^I the concentration of species i in region I (corresponding to all species present, that means the reactant molecule to be dehydrogenated, the dehydrogenated product molecule, the hydrogen, and the inert). D_i^I is an effective bed radial diffusivity, ρ_b the reactor bed density (kg/m^3), α_i the stoichiometric coefficient (equal to zero for inerts or diluents), and φ^I the reaction rate (mol/kg·s). The superficial fluid velocity U_I is considered constant by Becker *et al.* [5.17], but this assumption is relaxed in the model of Tayakout *et al.* [5.18, 5.19].

In the reactive membrane (region II)

$$D_i^{II} \frac{1}{r} \frac{\partial}{\partial r} \left(r \frac{\partial C_i^{II}}{\partial r} \right) = \rho_m \alpha_i \varphi^I \tag{5.2}$$

$r_i \leq r \leq r_m$

where ρ_m is the membrane density, D_i^{II} the membrane diffusivity, and φ^{II} the reaction rate in region II.

In the macroporous support (zone III)

$$D_i^{III} \frac{1}{r} \frac{\partial}{\partial r} \left(r \frac{\partial C_i^{III}}{\partial r} \right) = 0 \qquad (5.3)$$

$r_m \leq r \leq r_s$

In the shellside (zone IV)

$$U_{IV} \frac{\partial C_i^{IV}}{\partial z} = D_i^{IV} \frac{1}{r} \frac{\partial}{\partial r} \left(r \frac{\partial C_i^{IV}}{\partial r} \right) - \rho_b \alpha_i \varphi^{IV} \qquad (5.4)$$

with $r_s \leq r \leq r_e$ and $0 \leq z \leq L$

In the above equations D_i^{III} is the membrane diffusivity in region III, D_i^{IV} is the effective bed radial diffusivity, and φ^{IV} the reaction rate in region IV. Equations (5.1) and (5.4) could be further augmented by a corresponding axial dispersion term, as it was done in the work of Tayakout *et al.* [5.18]. Assuming the axial and radial diffusivities to be equal the equation in region IV, for example, Equation (5.4) then becomes

$$U_{IV} \frac{\partial C_i^{IV}}{\partial z} = D_i^{IV} \left(\frac{\partial^2 C_i^{IV}}{\partial z^2} + \frac{1}{r} \frac{\partial}{\partial r} \left(r \frac{\partial C_i^{IV}}{\partial r} \right) \right) - \rho_b \alpha_i \varphi^{IV} \qquad (5.5)$$

The above equations (5.1) to (5.5) are complemented by a set of corresponding boundary and initial conditions

At Z = 0 (reactor inlet)

$$C_i^I = C_{i0}^I \qquad (5.6)$$

$$C_i^{IV} = C_{i0}^{IV} \qquad (5.7)$$

At r = 0 (symmetry condition)

$$\frac{\partial C_i^I}{\partial r} = 0 \qquad (5.8)$$

At r = r_e (wall no-flux condition)

$$\frac{\partial C_i^{IV}}{\partial r} = 0 \qquad (5.9)$$

At the interfaces between the various regions relationships describing the continuity of fluxes and concentrations apply. For example, between regions I and II the continuity of fluxes is described by the following equation:

$$D_i^I \left(r \frac{\partial C_i^I}{\partial r} \right)_{r=r_i-} = D_i^{II} \left(r \frac{\partial C_i^{II}}{\partial r} \right)_{r=r_i+} \tag{5.10}$$

with similar types of conditions applying in the other interfacial regions.

If axial dispersion in the tubeside and shellside are to be taken into account the initial conditions (5.6) and (5.7) must be replaced by corresponding conditions, which account for these effects. Condition (5.6), for example then becomes

$$\left(\overline{U}_I \cdot C_i^I \right)_0 = -D_i^I \frac{\partial C_i^I}{\partial z} + U_I \left(0^+, r \right) C_i^I \tag{5.11}$$

where $\overline{U}_I$ is the average fluid velocity entering the inner compartment. A similar equation is obtained for the shellside. In the paper of Tayakout *et al.* [5.19] the diffusion coefficient of species i in the mixture was estimated using the binary diffusivities (D_{ij}) and the Stefan Maxwell formulation of transport [5.24]. The Knudsen diffusion equation was used for the mesoporous active membrane with the effective diffusion coefficients estimated based on a comparison between the theoretical and the experimental values. The reaction kinetics were estimated experimentally.

The above system of equations can be made dimensionless by defining appropriate dimensionless variables, for example, dimensionless concentrations $\Psi_i^I = \dfrac{C_i^I}{\left(C_i^I \right)_0}$, radial

$\Theta = r/r_i$, and axial coordinates, $\xi = z/L$, etc. The resulting systems of dimensionless equations contain a number of dimensionless groups, which typically represent the ratios of characteristic times for the various processes, which occur in the various reactor compartments. For a CMR (i.e., with only the membrane being catalytically active and no catalyst being present) Tayakout *et al.* [5.18, 5.19] define two dimensionless groups a Peclet (Pe) number (the ratio of characteristic time for radial diffusion to the characteristic time for flow in the tubeside), and the Thiele (Φ) modulus (the ratio of characteristic time for radial diffusion to the characteristic time for reaction in the membrane), described by the following equations:

$$Pe = \frac{\overline{U}_I L}{D_i^I} \tag{5.12}$$

$$\Phi^2 = \frac{k^{II} \left(r_m - r_i \right)^2 RT}{D_i^{II} P_0} \tag{5.13}$$

where $k^{II} = \dfrac{\varphi^{II} \cdot \rho_m}{P_c} P_H$ (5.14)

The dimensionless equations are solved by discretizing the first and second order derivatives, thus, reducing the system of differential equations into a system of algebraic equations that can be solved by using DASSL or ACSL numerical packages.

The simulations of Becker *et al.* [5.17] showed good agreement with the experimental results. Figure 5.3 shows the simulated ethylbenzene concentration profiles along the reactor length for a PBMR (inert membrane). The ethylbenzene concentration decreases sharply near the inlet in the tubeside; this decrease is the result not only of the reaction, but also of the permeation of the hydrocarbon to the shellside. There is, as a result, a parallel increase of the ethylbenzene concentration in the shellside. Interestingly enough, there is a maximum in the ethylbenzene concentration in the shellside, which is indicative of the fact that there is some back-diffusion of the ethylbenzene concentration from the shellside to tubeside, as its concentration in the tubeside is depleted due to the reaction. These phenomena are indicative of the fact that the membrane used in the simulations is mesoporous (the mean pore radius of the separative layer is 40 Å), and allows significant and rapid back and forth diffusion of the ethylbenzene. Reactant loss to the shellside is, of course, detrimental to the reactor conversion and yield. This and other simulations show the importance for membrane reactor performances of using a membrane with a good permselectivity. Sometimes, when using poorly permselective membranes, the detrimental effect of reactant loss to the shellside overtakes the conversion enhancement obtained by the equilibrium displacement, with the result that for this particular case the maximum conversion obtained is comparable or even lower than that of the corresponding conventional packed-bed reactor.

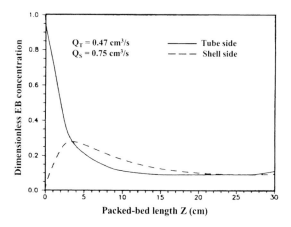

Figure 5.3. Ethylbenzene concentration axial profiles in a PBMR at 891 K. From Becker *et al.* [5.17], with permission from Elsevier Science.

The effect of reactant loss on membrane reactor performance was explained nicely in a study by Harold *et al.* [5.25], who compared conversion during the cyclohexane dehydrogenation reaction in a PBMR equipped with different types of membranes. The results are shown in Fig. 5.4, which shows the cyclohexane conversion in the reactor as a function of the ratio of permeation to reaction rates (proportional to the ratio of a characteristic time for reaction in the packed bed to a characteristic time for transport through the membrane). Curves 1 and 2 correspond to mesoporous membranes with a Knudsen (H_2/cyclohexane) separation factor. Curves 3 and 4 are for microporous membranes with a separation factor of 100, and curves 5 and 6 correspond to dense metal membranes with an infinite separation factor. The odd numbered curves correspond to using an inert sweep gas flow rate equal to the cyclohexane flow, whereas for the even numbered curves the sweep to cyclohexane flow ratio is 10.

As can be observed in Figure 5.4, when porous membranes are used, the conversion passes through a maximum. This is because at high permeation/reaction rate ratios the yield is negatively impacted by reactant losses. Clearly this effect is more significant for the mesoporous membranes, with the Knudsen like diffusion, than for the microporous membranes, which typically have higher permselectivity factors (assumed to be 100 in these calculations). In the case of ideal dense membranes, which in this example are assumed to be only permeable to hydrogen, no maximum in conversion is observed (no reactant loss is present to negatively impact conversion), and better performances are obtained at higher permeation/reaction rate ratios. For all membranes, when the ratio of inert sweep gas flow/cyclohexane flow increases, the maximum cyclohexane conversion increases.

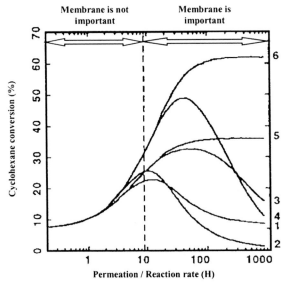

Figure 5.4. Cyclohexane conversion vs. the (permeation/reaction rate) ratio for the cyclohexane dehydrogenation reaction at 477 K and a feed pressure of 100 kPa. Adapted from Harold *et al.* [5.25].

This effect is more pronounced for the dense membranes. For porous membranes, at the very high permeation/reaction rate ratios, the conversion decreases for the higher sweep ratios, since the positive effect of increasing the transmembrane pressure gradient is counterbalanced by the negative impacts of enhanced reactant losses.

Figure 5.5 shows the simulations of Tayakout *et al.* [5.19], showing the influence of the Peclet number and Thiele modulus on the conversion (X) during cyclohexane dehydrogenation in a CMR. The conversion decreases as the Peclet number increases. This is because for larger Pe the axial flow dominates when compared with the radial diffusion. Since radial diffusion is the primary means by which the reactant (cyclohexane) reaches the catalytic membrane, a large Pe means that a large fraction of it washes out of the reactor before reaching the membrane. The Pe effect is more significant for small values of Thiele modulus. For the same value of Pe, reactor conversion increases as the Thiele modulus increases. Figure 5.6 shows the simulated cyclohexane concentration profiles of Tayakout *et al.* [5.19] as a function of the radial and axial coordinates for a CMR. The results in the figure are for a high Pe value of 1000, and a low Thiele modulus value of 1. In the reactor configuration of Figure 5.6 the hydrocarbon is fed in the tubeside in contact with the catalytic mesoporous membrane. Significant cyclohexane axial and radial concentration gradients are observed in the feed compartment, and as also noted with the results of Becker *et al.* [5.17] reactant loss is a concern. Tayakout *et al.* [5.19] also studied a second reactor configuration, where the cyclohexane was fed in the shellside in contact with the membrane support. In the latter case the diffusion of cyclohexane through the support limited its availability to the reactive membrane. In this case the hydrocarbon conversion obtained is lower than the calculated equilibrium value.

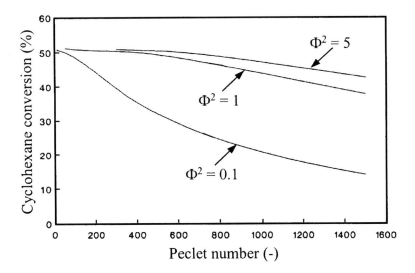

Figure 5.5. Cyclohexane conversion as function of the Peclet number and Thiele modulus for the cyclohexane dehydrogenation in a CMR at 480 K. Adapted from Tayakout *et al.* [5.19].

A similar model was utilized by Hermann *et al.* [5.26] to study the catalytic dehydrogenation of ethylbenzene to styrene in a tubular palladium membrane reactor using a commercial catalyst. The mathematical model of the membrane reactor again takes into account the different mass transport mechanisms prevailing in the various layers of the membrane, that is, multicomponent diffusion in the stagnant gas films on both faces of the membrane, combined multicomponent molecular diffusion, effective Knudsen diffusion, and viscous flow in the macroporous support, Knudsen diffusion in the microporous intermediate layer, and Sieverts' law of hydrogen transport through the Pd-film. A kinetic model from the literature was adjusted to match conversion and selectivity observed during experiments with a commercial catalyst in a laboratory fixed-bed reactor. Simulation calculations based on the resulting effective kinetics were carried out for industrially relevant operating conditions and various process configurations, which included the use of inert sweep gas, evacuation of the permeate gas, and oxidation of the permeated hydrogen. The results demonstrated that under typical process conditions removal of hydrogen through the membrane gives only a small increase in styrene yield. However, the model predicts that by increasing the reaction pressure in the membrane reactor the kinetic limitation can be overcome, and ethylbenzene conversion can be increased to above 90 %, without markedly decreasing styrene selectivity.

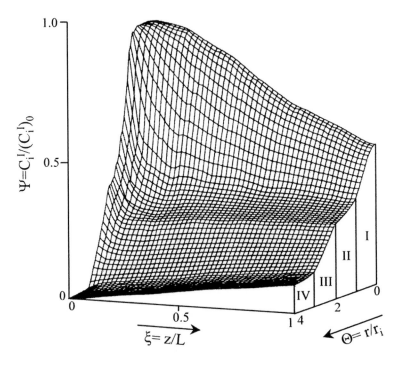

Figure 5.6. Cyclohexane axial and radial profiles in a CMR. Adapted from Tayakout *et al.* [5.19].

A number of studies involving the modeling of membrane reactors incorporating solid oxide membranes have also been reported. Wang and Lin [5.27] and Lu *et al.* [5.28] have presented models of membrane reactors utilizing solid oxide membranes for direct CH_4 activation. Nozaki and Fujimoto [5.29] developed a model to fit their data of selective oxidative methane coupling in a membrane reactor utilizing a PbO oxide membrane impregnated on a porous silica-alumina tube. More recently a model of the same reaction in a PBMR was presented by Jin *et al.* [5.30]. In their experiments a dense perovskite membrane was used, and the oxygen transfer through the membrane was measured experimentally; an empirical equation taking into account the temperature and oxygen partial pressures was established and utilized in the model. A good agreement between the experimental data and the model was obtained. Dixon and coworkers [5.23] have also presented a modelling study of the application of solid oxide membrane reactors in the area of environmentally benign processing, involving the study of CO_2 and NO decomposition, and the use of perovskite membranes with a high oxygen flux.

5.1.2   Thermal Effects

Many of the modelling studies to date have considered the membrane reactors to operate isothermally. However, as with conventional reactors temperature gradients in membrane reactors could have a significant effect on reactor performance. Accounting for temperature non-uniformities and radial and axial temperature profiles is of added significance for membrane reactors utilizing non-metallic inorganic membranes, for which the low thermal conductivity of such membranes makes it difficult to heat and operate isothermally. One experimental strategy to avoid membrane reactor temperature non-uniformities, at least in laboratory reactors, involves using a longer heating furnace region than the membrane reactor itself, long enough that it guarantees an efficient preheating of feed gases into the reactor [5.19]. However, this type of strategy may still be inefficient in the case of low reactant and product dilution ratios or highly endothermic or exothermic gaseous reactions, because the thermal conductivities of gases, and of the ceramic porous membranes are very low. As noted in Section 5.1.1, Becker *et al.* [5.17] measured in their experiments the temperature profile in the membrane reactor. They, then, used the obtained experimental temperature profiles during their isothermal simulations in the calculation of the kinetic rate constants and transport coefficients. This approach, though proven successful in fitting their experimental data, is, nevertheless, long and tedious, and difficult to generalize. It, furthermore, lacks the predictive capabilities, which may be one of the key reasons for carrying out reactor simulations in the first place.

As was mentioned earlier, Tsotsis and coworkers developed a few years back [5.6] a general model for the PBCMR reactor, which accounts for the non-isothermal effects, by coupling for this purpose the energy and mass balance equations. The complete set of equations with the corresponding boundary conditions is given in detail in the paper of Tsotsis *et al.* [5.6]. In their original publication they discuss an interesting modelling study by Itoh [5.31], who studied a Pd/PBMR in which the cyclohexane dehydrogenation reaction in the tubeside was coupled to the oxidation reaction of the permeated hydrogen gas in the shellside. Some of the simulation results are shown in Figure 5.7. The parameter H in this figure is defined by Tsotsis and coworkers [5.6] as:

$$H = \frac{2\pi \lambda_{e,m} L}{F_0^t \ln(1 - \varepsilon) C_{pA}}$$

(5.15)

where F_0^t is the total molar feed rate, $\lambda_{e,m}$ is the effective membrane conductivity, C_{pA} is the molar heat capacity of purge gas A, and ε is the catalyst bed void fraction. The results of the non-isothermal simulations are often surprising as can be seen in Figure 5.7, where increasing H for the same dimensionless reactor length seems to lead to a decrease in reactor conversion. Since heat is being produced in the shellside a diminished H would imply a decreased transfer of this heat to the tubeside, which will negatively impact on the performance for the catalytic dehydrogenation reaction. On the other hand, this means a increase in the temperature in the shellside, which implies a faster hydrogen oxidation, and an increased transport through the membrane, which would tend to favorably impact on the rate of the dehydrogenation reaction. Higher temperatures in the shellside would also tend to compensate for the effect of lower H. The need to take into account thermal effects was also emphasized in a recent study by Koukou *et al.* [5.32], who developed a detailed model of a non-isothermal membrane reactor, and solved it by finite volume techniques. Their results show that the assumption of an isothermal membrane reactor leads to significant errors in the prediction of reactor conversion.

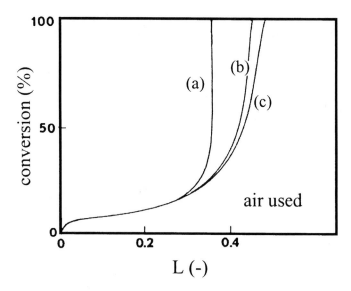

Figure 5.7. Conversion as a function of dimensionless reactor's length at different H. (a) H=100, (b) H=500, (c) H=10^4. Adapted from [5.31].

Moon and Park [5.33] carried out recently a modelling study of a non-isothermal two-dimensional model of a PBMR. In the study of Moon and Park the packed-bed of catalysts was placed in the shellside, whereas the tubeside is empty. The reaction studied was the cyclohexane dehydrogenation. The hydrocarbon was fed in the shellside diluted with nitrogen, which was also used as a sweep gas in the tubeside. They analyzed an isothermal PBMR model, and a non-isothermal PBMR, in which the reactor temperature was assumed to follow a parabolic profile inside the reactor, as was observed experimentally by Dixon and coworkers [5.17]. Two key dimensionless groups characterize the membrane reactor behavior. They are the Peclet number (Pe), and the Damköhler (Da) number in the shellside. Pe represents the ratio of a characteristic time of transport through the membrane with respect to a convective transport characteristic time. Da is the ratio of the convective transport characteristic time with respect to the reaction characteristic time. The key simulation results are presented in Fig. 5.8 in terms of iso-conversion lines in the log(Da)-log(Pe) parameter space. As can be seen in Fig. 5.8c for the membrane reactor characterized by the parabolic temperature profile, one can distinguish various regions of distinct behavior. Note that in region 1 of low Da values, the conversion is controlled by the reaction kinetics, and is not strongly affected from the changes in the Pe number. In region 3, on the other hand, the conversion is controlled by the permeation rate, and small changes in the Pe number have a strong effect on conversion. Comparing the conversions obtained for isothermal conditions (Figure 5.8a) with those obtained for non-isothermal reactor conditions (Figure 5.8b) one can observe that the highest conversions are obtained for isothermal conditions; this result is in good agreement with the conclusions obtained previously by Koukou *et al.* [5.32]. Moon and Park note another important difference between the two types of operation. Under non-isothermal operation in the region of large Da and Pe numbers, the exit conversion diminishes near the exit of the reactor; this result may be due to an increase in the rate of the reverse reaction as a result of the temperature decrease at the reactor exit.

A comprehensive theoretical model of a non-isothermal CMR utilizing a multilayer mesoporous membrane was recently developed by Brinkman *et al.* [5.34]. The model assumes plug-flow conditions prevailing in the tubeside and shellside, and utilizes the dusty gas model (see further discussion in Section 5.1.4) to describe transport in the three support membrane layers, and transport and reaction in the catalytically active top layer. Differential energy balances describe the temperature profiles in the reactor tubeside and shellside, and in the various membrane layers. Results of the theoretical investigation were compared with experiments with the oxidative methanol dehydrogenation. The model provided an accurate description of the measured temperature profiles but performed poorly in describing the hydrogen yield and selectivity. The authors attributed the discrepancies to inaccurate kinetic information concerning side reactions, including methanol combustion to CO_2 and water.

Non-isothermal 1-D models for adiabatic PBMR and FBMR reactors utilizing Pd tubular membranes have been developed by Elnashaie *et al.* [5.35], and applied to the catalytic ethylbenzene dehydrogenation reaction. In contrast to many other modelling studies their model takes into account intraparticle diffusional limitations. The catalyst particles are modeled as 1-D slabs using the dusty gas model. The kinetic data utilized in the model are extracted from industrial fixed bed reactor data, and from laboratory experiments using a detailed kinetic scheme. The membrane reactors are shown to significantly outperform the conventional reactors. This is particularly true for the FBMR.

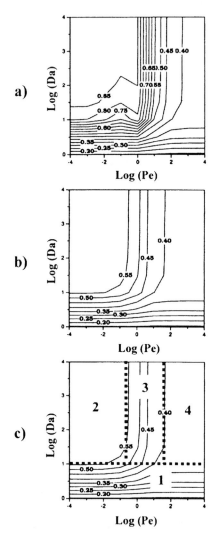

Figure 5.8. Cyclohexane conversion as function of Da and Pe numbers at 493 K. The sweep gas/feed gas ratio is 4.23. (a) Isothermal operation, (b) Non-isothermal operation, with a parabolic temperature profile between 463 K (reactor's ends) and 493 K (center), (c) different regions observed. The selectivity considered was ten times the Knudsen. From Moon and Park [5.33], with permission of Elsevier Science.

The consideration of thermal effects and non-isothermal conditions is particularly important for reactions for which mass transport through the membrane is activated and, therefore, depends strongly on temperature. This is, typically, the case for dense membranes like, for example, solid oxide membranes, where the molecular transport is due to ionic diffusion. A theoretical study of the partial oxidation of CH_4 to synthesis gas in a membrane reactor utilizing a dense solid oxide membrane has been reported by Tsai *et al.* [5.22, 5.36]. These authors considered the catalytic membrane to consist of three layers: a macroporous support layer and a dense perovskite film ($La_{1-x}Sr_xCo_{1-y}Fe_yO_{3-\delta}$) permeable only to oxygen on the top of which a porous catalytic layer is placed. To model such a reactor Tsai *et al.* [5.22, 5.36] developed a two-dimensional model considering the appropriate mass balance equations for the three membrane layers and the two reactor compartments. For the tubeside and shellside the equations were similar to equations (5.1) and (5.4), without the reaction term. For the reactive membrane and macroporous support regions the corresponding mass balance equations were similar with equations (5.2) and (5.3) respectively. The dense perovskite layer allows only oxygen to transport through; the authors assumed that the oxygen transfer was by a Fickian-like diffusion with an activated diffusion coefficient. The equation they utilized to describe the oxygen flux through the perovskite layer in cylindrical coordinates was:

$$F_{O_{2,r}} = \frac{A \exp\left[-\dfrac{E}{RT}\right] T}{r \ln\left(\dfrac{r_m}{r_i}\right)} \ln\left(\frac{P_{O_{2,r_i}}}{P_{O_{2,r_m}}}\right) \tag{5.16}$$

In the above equation, r is the radial position within the perovskite layer, which is located between position r_m, at the catalytic layer side and r_i at the support side, E is the activation energy, A is a pre-exponential factor (kgmol/m·s·K) whereas the P_{O_2} are the corresponding oxygen pressures at these positions. Tsai *et al.* [5.22, 5.36] also considered the energy balance equations in every region accounting for energy changes due to the transport of the various species, and due to energetic effects associated with the reaction. The CH_4 partial oxidation to syngas was described to be due to the total oxidation of CH_4 coupled with its CO_2 and steam reforming.

The consideration of thermal effects is also of significance for the modelling of membrane reactors involving exothermic reactions, like hydrocarbon partial oxidation. During such reactions reactor hot-spots and runaway phenomena could lead to dramatic decreases in the product yield and potentially dangerous conditions. As noted in Chapter 2, the use of a membrane in order to distribute the oxygen uniformly all along the reactor length is a promising concept. Theoretically the decrease in the oxygen partial pressure in the feed promises to increase the selectivity towards the desired intermediate products, in addition to making the reactor an inherently safer device, avoiding hydrocarbon/oxygen ratios near the explosion limits. Dixon *et al.* [5.8, 5.23] modelled the catalytic oxidation of *o*-xylene to pthalic anhydride using a non-isothermal PBMR. Figure 5.9 provides a comparison between the calculated temperature profiles in the PBMR and in a conventional packed bed reactor. The simulation results indicate that a hot-spot develops in the classical packed bed reactor, represented by a sharp temperature peak near the reactor entrance, where the oxygen concentration is the greatest. This phenomenon is avoided in the PBMR as a result of the controlled introduction of the oxygen, which provides a better control of the reaction temperature.

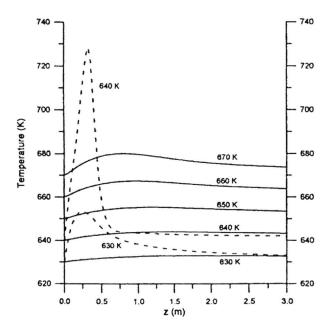

Figure 5.9. Calculated temperature profiles in a PBMR (solid lines) and PBR (dashed lines). From Dixon [5.8], with permission from the Royal Society of Chemistry.

A simulation study of the same scope was reported by Tellez *et al.* [5.37, 5.38] who simulated a membrane reactor for the oxidative dehydrogenation of butane over V/MgO catalysts for which the membrane was used again as a means to distribute the oxidant. The model of Tellez *et al.* [5.38] for the first time, took into consideration the effect of oxygen distribution through the membrane on the state of the catalyst, by means of kinetic parameters that relate the activities of selective and non-selective sites on the catalyst surface. This allowed them to explain the differences observed experimentally in the conversion-selectivity behavior of fixed beds and membrane reactors. In a more recent study by the same group Pedernera *et al.* [5.39, 5.40] modelled the selective oxidation of butane to maleic anhydride (MA) in a CNMR for two different configurations, one in which the catalyst was packed in the inside of the membrane and oxygen was fed in the shellside (termed the IMR configuration), and the other in which the catalyst was packed in the shellside and the oxygen was fed in the tubeside (termed the OIFMR configuration). These two configurations had been studied experimentally previously by Mallada *et al.*

[5.41]. The OIFMR configuration (for the same level of conversion) was shown to exhibit superior MA selectivity and yield. Their model ignores interparticle/intraparticle mass transport limitations, and again utilizes a Mars van Krevelen type kinetic mechanism with two types of catalytically sites; these include selective sites for the production of MA and unselective sites, which give CO and CO_2. The kinetic constants of the model were fitted to the data by the same group. The two-dimensional, non-isothermal model accounts for convective and diffusive transport in the radial direction, convective transport in the axial direction, and utilizes the Dusty Gas model to describe transport of oxygen in the inert membrane. The pressure of oxygen was considered constant while the pressure on the re-action side was described by the Ergun equation. The reason for using the two-dimensional model is that although the 1-D model gave satisfactory agreement with the conversion results it was unable to predict satisfactorily the differences between the two reactor configurations (as previously noted, the reactor with the catalyst in the shellside was shown by Mallada *et al.* [5.41] to exhibit a somewhat larger yield). The two-dimensional model is capable of accounting for gradients in the radial direction of both temperature and active site concentration, and to qualitatively predict the differences in observed yields for the two types of reactors.

Hou *et al.* [5.42] have also presented a model taking into account radial concentration and temperature profiles, which applies to a membrane reactor with a fixed-bed of cata-lysts inside the membrane and with distribution of one of the reactants through the mem-brane. They have applied this model to the oxidative dehydrogenation of propane using kinetic data previously published by the group (Ramos *et al.* [5.43]). The model assumes that within the tubeside the gas velocity has two components: an axial component, which was assumed to not depend on the radial position, and a radial component, which was as-sumed to vary linearly with radius. Transport in the radial direction is by diffusion and convection, while transport in the axial direction is only by convection. The energy equa-tion allows for transport by conduction and convection in the radial direction, and by con-vection in the axial direction. At the membrane wall the concentration of all species, with exception of O_2, was assumed to be zero, and the flux of oxygen was assumed to have ei-ther a constant value or to vary linearly with position. Model predictions were compared with the experimental results, and the effect of an increase in the reactor diameter and other changes in the operation conditions were studied. Increasing the reactor diameter was shown, for example, to lead to reduced selectivities. Radial temperature gradients (~15K) were observed for low values of radial thermal conductivity even for the small di-ameter (0.7 cm) membrane tubes that they utilized in their study.

Dixon recently [5.44] provided a comparative study of the yield to intermediate prod-ucts for partial oxidation for a distributed-feed membrane reactor, and the conventional cooled-tube fixed-bed reactor (FBR). They studied a generic parallel/consecutive partial oxidation reaction scheme (reactions 5.25 and 5.26 in Section 5.1.4 with all the stoichiometric coefficients being equal to 1). The reactor resembled that shown in Figure 5.1 but with catalyst placed only in the shellside. The hydrocarbon (B) was fed in the shellside while the oxygen (A) was fed in the tubeside and permeated (in the case of the MR) through the membrane, which is assumed to only permeate A. For the FBR both reactants are fed in the shellside and the membrane is replaced by an impermeable tube. A non-isothermal, non-adiabatic, one-dimensional pseudo-homogeneous reactor model was used without axial dispersion, and with no pressure drops. In the analysis the FBR was optimized first with respect to the feed of reactant A. Optimal conditions for the FBR usually lie on or close to the boundaries between either pseudo-adiabatic operation or hot-spot operation and runaway operation. The MR was then constrained to use the same amount of catalyst, equal feed B, and equal heat-removal rate. It was not required to use the same amount of A (oxygen), as A was considered to result from air through oxygen diffusion through the membrane and, therefore, to be inexpensive. Each reactor was allowed to utilize A the best way it could. The range of membrane permeabilities was then determined that allowed an increase in yield with no worse temperature rise. The results of the analysis were presented in terms of dimensionless groups, like the Damköhler and Stanton numbers, and the dimensionless adiabatic temperature rise in order to delineate the behavior in various regions of the parameter space. Results show that the MR performance is better than that of the FBR only over restricted regions of parameter space, at higher Da, and under conditions of high heat release or lower cooling rates. This gain in performance comes at a price in that the MR utilizes a much higher feed of A. The MR performance can be enhanced by a mixed-feed strategy, in which part of A is fed in the shellside, or by a slightly increased inlet temperature. The rationale for the latter is that temperature profiles along the length of the membrane reactor are generally lower, because the distribution of the reactant leads to a more even heat release, and the membrane reactor shows no hot-spots for an extended range of feed temperatures and operating pressures.

5.1.3   The Catalytic Membrane Reactor vs. Conventional Reactors

As it may have become apparent to the reader from some of the relevant discussions in the previous chapters, and will be discussed in greater detail in Chapter 6, proving whether a membrane reactor based process is feasible technically and economically is a complex undertaking involving many other factors beyond a straightforward comparison between the yield and selectivity of the membrane reactor, and any of the competing conventional reactor systems. Cost of the membranes and membrane robustness, for example, sway more weight in the decision process, and often the separation aspects of the membrane-based process (particularly for membrane bioreactors) may be more critical in determining process success rather than improvements in selectivity and yield. Nevertheless, particularly in the field of catalytic membrane reactors, many authors have often attempted in the past to provide a direct comparison in the performance of their membrane reactor systems with that of more conventional reactors (PFR, CSTR, etc.) in terms of the attained yields and selectivity. There are a number of interesting technical issues that arise, when attempting such a side-by-side comparison. Often what appears to be a rather simple task in comparing the conversion and yield, say of a packed-bed reactor (PBR) or a plug-flow reactor (PFR) to that of PBMR, turns out to be a much more complex undertaking, as Gokhale *et al.* [5.45], for example, have noted. One must be careful in choosing an appropriate "yardstick" in order to compare performance for both types of reactors on equitable terms, since an incorrect analysis may turn out to provide an erroneous advantage of the PBMR over the PFR and vice versa. For laboratory catalytic membrane reactors, for example, and for equilibrium limited reactions in order to improve the driving force for the species to be removed (e.g., hydrogen) across the membrane the shellside of the membrane reactor is typically swept with an inert gas. For porous membranes some of the sweep gas back-diffuses into the tubeside diluting the reacting stream there; it is then difficult to attribute with any degree of certainty the gains in conversion and yield for the membrane reactor to the shift in equilibrium, due to the removal of the product, or due to the presence of the dilutant. From a practical standpoint, of course, the presence of the inert gas in the hydro-carbon stream in the tubeside or the loss of useful products (e.g., hydrogen) in the shell-side has a substantial impact on process economics.

The technical characteristics of inert swept catalytic membrane reactors and the issues related to the equitable comparison between their yield and that of conventional reactors have been discussed by a number of authors (Tayakout *et al.* [5.19], Itoh [5.46], and Moon and Park [5.33]). Reo *et al.* [5.47, 5.48] have nicely summarized these matters in two theoretical papers that were published in 1997. In their analysis they attempted to compare the performance of an isothermal PBMR (with a porous membrane obeying Fickian diffu-sion and no overall pressure drop between the shellside and tubeside) with that of a PFR for an equilibrium limited reaction. In order to accomplish this task equitably, Reo *et al.* [5.47, 5.48] imposed a set of technical constraints. Specifically, they set the feed to both reactors to be the same, and assumed identical pressure, temperature, and reactor volumes (or catalyst amounts) for both reactors. In order to address the issue of conversion en-hancement due to dilution rather than to equilibrium shift due to product removal, in their analysis they allowed for the possibility for the PFR to be also diluted by an inert; they re-quired, however, that the amount of dilutant introduced into the PFR should be either the same or smaller than the quantity introduced into the PBMR as a sweep gas. Key pa-rameters in their analysis of the PBMR performance are the Pe and Da dimensionless numbers, as were defined above. Pe is the ratio of a characteristic time for transport through the membrane to a characteristic time for flow in the reactor. Da is defined as the ratio of a characteristic time for flow for the reactor to the characteristic time for reaction. In their simulation they plot the olefin yield of the PBMR as a function of DaPe (which is the ratio of characteristic time for permeation to the characteristic time for reaction) for various values of Da. They note (as did Harold *et al.* [5.25] before them, see Figure 5.4) that there is a optimal value of DaPe that maximizes olefin yield. To compare the PBMR (co-current configuration) with the PFR, the Da was fixed for both reactors, and then the PBMR was optimized with respect to the DaPe (equivalent to choosing an optimal mem-brane permeance). The simulation results of olefin yield as a function of Da are shown in Figure 5.10 for three different reactors, the optimized PBMR, a PFR with the same Da, and a diluted PFR also with the same Da; the latter was optimized with respect to the ad-dition of inert, care being taken that the amount of inert gas added does not exceed the sweep gas added to the PBMR. The olefin yield in the PBMR is distinctly higher (for large Da values) than that of the conventional PFR, but approaches in that region of Da values the olefin yield of the so-called diluted PFR (it is unlikely, of course, that such a reactor would be really competitive with either the undiluted PFR or the PBMR, as the olefin and hydrogen must be separated from the inert). The differences in performance are a strong function of the membrane's permselectivity, with more permselective membranes tending to provide superior performance.

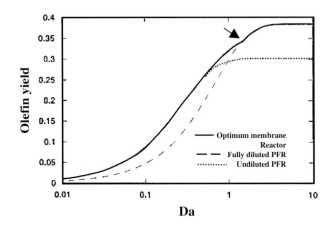

Figure 5.10. Olefin yield from a PBMR using the optimum DaPe. From Reo *et al.* [5.47], reproduced with permission from the American Institute of Chemical Engineers.

Reo *et al.* [5.48] also studied the behavior of a PBMR, for which a total pressure difference exists across the membrane in between the tubeside and the shellside compartments. For such reactor, in principle, there is no need to utilize a sweep gas in the shellside, as the transmembrane pressure difference may be sufficient to provide the driving force for transporting the product (e.g., hydrogen) across the membrane (since alkane dehydrogenation is negatively impacted by pressure it is unlikely that such a PBMR would be economic, unless the downstream processing of the alkene necessitated the use of high pressures in the first place). Again an optimal value of DaPe exists, which maximizes PBMR performance, and the olefin yield in the PBMR is higher than the one obtained in the PFR operating at the same pressure. For comparison purposes Reo *et al.* [5.48] considered a PFR network consisting of two reactors in series (see Figure 5.11) the first operating at a pressure equal to the tubeside pressure of the PBMR, and the second at a pressure equal to the shellside pressure. In the PFR network a part of the exit stream from the high pressure PFR is diverted into the low pressure PFR, where it undergoes further reaction.

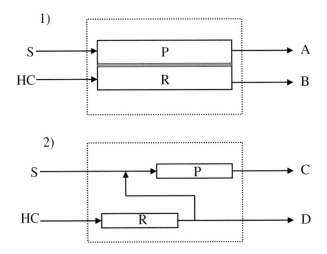

Figure 5.11. Schematic representation of the 1) PBMR and 2) PFR network. P: permeate side, R reaction side, S: inert sweep gas, HC: hydrocarbon, A, B, C, and D: a mixture of hydrocarbon, sweep gas, olefin and hydrogen. Adapted from Reo *et al.* [5.48].

By varying the amount of the side stream into the low pressure PFR one can optimize the performance of the two PFR network. The optimized reactor network showed an olefin yield close to the PBMR, except in the region of moderate to low conversions, where the PBMR offered modest yield advantages (less than 20 %) over the PFR network. As Reo *et al.* [5.48] note, assuming that the high pressure PBMR was economic (which would imply the need for a high pressure olefin product), it is unlikely that the two PFR network would provide an advantage in performance, as most of the alkene would have to be delivered at the lower shellside pressures. Despite the clear advantage in performance pressure driven PBMR may offer over inert swept PBMR, their applicability is closely tied to the need to deliver the olefin at the operating high pressure. One must also be aware of the fact that, when a transmembrane pressure is applied, non-separative convective flow may become of significance through the omnipresent macroporous membrane defects. From a modelling standpoint other types of models, like the Dusty Gas Model [5.49] must then be used to describe gas transfer through the membrane; the application of this model will be discussed in the next section for the case of non-permselective membrane reactors.

5.1.4 Non-permselective Catalytically Active Membranes

As discussed in Chapter 2, in a number of membrane reactor applications the membrane is non-permselective, and it simply acts as a contactor device (when it is catalytic), or simply as a means to distribute one of the reactants in a more uniform manner (when it is inert). In modeling such reactors one must take into consideration, in addition to Knudsen diffusion, the presence of molecular diffusion and convective transport. The Dusty Gas Model

(DGM) of transport is, typically, utilized in the description of such reactors. The same model is probably also applicable to membrane reactors utilizing mesoporous membranes and operating under a significant transmembrane pressure gradient. The isothermal models describing such reactors consist of simple mass balance equations in every membrane reactor region. For regions where reaction occurs, for example, the mass balance equation is as follows

$$\nabla J_i = r_i \tag{5.17}$$

while in non-reactive regions

$$\nabla J_i = 0 \tag{5.18}$$

where J_i is the flux for species i and r_i the corresponding reaction rate. J_i is described, in turn, according to DGM as follows:

$$\sum_{j=1,J\neq i}^{n} \frac{x_i J_j - x_j J_i}{P\ D_{ij}} - \frac{J_i}{P\ D_{iK}} = \frac{1}{RT}\nabla x_i + \frac{J_i}{PRT}\left(\frac{B_0 P}{\mu D_{iK}}\right)\nabla P \ ; \ (i=1......n) \tag{5.19}$$

where D_{ij} is the effective binary diffusion coefficient of components i and j (already includes the porosity ε and the tortuosity τ of the membrane), D_{iK} is the corresponding effective Knudsen diffusion coefficient, x is the molar fraction. B_0 corresponds to the permeability coefficient for convective flow in m^2, and μ is the viscosity in $N \cdot s/m^2$. One of the difficulties with the DGM is that D_{ij}, D_{iK}, and B_0 are all empirical parameters, which depend on the structure of the porous (membrane) region. One of the simplest ways to describe such structure is through the "bundle of parallel tubes" model, which envisions the porous medium to consist of straight, non-intersecting parallel pores with the same average size. For such model B_0 is typically described by the following equation:

$$B_0 = \frac{\varepsilon r_p^2}{8\tau} \tag{5.20}$$

where r_p is the mean pore radius. Van Swaaij and coworkers [5.50] have pioneered the use of the DGM models to membrane reactors. They have applied these models to the CNMR studies of reactions, which require strict stoichiometric ratios. As discussed in Chapter 2, these include the Claus desulfurization reaction [5.51, 5.52], the catalytic NO_x reduction by NH_3 [5.53], and the combustion of CO and other hydrocarbons. In the membrane reactors studied by van Swaaij and coworkers [5.51] the reactants are fed in different compartments (regions I and IV in Figure 5.12) of the membrane reactor, while the reaction takes place within the active membrane, itself, which is not very selective (a macroporous or mesoporous one). In one of their early studies [5.50] they considered a model reaction of the type A + B → 2C, following mass action kinetics. They considered an isothermal reactor with well-stirred compartments I and IV, without any external gas-to-membrane resistances. They solved equations (5.17) to (5.20) together with the corresponding boundary conditions by a finite differences method.

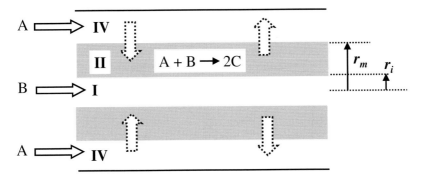

Figure 5.12. Catalytic non-permselective membrane reactor for the reaction A + B → 2C.

For the reaction A + B → 2C no net change occurs in the number of moles and, therefore, no pressure gradient develops through the membrane, unless one is imposed externally. Figure 5.13 shows some of their simulation results in terms of the mole fraction profiles as function of dimensionless position ζ within the catalytic membrane layer, defined as

$$\zeta = \frac{r - r_i}{(r_m - r_i)} \tag{5.21}$$

for different values of the modified Thiele modulus, defined as:

$$\Phi'^2 = \frac{k^{'II}(r_m - r_i)^2}{D_{iJ}^{II} RT} P_0 \tag{5.22}$$

where $k^{'II}$ is the reaction rate constant, which is assumed to be given here by:

$$k^{'II} = \frac{\varphi^{II} \rho_m^{II}}{P_A P_B} \tag{5.23}$$

For low values of the Thiele modulus, reaction is slow and the whole process is controlled by kinetics. As shown in Figure 5.13a the product yield in this case is also low, and the reactants are not completely converted, and, therefore, reactants slip is observed. For moderate values of the Thiele modulus (Φ'^2=1) the reaction is faster; for these conditions the reactants are completely converted (solid lines in Figure 5.13b) within the membrane region, the molar concentration of the product (C) increases inside the membrane, and a reaction zone develops. This reaction zone can be sustained within the membrane by maintaining the right balance between the membrane permeability and the reaction rate. Under such conditions both reactants can be completely consumed within the catalytic membrane, thus, avoiding the need for further separation. At higher values of the Thiele modulus ($\Phi'^2=10^3$) the reaction is instantaneous (this is generally the case in combustion), and a reaction plane is established (broken lines in Figure 5.13b).

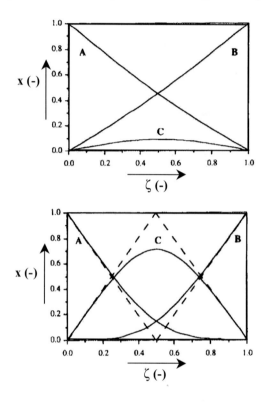

Figure 5.13. Calculated mole fraction profiles at different modified Thiele modulus. (a) top for small values of Φ', (b) bottom moderate and large values of Φ'. From Veldsink *et al.* [5.50], with permission from Overseas Publishers Association.

Veldsink *et al.* [5.50] also studied the influence of the dimensionless transmembrane pressure $\Delta\Pi$, defined as

$$\Delta\Pi = \frac{\Delta P}{P_0} \tag{5.24}$$

on the conversion and molar concentration profiles. In their analysis reactant B was fed into a higher overall pressure compartment, which means that its transport is, at least partially, due to convective flow. The simulation results are presented in Figure 5.14 for the case of instantaneous reaction ($\Phi'^2 = 10^3$). Figure 5.14 shows the molar concentration profiles for two different transmembrane pressures. One observes that when the transmembrane pressure increases the reaction zone shifts towards the lower pressure side, while simultaneously an expansion of the zone is also observed. This last observation is attributed to the increase in the transport rate, which in this case limits the overall process. The model of Veldsink *et al.* [5.50] was tested with experimental data from the Claus reaction; conversion simulations were in general in good agreement with the experimental values [5.52].

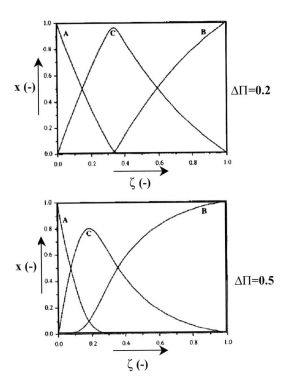

Figure 5.14. Calculated mole fraction profiles at different modified transmembrane pressures. From Veldsink *et al.* [5.50], with permission from Overseas Publishers Association.

A model for a similar membrane reactor, shown schematically on the top of Figure 5.15, has been developed by Harold *et al.* [5.54] to simulate reactant and products concentrations in parallel-consecutive reaction networks. The membrane reactor compartments on either side of the membrane were assumed to be completely stirred with no pressure gradient across the membrane. The model reaction studied is of relevance to hydrocarbon partial oxidation reactions, where the intermediate oxidation product can further react with oxygen to produce the undesirable total combustion products. Here the goal is to maximize the yield of the intermediate desired product. The calculation results indicate that the intermediate product yield increases with the Thiele modulus when the reactants are fed separately (completely segregated). This result indicates that this concept is attractive when the system is globally controlled by mass transport [5.54]. Harold and Lee [5.55] studied the membrane reactor of Figure 5.15 under non-isothermal conditions and for an integral reactor operation. The reaction network studied is described by the following set of equations:

$$A + \alpha_B\, B \rightarrow \alpha_{C1}\, R \tag{5.25}$$

$$A + \alpha_{C2}\, R \rightarrow \alpha_x\, X + \alpha_y\, Y \tag{5.26}$$

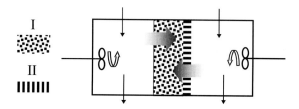

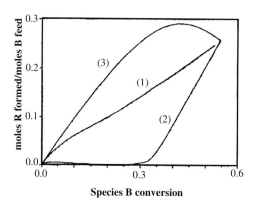

Figure 5.15. The use of CNMR in in partial oxidation reactions. Upper figure represents the schematic of the concept. Upper figure I: porous support, II: active layer. Lower figure gives the calculated yield of species R formed as function of B conversion, for (1) a mixed feed, (2) a segregated feed with A mainly supplied in the active layer, (3) a segregated feed with A mainly fed into the support layer. $\Phi'^2 = 5$. Adapted from [5.55], with permission from Elsevier Science.

In their study the model is taken to represent the oxidation of ethylene (B) by molecular oxygen (A) in order to produce acetaldehyde (R), an intermediate product for this reaction, which can further react under these conditions towards the products of total oxidation CO_2 (X) and H_2O (Y). Some of the simulation results are presented in the bottom part of the Fig. 5.15. In this figure the yield of R (in terms of moles of R formed per mole of B fed to the reactor) is plotted as a function of the conversion of B (which increases with reactor contact time). In the model of Harold and Lee the feed consists of an equimolar mixture of A and B, which means that the limiting reactant (A) is completely consumed and the reaction stops before the excess reactant (B) approaches the 100 % level. In the case of a mixed feed of A and B that is fed in both compartments, the yield of R increases monotonically with the conversion of B. This behavior is explained by the fact that A is completely consumed, before the consumption of R through the undesired sequential reaction (5.26) becomes important. When A is supplied mainly from the active layer side (curve (2)) a detrimental effect on the conversion is observed at low contact times (small B conversions) because the supply of B is limited by the diffusion through the support layer. In this case the secondary reaction is favored by the relative high A concentration found within the active layer. At higher contact times, at the level when a 30 % conversion of B is attained (up to that point resulting mostly in total oxidation products from reaction (5.26)) the rest of B begins to transform into to the desired intermediate products, without substantial further oxidation, because by that point the concentration of A is very low, and the extent of secondary reactions is very limited. The most favorable case is obtained when the oxidant (A) is supplied from the support side (curve 3). In this case the transport of the oxidant into the catalytic layer is controlled by the diffusion through the support. As a result its concentration levels are maintained low enough in the catalytic layer and total oxidation is largely avoided. The simulation results of Harold and Lee [5.55] point out the importance of choosing the right strategy for feeding the reactants into the membrane reactor, in terms of obtaining an advantage in the CNMR with respect to the conventional catalytic reactor.

More recently Neomagus *et al.* [5.56] utilized a similar model to study methane combustion in a CMR utilizing a catalytic porous alumina membrane activated with platinum. As in the previous studies the oxidant and the hydrocarbon were fed in different compartments separated by the porous, catalytically active membrane. In their study Neomagus *et al.* [5.56], however, took into consideration thermal effects by writing a corresponding energy balance equation coupled to the mass balance equations. They simulated the concentration and temperature profiles through the membrane, and compared the simulation results with experimental results of methane conversion. The authors report that in the transport controlled regime no slip of reactants is observed, easy controllability of flow rates and pressures is obtained, and thermal runaways is avoided, since transport is much less sensitive to temperature than reaction kinetics. Saracco and Specchia [5.57] utilized a similar type of model for the propane oxidation reaction. Their model takes into account ignition, extinction and multiple steady state phenomena. In their model the effect of the catalyst distribution in the membrane layer, and external and internal (inside the membrane) heat and mass transfer resistances on reactor performance were also evaluated. Their results demonstrate that in the case of moderate to high (above 6.3% in volume) propane feed concentrations a monodispersed catalyst model was not able to simulate the experimental results. Assuming a non-uniform catalyst distribution the simulations matched well with the ignition and extinction phenomena observed. A number of additional studies detailing the effect of catalyst distribution on CMR performances will be discussed further below.

5.1.5 Three-Phase Catalytic Membrane Reactors

In Chapter 2 we discussed a number of studies with three-phase catalytic membrane re-actors. In these reactors the catalyst is impregnated within the membrane, which serves as a contactor between the gas phase (B) and liquid phase reactants (A), and the catalyst that resides within the membrane pores. When gas/liquid reactions occur in conventional (packed, -trickle or fluidized-bed) multiphase catalytic reactors the solid catalyst is wetted by a liquid film; as a result, the gas, before reaching the catalyst particle surface or pore, has to diffuse through the liquid layer, which acts as an additional mass transfer resistance between the gas and the solid. In the case of a catalytic membrane reactor, as shown schematically in Fig. 5.16, the active membrane pores are filled simultaneously with the liquid and gas reactants, ensuring an effective contact between the three phases (gas/liquid, and catalyst). One of the earliest studies of this type of reactor was reported by Akyurtlu *et al.* [5.58], who developed a semi-analytical model coupling analytical results with a numerical solution for this type of reactor. Harold and coworkers (Harold and Ng [5.59] and Cini and Harold [5.60]) also studied this type of reactor. In their proposed model a catalytically impregnated symmetric membrane is used as a contactor between a gaseous reactant A (a mixture of A in an inert), which flows through the tube-side and a liquid phase reactant B circulated co-currently in the shellside. The product of the reaction C is also in the liquid phase (which, in addition, contains a solvent). Harold and coworkers assume that the liquid fills the membrane pores, where the reaction occurs. The CMR per-formance was compared with a model trickle-bed reactor, which consisted of a string of catalyst particle. The simulation results indicate that, in general, the CMR exhibits a better efficiency. The authors list as potential advantages for the CMR better temperature con-trol, reduced transport limitations, and a well-defined reaction interface.

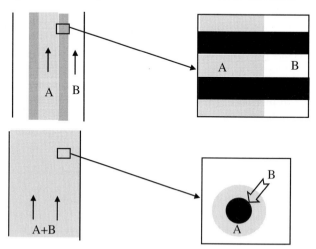

Figure 5.16. Schematic of the three-phase CMR (top) and comparison with a classical three-phase reactor (bottom).

Torres *et al.* [5.61] have also studied a similar membrane reactor system. Their model is similar to that of Harold and coworkers [5.59, 5.60], but the membrane of Torres *et al.* [5.61] consists of two layers, an active thin layer (mesoporous γ-alumina) where the catalyst (Pt) is placed, and a macroporous support (α-alumina). Two membrane reactor configurations were studied: in the first one (configuration 1) the gas phase reactant (hydrogen) is fed in the tubeside (the side of the catalytically active layer) whereas the liquid phase (nitrobenzene) was fed in the shellside (the macroporous membrane side). In the second configuration the hydrogen was fed in the shellside, while the nitrobenzene was fed in the tubeside. The results of the model were validated with experiments with the nitrobenzene hydrogenation reaction. The mesoporous γ-alumina membrane layer with its high specific surface area ($150 \text{ m}^2 \cdot \text{g}^{-1}$) turned out to be a very effective contactor resulting in a rapid hydrogen saturation of the liquid phase. The simulation results of the hydrogen concentration profiles as a function of the radial position in the membrane for two values of Thiele modulus for configuration 1 are shown in Fig. 5.17. The Thiele modulus here is defined as

$$\Phi_H^2 = \frac{k'^{II}(r_m - r_i)^2 C_{Heq}}{D_H^{II}} \tag{5.27}$$

where C_{Heq} is the hydrogen equilibrium molar concentration. From Figure 5.17 one can conclude that for large values of Thiele modulus (i.e., large reaction rates), almost all the hydrogen arriving in the reactive membrane is consumed. In this case the system is controlled by the slow diffusion of hydrogen through the membrane. In addition one observes that the hydrogen concentration gradient in the support is also very significant, due to the slow diffusion of hydrogen in the support layer which is thick and filled with liquid. These effects have been shown to be even more pronounced for nitrobenzene and aniline, because these molecules have lower diffusion coefficients than hydrogen. For intermediate values of the Thiele modulus, consistent with the experimental reaction rates, Torres *et al.* [5.61] observed that the reactor operates in an intermediate regime affected by both mass transfer and chemical reaction. Taking into consideration the slow fluxes obtained in the liquid-filled support, they concluded that the better reactor efficiency is attained in the configuration where gaseous hydrogen is fed on the support side.

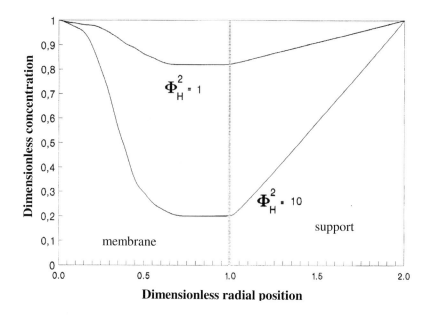

Figure 5.17. Simulated hydrogen concentration profiles at different values of the Thiele modulus. From [5.61], with permission from the American Chemical Society.

5.1.6   Other Modelling Aspects of Catalytic Membrane Reactors

As was noted previously, Saracco and Specchia [5.57] have studied in their paper the effect that a non-homogeneous distribution of the catalyst in the active membrane layer has on reactor performance. Other investigators have also investigated the effect that distributing the catalyst non-uniformly in the membrane layer has on membrane reactor performance. Varma and coworkers [5.62, 5.63] have systematically investigated these issues for the case of a simple first order reversible reaction, A ⇌ B. They have analyzed the effect of non-uniform catalyst distribution for three types of reactors: a PBMR, a CMR (both with well mixed tubeside and shellside regions), and a conventional packed bed CSTR (PBR). They compared the behavior between these reactors both for uniform and non-uniform catalyst distributions (Dirac delta catalyst activity profiles) in the catalyst pellets (for the PBMR and PBR) or in the membrane (CMR). In their model they assumed negligible external mass transfer resistances, and identical species diffusivities in both the catalyst pellets and the membrane. Figure 5.18 shows the total reactor conversion (for the membrane reactors based both on the feed and the permeate sides) as a function of a dimensionless residence time, defined by equation (5.28).

$$\theta = \frac{D_A V P_0}{RTQ_0(r_m - r_i)^2}$$

(5.28)

where D_A is the effective diffusion coefficient of A, V is the reactor volume, and Q_0 the volumetric flow rate.

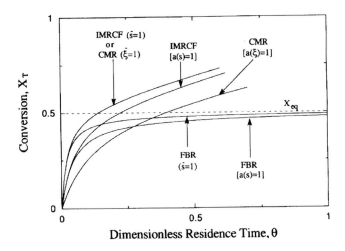

Figure 5.18. Total conversion as function of the dimensionless residence time θ for a CMR, PBMR and PBR with uniform and Dirac delta catalyst activity located at the feed side (CMR) or on the surface of catalyst pellets (PBMR, PBR) at identical Thiele modulus ($\Phi=5$) for membrane and catalyst pellets. From [5.62], with permission from Elsevier Science.

In Fig. 5.18 IMRCF (s=1) corresponds to the PBMR with catalyst pellets with a non-uniform catalyst distribution, while CMR ($\xi=1$) is the CMR with the catalyst placed non-uniformly on the membrane surface in contact with the catalytic reactor feed. IMRCF (α(s)=1) and CMR (α(ξ)=1) correspond to the PBMR and CMR with uniform catalyst distributions. The conventional packed-bed reactor (FBR in Figure 5.18) exhibits conversions, which are below the equilibrium conversion, and for large residence times are lower than those exhibited by the CMR and the PBMR. The highest conversions are obtained with the non-uniform activity (Dirac delta case) profiles. This result was explained on the basis that the access of the reactants to the active catalytic sites was not limited by diffusion. When the catalyst is uniformly distributed the PBMR exhibits better performances than the CMR. It is interesting to note that at low residence times the packed-bed reactor conversion is higher than that of the PBMR with a uniformly distributed catalyst; this is because in this case for the PBMR the reactants are only partially in contact with the catalyst due to diffusional limitations.

An interesting two-membrane reactor system concept has been suggested by Tekic *et al.* [5.64] for the following reaction scheme:

$$A \rightleftharpoons \alpha_B B + \alpha_C C \qquad\qquad (5.29)$$

Their membrane reactor system, shown schematically in Fig. 5.19, is equipped with two membranes, which are individually 100% permselective to each reaction product. The reactor consists of a reaction chamber and two different permeate chambers, each corresponding to one of the two membranes.

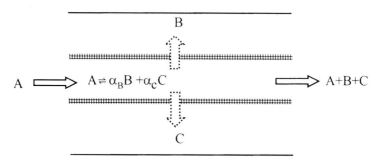

Figure 5.19. Schematic of the two-membrane reactor system. Adapted from [5.64].

Tekic *et al.* [5.64] use a simple, one-dimensional isothermal model of their proposed membrane reactor. Their model demonstrates that this membrane reactor configuration gives better conversions than the conventional one-membrane reactor system. The same group extended, in a later study [5.65], their model in order to account for non- isothermal effects. They have more recently theoretically examined the application of this concept to the thermal splitting reaction of water [5.66]. In this study the reaction was assumed to take place at 2000 K, and the two membranes were assumed to be individually 100 % permselective to hydrogen and oxygen, the products of the water splitting reaction. The simulation results indicated that conversions up to 99 % are possible in the double membrane reactor without the need to use a sweep gas. A two-membrane reactor system was also considered by Lu *et al.* [5.67] for the methane oxidative coupling reaction. In their study one membrane was considered capable of supplying the oxygen along the reactor length, whereas the second one was considered to be 100 % selective towards the desired C_2 products. The authors reported that higher C_2 yields could be obtained with respect to the co-feed fixed-bed reactor at high contact times. In spite of the advantages suggested by these theoretical studies, the practicality of the two-membrane reactors remains doubtful. There are no membranes and seals, for example, which are currently available that can withstand the high temperatures (~2000 K) required for the water splitting reaction. And no C_2 permselective membranes are currently available that can operate at the OCM temperatures.

A different type of a multiple membrane reactor system was proposed and modeled by Kim and Datta [5.68]. Their membrane reactor consists of liquid-phase catalytic layer supported on a porous matrix, which is sandwiched in between two different membranes. They considered a simple irreversible A→B reaction. The membrane, which is in contact with the feed stream of reactant A, is considered to be permeable to the reactant, but to provide a barrier for the product B. The second membrane, which was placed on the opposite side of the porous matrix, was considered to be permeable to both species A and B. The porous matrix consisted of a porous support, which was coated with a thin surface layer of the liquid catalyst, but which left enough of pore space for the reactant and product species to diffuse through. The feed and permeate compartments were considered well-mixed, whereas the porous matrix with the impregnated liquid catalyst was modeled as a one-dimensional packed-bed. The model was validated with experimental results for the ethane hydroformylation reaction to propionaldehyde. The authors report that a good product separation is obtained, when the transport resistance of membranes is higher than the one provided by the catalyst region.

In addition to the liquid phase membrane reactor models previously discussed, there have been a number of other interesting modeling investigations involving liquid reactants and products. Garayhi *et al.* [5.69] have reported a general isothermal model for a CMR where a liquid phase reaction takes place. The model does not consider the separation of species by the membrane.

Park *et al.* [5.70] developed a model of a liquid-phase membrane reactor, which utilizes a reaction scheme (shown below) and rate data applicable to the hydrocracking of California Hondo crude derived asphaltenes.

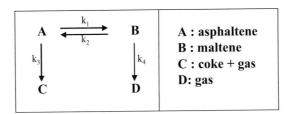

The model was applied to an isothermal membrane reactor, shown schematically in Fig. 5.20. Two reactor configurations have been investigated. In the first reactor configuration (configuration A) the shellside reactor volume is kept constant by continuously pumping into the reactor pure solvent. In the second reactor configuration (configuration B) no additional solvent is added to the reactor shellside, whose volume, as a result, decreases as the solvent and solutes diffuse across the membrane from the shellside to the tubeside under unsteady state conditions. For the calculations the asphaltene and maltene molecules were assumed to be single spherical molecules with weight-averaged MW of 9791 and 500, respectively, and corresponding equivalent Stokes-Einstein diameters of 13.4 and 8.2 Å. These are comparable to the size of membrane pores, and as a result transport through the membranes is by configurational diffusion [5.71]. The membrane in the theoretical analysis consists of four layers. Three of these layers are macroporous, while the fourth and top layer is micro/mesoporous. The micro/mesoporous layer for most of the calculations was also assumed to be monodisperse. For one set of calculations (in which the interest was in the effect of pore size distribution parameters) the layer was assumed to be polydisperse, i.e., having a pore size distribution.

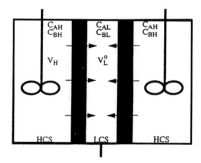

Figure 5.20. Schematic of the membrane reactor modelled by Park *et al.* [5.70], with permission from the American Chemical Society.

The model is rather straightforward and involves differential equations for the unsteady-state mass balance equation for every reactor compartment. The analysis of transport through the membrane involved the solution of the hindered convection/diffusion equations in every membrane layer, while accounting for the continuity of fluxes and concentrations at the interfaces between the various layers. The pore structure of the membrane layers is represented by the "bundle of parallel pores" model, and asphaltene and maltene molecules are represented as spheres, each with a unique equivalent Stokes-Einstein diameter. The effect of transmembrane pressure gradient and membrane pore size (for these calculations the microporous layer is considered monodisperse) on reactor conversion and maltene yield (after 200 h of reaction time) are shown in Fig. 5.21. For membranes with pore sizes between 9 and 13 Å there is no asphaltene loss through the membrane, and hence higher pressure gradients lead to higher conversions and yields. For large pore sizes (>18 Å), there is significant asphaltene loss through the membrane and the membrane has a negative effect on the asphaltene conversion and maltene yield. For this case increasing pressure gradients lead to decreasing conversions and yields. For intermediate pore sizes (~16 Å) the maltene yield shows a maximum at around 5 atm; at higher pressure gradients, the yield decreases due to the loss of asphaltenes through the membrane. The effect of pore size distribution on reactor behavior is shown in Fig. 5.22. For these calculations the structure of the microporous top layer is now assumed to follow a normal-Gaussian pore size distribution. In the figure the effect of varying the standard deviation of distribution (while keeping the mean pore diameter constant at 12 Å) on the asphaltene conversion and maltene yield is shown. For low standard deviations and a mean pore size of 12 Å, there is no asphaltene loss through the membrane. Hence, the asphaltene conversion and maltene yields, when plotted as a function of pressure gradient, reach an asymptotic value, and conversions are higher at higher pressure gradients. Increasing the standard deviation up to 0.1 does not appear to have a significant detrimental effect on the maltene yields. Pore size distributions with standard deviation larger than 0.2 result in negative effects in the membrane reactor in terms of both the maltene yield and asphaltene conversion. This is due to the significant asphaltene loss by leakage of asphaltene through the larger membrane pores. Asphaltene loss increases at higher pressure gradients. The maltene yield goes through a maximum and declines to values below that of the conventional reactor at higher pressure gradients. We note in passing, here, that even for gaseous reactants the "average" pore size of microporous (e.g., zeolite, carbon) membranes is close to the size of the transported molecules. Under such conditions the fluid is no longer a continuum, and the utility of models like the Dusty Gas Model comes into question. Non-equilibrium Molecular Dynamics techniques are finding increased utilization in this area [5.72, 5.73].

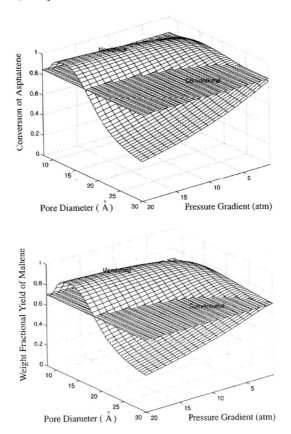

Figure 5.21. Effect of pressure and pore size on the asphaltene conversion and maltene weight fraction yield for membrane reactor configuration A at 350 °C. The flat surface is for the conventional reactor. From [5.70], with permission from the American Chemical Society.

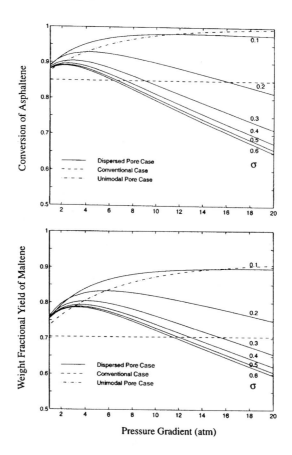

Figure 5.22. Effect of standard deviation of the pore size distribution on the asphaltene conversion and maltene weight fraction yield for reactor configuration A. From [5.70], with permission from the American Chemical Society.

Langhendries *et al.* [5.74] analyzed the liquid phase catalytic oxidation of cyclohexane in a PBMR, using a simple tank-in-series approximate model for the PBMR. In their - reactor the liquid hydrocarbon was fed in the tubeside, where a packed bed of a zeolite supported iron-pthalocyanine catalysts was placed. The oxidant (aqueous *t*-butyl-hydroperoxide) was fed in the shellside from were it was extracted continuously to the tubeside by a microporous membrane. The simulation results show that the PBMR is more efficient than a co-feed PBR in terms of conversion but only at low space times (shorter reactors). A significant enhancement of the organic peroxide efficiency, defined as the amount of oxidant used for the conversion of cyclohexane to the total oxidant converted, was also observed for the PBMR. It was explained to be the result of the controlled addition of the peroxide, which gives low and nearly uniform concentration along the reactor length.

Yawalkar *et al.* [5.75] have also developed a mathematical model to describe the oxidation of alkenes to epoxides in a CMR. In the model the emphasis is on the effect of the membrane characteristics. The tubeside and shellside are assumed to be well-stirred and operate under differential conditions. The hydrocarbon is fed in one side of the membrane, while the liquid peroxide oxidant is fed in the other. The membrane is modelled as a homogeneous polymeric phase, in which cubic zeolite catalyst particles are dispersed uniformly at equal distances from each other. Alkene epoxidation is taken to take place in two steps. The epoxidation reaction, itself, between the hydrocarbon and the oxidant (following mass action kinetics) to produce the epoxide, and a side undesirable reaction in which the peroxide oxidant decomposes into alcohol and oxygen (taken to be second order in peroxide concentration). The concentration of the hydrocarbon within the membrane and catalyst was considered to be in excess. Equations are, therefore, only presented for the transport and reaction of the oxidant. Reaction is assumed to occur only within the catalyst particles. A 1-D diffusion equation applies to the transport of the peroxide in the polymeric phase, and a 1-D diffusion and reaction equation applies to the catalytic phase. Continuity of fluxes and Henry's law applies across the membrane/catalyst interface. Henry's law also applies between the liquid oxidant and membrane phase. Yawalkar *et al.* [5.75] studied the effects of various parameters, such as peroxide and alkene concentration on either side of the membrane, sorption coefficient of the membrane for peroxide and alkene, membrane-catalyst Henry distribution coefficient for peroxide and alkene, and catalyst loading. The model predictions show that in the CMR with an organophilic membrane phase, peroxide concentration remains low at the catalyst active sites, thus, reducing the rate of peroxide decomposition. By proper selection of the various parameters, high peroxide efficiency and significant rate of generation of the epoxide can be obtained.

Sousa *et al.* [5.76, 5.77] modeled a CMR utilizing a dense catalytic polymeric membrane for an equilibrium limited elementary gas phase reaction of the type $\alpha_a A + \alpha_b B \rightleftharpoons \alpha_c C + \alpha_d D$. The model considers well-stirred retentate and permeate sides, isothermal operation, Fickian transport across the membrane with constant diffusivities, and a linear sorption equilibrium between the bulk and membrane phases. The conversion enhancement over the thermodynamic equilibrium value corresponding to equimolar feed conditions is studied for three different cases $\Delta n > 0$, $\Delta n = 0$, and $\Delta n < 0$, where $\Delta n = (\alpha_c + \alpha_d) - (\alpha_a + \alpha_b)$. Souza *et al.* [5.76, 5.77] conclude that the conversion can be significantly enhanced, when the diffusion coefficients of the products are higher than those of the reactants and/or the sorption coefficients are lower, the degree of enhancement affected strongly by Δn and the Thiele modulus. They report that performance of a dense polymeric membrane CMR depends on both the sorption and diffusion coefficients but in a different way, so the study of such a reactor should not be based on overall component permeabilities.

Tan and Li [5.78] and Li and Tan [5.79] developed theoretical models to describe the hollow-fiber membrane reactors (Shanbhag *et al.* [5.80, 5.81]), discussed in Chapter 2, which are applicable to water treatment processes, in which the membrane functions as an efficient reactant (e.g., ozone, hydrogen) distributor, and the chemical reaction takes place either in the hollow-fiber lumen or in the shellside. The model assumes a constant density for the liquid stream, fully developed flow and radial diffusion (when the liquid is introduced in the lumen side) or plug flow behavior and axial dispersion (when the liquid is introduced in the shellside). Reaction only occurs in the liquid phase, and follows mass action kinetics. The effects of various design and operating parameters, and membrane physical properties on the performance of the reactors were studied. The studies indicate that axial dispersion has to be considered, when the chemical reaction takes place in the shellside and Pe < 40. The model predicts that the mass-transfer direction of the pollutant through the membrane may be reversed, when the chemical reaction in the liquid is extremely fast, because under such conditions the pollutant that may have diffused in the gas phase can back-diffuse into the liquid phase. The results of the model compared favorably with experimental data from an ozonation MR (Shanbhag *et al.* [5.80]) and a hybrid UV-MR system for dissolved oxygen (DO) removal (Sinha [5.82]).

Li *et al.* [5.83] developed a model to describe the non-oxidative conversion of CH_4 to useful C_2–C_{10} hydrocarbons using hydrogen permeable solid oxide membranes. A schematic of the proposed experimental system and other experimental details are provided in Chapter 2. A distinct feature of the model of Li *et al.* [5.83] is that it utilizes detailed kinetic models to describe the homogeneous and heterogeneous processes that occur during the conversion of CH_4 into higher hydrocarbons. In their analysis Li et al. [5.83] consider all hydrocarbons larger than naphthalene ($C_{10}H_8$) to be undesirable polynuclear aromatic hydrocarbons (PAH) and monitor them as a lump, though they consider their processes of formation from naphthalene and benzene rigorously without the use of any lumping approximations. The basic homogeneous pathway consists of 44 steps and 25 species, and was shown previously (Dean [5.84]) to describe methane pyrolysis at low conversions < 1 %. To this pathway Li *et al.* [5.83] have added eight elementary steps in order to describe methane pyrolysis at higher conversions and to account for species formation higher than naphthalene, bringing the total number of steps used to 52 and the total number of species to 33. The homogeneous mechanism is complimented with an additional heterogeneous step to describe methyl radical formation by direct activation on catalytic sites. The 1-D isothermal membrane reactor model was simple, assuming plug-flow conditions prevailing in the feed and permeate side, and Fickian diffusion of H_2 through the membrane. Together with the kinetic pathways it was used to explore thermodynamic and kinetic barriers, and the effects of continuous hydrogen removal and the presence of catalytic sites on attainable reactor yields. The H_2 formed during the reaction decreases the CH_4 pyrolysis rates and equilibrium conversions, and favors the formation of lighter products. Its removal by the membrane increases reaction rates and equilibrium CH_4 conversions. Calculated C_2–C_{10} yields reach values greater than 90 % at an intermediate range of values of δ (1-10), which is defined as the ratio of characteristic times for CH_4 reaction and hydrogen permeation through the membrane. Homogeneous reactions alone require impractical residence times, even with H_2 removal, due to the large characteristic times, because of the slow initiation (the formation of methane and hydrogen radicals) and chain transfer (the formation of products from such radicals without their net consumption) rates. Using catalytic sites that form methyl radicals eliminates the induction period without influencing the homogeneous product distribution. Methane conversion, however, occurs predominately in the chain transfer regime, within which individual transfer steps and the formation of C_2 intermediates become limited by thermodynamic constraints. Catalytic sites alone cannot overcome these limitations. Membrane reactors with continuous H_2 removal, on the other hand, overcome these thermodynamic constraints on C_2 formation, and favorably impact the required residence times. Reaction rates then become limited by homogeneous reactions of C_2 products to form C_{6+} aromatics. Here, the use of a bifunctional catalyst like Mo/H-ZSM5, which may increase the rate of cyclization and oligomerization reactions by acid sites within a shape-selective environment, may turn out to be of value. Higher δ values, outside the optimal range, lead to conversion of the desired C_2–C_{10} products to larger polynuclear aromatics. As previously noted in Chapter 2, the same group is experimentally implementing the concept of the direct conversion of methane to higher hydrocarbons via hydrogen removal using shape-selective, bifunctional pyrolysis catalysts, and hydrogen permeable perovskite membrane.

5.2   Modelling of Pervaporation Membrane Reactors

As noted in Chapter 3, a class of liquid phase membrane reactors which are attracting significant attention are pervaporation membrane reactors. There are several published studies modelling the behavior of such reactors, some of which have already been briefly mentioned in Chapter 3. The early PVMR models described reaction kinetics and membrane permeation in terms of concentrations of the reacting species. For thermodynamically non-ideal mixtures, however, activities are needed in the description of transport (pervaporation) by a solution diffusion mechanism through the membrane. As detailed in the very comprehensive treatment of pervaporation by Heintz and Stephan [5.85, 5.86], proper care must be taken for describing the non-ideal effects through the calculation of activity coefficients. For non-ideal mixtures, furthermore, expressing the reaction rates in terms of concentrations results in reaction rate constants, which often depend on concentrations (see Froment and Bischoff [5.87] for further detailed discussions). Using activities to describe reaction rates not only rectifies this problem, but also provides a unified approach in describing both reaction kinetics and thermodynamic equilibrium (Venkateswalu *et al.* [5.88], Keurentjes *et al.* [5.89]).

The first pervaporation membrane reactor model which takes into account solution non-idealities was developed and validated experimentally by Zhu *et al.* [5.90]. Prior studies [5.89, 5.91] also made note of such non-idealities, but offered no unified means for accounting for these phenomena in the description of PVMR. Since the model of Zhu *et al.* [5.90] appeared, other groups have also utilized similar models [5.92]. A more comprehensive analog of this model was, for example, recently presented and validated experimentally by Park [5.93], and by Lim *et al.* [5.94]. Zhu *et al.* [5.90] analyzed a tubular PVMR, in which the homogeneously catalyzed esterification reaction of acetic acid with ethanol to produce ethyl acetate and water took place. The reaction can be expressed generally as:

$$A + B \leftrightarrow E + Y \tag{5.30}$$

where A represents the acetic acid, B the ethanol, E the ethyl acetate, and W the water, correspondingly. The reaction kinetics have been studied both in a plug flow [5.90] and in a well-stirred batch reactor [5.93], and have been shown to obey the following rate expression in terms of activities, α_i,

$$r = k\left[\alpha_A \alpha_B - \alpha_E \alpha_W / K_{eq}\right] \tag{5.31}$$

where r is the reaction rate, k is the forward reaction rate constant, and K_{eq} the equilibrium constant. In the above expression the activities of the various chemical species i can be replaced by the activity coefficient, γ_i and the mole fractions X_i:

$$r = k\left[\gamma_1 \gamma_2 X_1 X_2 - \gamma_3 \gamma_4 X_3 X_4 / K_{eq}\right] \tag{5.32}$$

For the calculation of the activity coefficients use is made of the UNIQUAC equation [5.95]. Further details about the use of this equation can be found elsewhere [5.90, 5.93]. The pervaporative transport of a component i through a dense polymeric membrane is generally described in the literature [5.85, 5.86] by the following equation

$$J_i = U_i A_m \left[\gamma_i X_i P_{is} - \varphi_i \psi_i P\right] \tag{5.33}$$

where J_i is the flux of i, U_i the permeance of i in the mixture, P_{is} the saturation pressure of i, φ_i its fugacity, ψ_i its mole fraction in the vapor phase, and P represents the total pressure on the permeate side. In laboratory experiments P is typically very small, and the gas phase is taken to be ideal. U_i, the membrane permeance for species i, is typically dependent on composition and temperature, and can be represented as:

$$U_i = \lambda_i(X_i, T) U_i^o \tag{5.34}$$

where $\lambda_i(X_i, T)$ is the function that describes the composition/temperature dependence, and U_i^o is the permeance of pure i. A composite polymeric-inorganic membrane was utilized in the experiments of Zhu *et al.* [5.90] and Park [5.93]. For this membrane the permeances for the various components of the multicomponent mixture were determined experimentally, as there are currently no predictive relationships. The transport characteristics of each membrane were determined *in situ* in the pervaporation membrane reactor, prior to the initiation of each reaction experiment [5.90]. The experimentally measured permeances, together with the reaction rate constants measured independently, were utilized in the validation of the membrane reactor models [5.93].

The following assumptions were made in developing the equations for the model for the plug-flow pervaporation membrane reactor (PFPVMR): The reactor behaves as an ideal, homogeneous plug-flow reactor. The membrane is completely unreactive. The main resistance to transport lies in the dense polymeric layer. Transport resistance in the inorganic support structure on the membrane permeate side, and concentration polarization effects on the membrane tubeside are considered negligible. This is, indeed, true for the laboratory reactors for low pressures on the shellside, and for relatively small residence times on the tubeside. For large-scale pervaporation membrane reactors both effects may turn out to be significant [5.96]. Generally, both effects tend to impact negatively on membrane performance, so reactor performance, when ignoring such effects, should be viewed as "the best case" scenario. Based on the above assumptions the equations describing the plug-flow pervaporation membrane reactor (PFPVMR) are as follows:

Tubeside

$$\frac{dF_{ir}}{dV} = \upsilon_i k \left[\gamma_1 \gamma_2 X_1 X_2 - \gamma_3 \gamma_4 X_3 X_4 / K_{eq} \right] - \alpha_m \lambda_i U_i^0 \left[\gamma_i X_i P_{is} \varphi_i \psi_i P \right] \tag{5.35}$$

at $V = 0$, $F_{ir} = F_{AR}^0 (\theta_i - f_0)$

Shellside

$$\frac{dF_{ip}}{dV} = \alpha_m \lambda_i U_i^0 \left[\gamma_i X_i P_{is} - \varphi_i \psi_i P \right] \tag{5.36}$$

at $V = 0$, $F_{i,p} = 0$ (for $I = A, B, E, W$); $F_{i,p} = F_{i,p0}$ (for $I = $ inert)

where α_m is the membrane area per unit reactor volume, F_{ir} the molar flow rate of species i in the reactor, and F_{ip} the molar flow rate of i in the membrane shellside. In equation (5.35) the assumption is made that the PFPVMR follows a conventional reactor (e.g., PFR). The inlet into the PFPVMR is then the outlet of the PFR. The molar flow rates of the various species into the PFR are represented by F_{iR}^0, and $\theta_i = F_{iR}^0 / F_{AR}^0$, A being the limiting reagent. The PFPVMR conversion f is defined in terms of the limiting reagent as

$$f = \frac{F_{AR}^0 - F_{AR}}{F_{AR}^0} \tag{5.37}$$

f_0 is the conversion at the inlet of the PFPVMR. When $f_0 = 0$, the F_{iR}^0 are the inlet molar flow rates into the PFPVMR. Lim *et al.* [5.94] define the following dimensionless variables and groups:

$$Y_{ir} = \frac{F_{ir}}{F_{AR}^0 (1 + \theta_B)} \qquad Y_{ip} = \frac{F_{ip}}{F_{AR}^0 (1 + \theta_B)} \qquad \varepsilon = \frac{P}{P_{ws}} \qquad \xi = \frac{V}{V_R} \qquad \alpha_i = \frac{U_i^0}{U_w^0} \qquad \delta = \frac{P_{is}}{P_{ws}}$$

$$D_w = \frac{\alpha_m V_R U_w^0 P_{ws}}{F_{AR}^0 (1+\theta_B)} \qquad Da = \frac{V_R k}{F_{AR}^0 (1+\theta_B)} \qquad \Omega = \frac{D_w}{Da}$$

In the above Da is a modified Damköhler number (ratio of a characteristic flow time to a characteristic reaction time), Dw is the ratio of a characteristic time for flow to the characteristic time for transport, and Ω is their ratio (the ratio of the characteristic time for reaction to that for transport). Equations 5.35 and 5.36 become dimensionless as:

Tubeside

$$\frac{dY_{ir}}{d\xi} = \upsilon_i Da[g_1 g_2 X_{1r} X_{2r} - g_3 g_4 X_{3r} X_{4r}/K_{eq}]\left[\frac{1}{\sum Y_{ir}}\right]^2 - \alpha_i \lambda_i D_w [\gamma_i \frac{Y_{ir}}{\sum Y_{ir}} \delta_i - \varphi_i \frac{Y_{ip}}{\sum Y_{ip}} \varepsilon]$$

at $\xi = 0$, $Y_{ir} = \dfrac{\theta_i - f_0}{1+\theta_B}$

$$(5.38)$$

Shellside:

$$\frac{dY_{ip}}{d\xi} = \alpha_i \lambda_i D_w \left[\gamma_i \frac{Y_{ir}}{\sum Y_{ir}} \delta_i - \varphi_i \frac{Y_{ip}}{\sum Y_{ip}} \varepsilon\right]$$

$$(5.39)$$

at $\xi = 0, Y_{ip} = 0$ $(1 = A, E, B, W); Y_{ip} = Y_{ip0}$ $(I = inert)$

Reactor conversion for the PFPVMR configuration as a function of the modified Damköhler number, Da, for various values of the parameter Ω, is shown in Fig. 5.23. The line for $\Omega=0$ represents the conversion of a conventional plug-flow reactor (PFR).

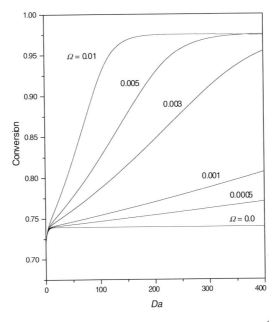

Figure 5.23. Conversion of PFPVMR versus Da for various Ω at constant $\alpha^{-1}_{EtOH} = 40$ and $f_0 = 0.72$. From [5.94].

As expected, the PFR conversion asymptotically reaches the equilibrium conversion for large values of Da. The PFPVMR conversions are much higher than the corresponding conversion of the conventional PFR. This is the result of the efficient water removal by the pervaporation membrane, which increases the forward reaction rate, and allows the esterification reaction to overcome its thermodynamic limitations. Ω has a significant effect on the reactor conversion. For a given Da, a higher value of Ω yields a higher conversion. This is true up to a point, beyond which reactant losses begin to have a negative impact on reactor performance. Lim *et al.* [5.94] have also modelled and analyzed the behavior of a number of other common PVMR configurations. The technical details for the models can be found in the original publication.

5.3   Modelling of Membrane Bioreactors

Membrane bioreactors have been modelled using approaches that have proven successful in the more conventional catalytic membrane reactor applications. The simplest membrane bioreactor system, as noted in Chapter 4, consists of two separate units, a bioreactor (typically a well-stirred batch reactor) coupled with an external hollow fiber or tubular or flat membrane module. These reactors have been modelled by coupling the classical equations of stirred tank reactors with the mathematical expressions describing membrane permeation. What makes this type of modelling unique is the complexity of the mechanisms of biological reactions. In addition, the rate equations associated with biological reactions are, in general, more complicated than the reaction rates, typically, encountered in catalytic applications. For example, metabolites that are produced often inhibit the reaction. The reaction law to describe such an inhibition action is often very complex. Studies have also appeared to analyze the performance of flow MBR, for which the presence of an external dilution rate [5.97] allows one to independently adjust the permeate flow rate, or the mode of reactor operation [5.98].

The MBR, which couple the membrane and reactor in the same unit present additional complications, especially when they use microorganisms as biocatalysts. As previously noted, bio-films that grow (either intentionally or unintentionally) on or within the membrane structure significantly affect the membrane flux, resulting on occasion in complete membrane plugging. Since most biological reactions take place in the liquid phase, one may be ill-advised to ignore the bulk-phase diffusional limitations. How one properly describes transport in the liquid phase, when biomass is present or within immobilized biofilms is of fundamental importance for this type of reactor, because the mass transport limitations in these complex media could have dramatic effects on reactor performance. This subject is still a matter of ongoing research [5.99, 5.100].

Some of the efforts, so far, to model such membrane bioreactors seem to not have considered the complications that may result from the presence of the biomass. Tharakan and Chau [5.101], for example, developed a model and carried out numerical simulations to describe a radial flow, hollow fiber membrane bioreactor, in which the biocatalyst consisted of a mammalian cell culture placed in the annular volume between the reactor cell and the hollow fibers. Their model utilizes the appropriate non-linear kinetics to describe the substrate consumption; however, the flow patterns assumed for the model were based on those obtained with an empty reactor, and would probably be inappropriate, when the annular volume is substantially filled with microorganisms. A model to describe a hollow-fiber perfusion system utilizing mouse adrenal tumor cells as biocatalysts was developed by Cima *et al.* [5.102]. In contrast, to the model of Tharakan and Chau [5.101], this model took into account the effect of the biomass, and the flow pattern distribution in the annular volume. These effects are of key importance for conditions encountered in long-term cell cultures, when the cell mass is very dense and small voids can completely distort the flow patterns. However, the model calculations of Cima *et al.* [5.102] did not take into account the dynamic evolution of the cell culture due to growth, and its influence on the permeate flow rate. Their model is appropriate for constant biocatalyst concentration.

A comprehensive model of a membrane bioreactor has been developed by Moueddeb *et al.* [5.103] for a simple irreversible reaction A→B. The goal of the model was to describe their experimental reactor system, which was described earlier in Chapter 4. The model equations were established by taking into account the effect of the biomass on the permeate flow rate in the annular volume. The mass balance equations for the substrate (A) and the product (B) in cylindrical coordinates, utilized by Moueddeb *et al.* [5.103] are given as:

$$D_A^{II}\left(\frac{\partial^2 C_A^{II}}{\partial r^2}+\frac{1}{r}\frac{\partial}{\partial r}\left(r\frac{\partial C_A^{II}}{\partial r}\right)\right)-\frac{Q}{2\pi\varepsilon L}-\varphi_A=\frac{\partial A}{\partial t} \tag{5.40}$$

$$D_B^{II}\left(\frac{\partial^2 C_B^{II}}{\partial r^2}+\frac{1}{r}\frac{\partial}{\partial r}\left(r\frac{\partial C_B^{II}}{\partial r}\right)\right)-\frac{Q}{2\pi\varepsilon L}-\varphi_B=\frac{\partial B}{\partial t} \tag{5.41}$$

In these equations D represents the corresponding diffusion coefficients, and Q the permeate flow rate. The first term of each equation gives the radial dispersion and the second one corresponds to the radial convection. The authors [5.103] used in their model, a biological kinetic rate expression (φ), which was obtained by independent experiments and analysis of a batch reactor, and also made an effort to account for and correlate the permeate flow decrease with the amount of produced biomass. The simulation curves obtained matched well the experimental results in terms of permeate flow rate evolution and product concentration. One of the important aspects of the model is its ability to theoretically determine the biomass concentration profiles, and the relation between the permeate flow rate and the calculated biomass concentration in the annular volume (Fig. 5.24). Such information is important since the biomass evolution cannot be determined by any experimental methodology.

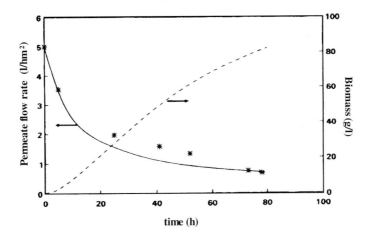

Figure 5.24. Evolution of permeate volumetric flow rate, calculated (——) , experimental (*) and calculated micro-organisms concentration profile in the annulus (-----). From Moueddeb [5.103], with permission from Elsevier Science.

Hollow fiber extractive membrane bioreactors (EMB), have been modelled by Pavasant *et al.* [5.104]. For this purpose the authors employed a diffusion-reaction model for the membrane to describe the dynamic biofilm growth. The wastewater and the biological treatment compartments were considered completely mixed. Pavasant *et al.* [5.104] report a good agreement between the simulation results and experimental data for the pollutant flux across the membrane, and the biofilm thickness evolution. The model describes successfully the radial concentration profiles in the membrane, but since the wastewater and the biological treatment compartments were considered completely mixed, the model cannot provide axial concentration profiles. EMB are generally operated in plug-flow mode, and, therefore, the calculation of the axial concentration profile is of importance in order to determine the degree of pollutant removal. In a second paper by the same group they reported a more complete model, which also takes into account the axial concentration profiles [5.105]. In another series of publications by the same group (Strachan and Livingston [5.106] and Livingston *et al.* [5.107]) they reported on the effect of membrane module configuration, and of other process design features on the performance of this type of membrane reactor. In the paper by Strachan and Livingston [5.106] they compared the behavior of three different membrane modules in a plug-flow or continuously stirred-tank reactor mode of operation for the continuous extraction of monochlorobenzene. The results showed that the CSTR mode of operation provides for better extraction performances than the plug-flow operation.

The modelling of enzymatic membrane reactors follows, in general, the same approach as described previously. In enzymatic membrane reactors the catalyst is a macromolecule (enzyme). It can be found either in a free form in the reactor or supported on the membrane surface, or inside the membrane porous structure by grafting it or in the form of a gel obtained by ultrafiltration. As in the case of the whole-cell membrane bioreactors discussed above, the proper calculation of the mass transfer characteristics is of great importance for the modelling of this type of reactor. One of the earliest models of enzymatic membrane bioreactors is by Salmon and Robertson [5.108]. These authors modelled an enzymatic membrane bioreactor, which was made of four coaxial compartments; the enzyme is confined within one of the compartments, and one of the substrates is fed in a gaseous form.

Steady state models of membrane bioreactors utilizing a multi-enzyme system, which in addition to the main reaction promotes the simultaneous regeneration of the co-factor (for further discussion see Chapter 4) have been developed by different groups in Japan [5.109, 5.110]. Several of these studies have also considered the effect of backmixing [5.111, 5.112]. A model of an enzymatic hollow fiber membrane bioreactor with a single enzyme, which utilizes two different substrates (reaction 5.42) has been developed recently by Salzman *et al.* [5.113].

$$A_1 + A_2 \rightarrow B_1 + B_2 \tag{5.42}$$

In the MBR of Salzman *et al.* [5.113] the soluble enzyme exists in the shellside, and is recirculated through en external reservoir to ensure its homogeneity. The substrates were fed in the tubeside and diffused through the hollow fibers to the shellside to react with the enzymes found there. The products that formed diffused back to the tubeside and left the reactor by convection. Back-mixing effects were considered only in the tubeside, whereas other potentially negative effects like side reactions or inhibition were not taken into account. To facilitate calculations the tubeside of the reactor was modelled as a series of n perfectly mixed compartments; the two limiting cases are for n=1 (completely stirred-tank) and n=∞ (plug-flow reactor). The effect of diffusional limitations that may prevail in this type of reactor, was studied with the aid of two dimensionless groups, the effectiveness factor η_f, and a group ϕ_{if} which gives a measure of the mass transport through the membrane, both calculated in the tubeside. In order to have uniform notation we redefine both dimensionless parameters as follows:

$$\eta_f = \frac{k_{n+1}^{IV}}{k_f^I} \tag{5.43}$$

$$\phi_{i,f} = \frac{V^{IV} \cdot k_f^I}{\kappa_i \cdot C_{i,f} \cdot L_f \cdot 2\pi \cdot r_m} \tag{5.44}$$

where f=1,2,...,n, i=A$_1$ or A$_2$, $C_{i,f}$ is the concentration of i in compartment f of the reactor shellside, κ_i is the mass transfer coefficient of the i component through the membrane, and V^{IV} corresponds to the total shellside and reservoir volumes. k_{n+1}^{IV} and k_f^I are the reaction rates in the presence and in the absence of diffusional limitations respectively, k_f^I calculated considering the reaction rate in the shellside of the reactor for substrate concentrations being equal with those in the lumen side. Figure 5.25(I) shows the relation between η_f and ϕ_{if}, and also defines the regions where the reactor may be either diffusionally or kinetically limited. For low values of the permeation modulus ϕ_{if} the value of the effectiveness factor is close to 1, and the reactor is under kinetic control. Mass transport control (diffusion) is encountered at low η_f and high ϕ_{if}, with an intermediate regime in between these two limiting cases. In Figure 5.25(II) the effectiveness factor is plotted at various locations along the reactor axis (i.e, f) for different enzymatic reactivities. Notice that the effectiveness factor increases along the length of the reactor. One can also observe that the effectiveness factor depends on the reaction rates; for high values of the maximal reaction rate the reactor is always under diffusion control.

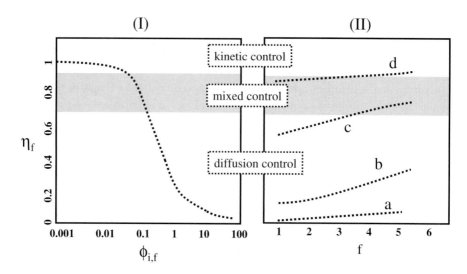

Figure 5.25. Effectiveness factor in a hollow fiber membrane reactor related to mass transport characteristics. (I): relation between the effectiveness factor (η_f) and the permeation modulus (ϕ_{if}). (II): η_f vs location section along the reactor axis at different maximal reaction rates (mol.s^{-1}): a: 10^{-1}, b: 10^{-2}, c: 10^{-3}, d: 10^{-4}. From reference [5.113], with permission from Elsevier Science.

5.4   References

[5.1]   W.N. Gill, E. Ruckenstein, and H.P. Hsieh, *Chem. Engng. Sci.* **1975**, 30, 685.

[5.2]   N. Itoh, *AIChE Journal* **1987**, 33, 9, 1576.

[5.3]   Y. Sun and S. Khang, *Ind. Eng. Chem. Res.* **1988**, 27, 7,1136.

[5.4]   N. Itoh, *AIChE Journal* **1987**, 33, 9, 1576.

[5.5]   P.M. Salmon and C.R. Robertson, *Chem. Engng. J.* **1988**, 38, 57.

[5.6]   T.T. Tsotsis, R.G. Minet, A.M. Champagnie, and P.K.T. Liu, "Catalytic Membrane Reactors", in: *Computer Aided Design of Catalysts*, E.R. Becker and C.J. Pereira (Eds.), Dekker, New York, pp. 471-551, 1993.

[5.7]   J. Sanchez and T.T. Tsotsis, "Current Developments and Future Research in Catalytic Membrane Reactors", in: *Fundamentals of Inorganic Membrane. Science and Technology*, A.J. Burggraaf and L. Cot (Eds.), Elsevier Science B.V., Amsterdam, 1996.

[5.8]   A. Dixon, "Innovations in Catalytic Inorganic Membraane Reactors", in: *Catalysis*, Vol. 14, The Royal Society of Chemistry, Cambridge, **1999**.

[5.9]   J. Shu, B.P.A. Grandjean, and S. Kaliaguine, *Appl. Catal. A.* **1994**, 119, 305.

[5.10]   G. Barbieri, V. Violante, F.P. Di Maio, A. Criscuoli, and E. Drioli, *Ind. Engng.Chem. Res.* **1997**, 33, 197.

[5.11]   J.S. Oklany, K. Hou, and R. Hughes, *Appl. Catal. A* **1998**, 170, 13.

[5.12]   J.S. Choi, I.K. Song, and W.Y. Lee, *J. Membr. Sci.* **2000**, 166, 175.

[5.13]   A. Trianto and T J. Kokugan, *Chem. Eng. Jpn.* **2001**, 34, 199.

[5.14]   T.T. Tsotsis, A.M. Champagnie, S.P. Vasileiadis, Z.D. Ziaka, and R.G. Minet, *Chem. Engng. Sci.* **1992**, 47, (9/11), 2903.

[5.15]   N. Itoh, W. C. Xu, and K. Haraya, *Ind. Eng. Chem. Res.* **1994**, 33, 197.

[5.16]   V.A. Papavassilion, J.A. McHenry, E.W. Corcoran, H.W. Deckman, and J.H. Meldon, *Proc. 1st Int. Workshop on Catalytic Membr.s*, September 1994, Lyon-Villeurbanne, France.

[5.17]   Y.L. Becker, A. G. Dixon, W. R. Moser, and Y.H. Ma, *J. .Membr. Sci.* **1993**, 77, 197.

[5.18]   M. Tayakout, B. Bernauer, Y. Toure, and J. Sanchez, *J. Simul. Pract. And Theory* **1995**, 2, 205.

[5.19]   M. Tayakout, J. Sanchez, D. Uzio, and J.A. Dalmon, *Proc. 1st Int. Workshop on Catalytic Membranes*, September 1994, Lyon-Villeurbanne, France.

[5.20]   M.K. Koukou, N. Papayannakos, N.C. Markatos, M. Bracht, H.M. van Veen, and A. Roskam, *J. Membr. Sci.* **1999**, 155, 241.

[5.21]   M.K. Koukou, N. Papayannakos, and N.C. Markatos, *Chem. Eng. J.* **2001**, 83, 95.

[5.22]   C.Y. Tsai, Y.H. Ma, W.R. Moser, and A.G. Dixon, in: Y.H. Ma, Ed. , *Proc. 3rd ICIM*, July 10-14, 1994, Worcester, MA, USA, 271.

[5.23]   A.G. Dixon, W.R. Moser, and Y.H. Ma, *Ind. Eng. Chem. Res.* **1994**, 33, 3015.

[5.24]   R.C. Reid, J.M. Prausnitz, and B.E. Poling, *The Properties of Ggases and Liquids*, Mac Graw Hill; Inc. NY, 1987.

[5.25]   M.P. Harold, C. Lee, A.J. Burggraaf, K. Keizer, V.T. Zaspalis, and R.S.A. de Lange, *MRS Bull.* **1994**, XIX, 4, 169.

[5.26]   C. Hermann, P. Quicker, and R. Dittmeyer, *J. Membr. Sci.* **1997**, 136, 161.

[5.27]   W. Wang and Y. S. Lin, in: Y.H. Ma, Ed., *Proc. 3rd ICIM*, July 10-14, 1994, Worcester, MA, USA, 259.

[5.28]   Y. Lu, A. Ramachandran, Y.H. Ma, W.R. Moser, and A.G. Dixon, in Y.H. Ma, Ed., *Proc. 3rd ICIM*, July 10-14, 1994, Worcester, MA, USA, 657.

[5.29]   T. Nozaki and K. Fujimoto, *AIChE J* **1994**, 40, 870.

[5.30]   W. Jin, X. Gu, S. Li, P. Huang, N. Xu, and J. Shi, *Chem. Engng. Sci.* **2000**, 55, 2617.

[5.31]   N. Itoh, *J. Chem. Engng. Jpn.* **1990**, 23,81.

[5.32]   M.K. Koukou, G. Chaloulou, N. Papayannakos, and N.C. Markatos, *Int. J. Heat Mass Transfer* **1997**, 40, 10.

[5.33]   W.S. Moon and S.B. Park, *J. .Membr. Sci.* **2000**, 170, 43.

[5.34]   T. Brinkmann, S.P. Perera, and W.J. Thomas, *Chem. Engng. Sci.* **2001**, 56, 2047

[5.35]   S.S.E.H. Elnashaie, B.K. Abdallah, S.S. Elshishini, S. Alkhowaiter, M.B. Noureldeen, and T. Alsoudani, *Catal. Today* **2001**, 64, 151.

[5.36]   C.Y. Tsai, Y.H. Ma, W.R. Moser, and A.G. Dixon, *Chem. Eng. Commun.* **1995**, 134, 107.

[5.37]   C. Tellez, M. Menendez, and J. Santamaria, *AIChE J.* **1997**, 43, 777.

[5.38]   C. Tellez, M. Menendez, and J. Santamaria, *Chem. Engng. Sci.* **1999**, 54, 2917.

[5.39]   M. Pedernera, M.J. Alfonso, M. Menendez, and J. Santamaria, Book of Abstracts, p. 125, *4th International Conference on Catalysis in Membr. Reactors ICCMR-2000*, July 3-5, 2000, Zaragoza, Spain.

[5.40]   M. Pedernera, R. Mallada, M. Menendez, and J. Santamaria, *AIChE J.* **2000**, 46, 2489.

[5.41]   R. Mallada, M. Pedernera, M. Menendez, and J. Santamaria, *Ind. Eng. Chem. Res.* **2000**, 36, 620.

[5.42]   K. Hou, R. Hughes, R. Ramos, M. Menendez, and J. Santamaria, *Chem. Engng. Sci.* **2001**, 56, 57.

[5.43]   R. Ramos, M. Menendez, and J. Santamaria, *Catal. Today* **2000**, 56, 239.

[5.44]   A.G. Dixon, *Catal. Today* **2001**, 67, 189.

[5.45]   Y.V. Gokhale, R.D. Noble, and J.L. Falconer, *J. Membr. Sci.* **1995**, 103, 235.

[5.46]   N. Itoh, *Catal.Today* **1995**, 25, 351.

[5.47]   C.M. Reo, L. A. Bernstein, and C.R.F. Lund, *AIChE J.* **1997**, 43, 495.

[5.48]   C.M. Reo, L. A. Bernstein, and C.R.F. Lund, *Chem. Engng. Sci.* **1997**, 52, 3075.

[5.49]   E.A. Mason and A.P. Malinauskas, *Gas Transport in Porous Media: The Dusty Gas Model*, Elsevier, Amsterdam, 1983.

[5.50]   J.W.Veldsink, R.M.J. van Damme, G.F. Versteeg, and W.P.M. van Swaaij, *Chem. Eng. Comm.* **1998**, 169, 145.

[5.51]   H.J. Sloot, G.F. Versteeg, and W.P.M. van Swaaij, *Chem. Engng. Sci.* **1990**, 45, 2415.

[5.52]   H.J. Sloot, C.A. Smolders, W.P.M. van Swaaij, and G.F. Versteeg, *AIChE J.* **1992**, 36, 887.

[5.53]   V.T. Zaspalis, W. van Praag, K. Keizer, J.G. van Ommen, J.R.H. Ross, and A.J. Burggraaf, *Appl. Catal.* **1991**, 74, 235.

[5.54]   M.P. Harold, V.T. Zaspalis, K. Keizer, and A.J. Burggraaf, *Chem. Engng. Sci.* **1993**, 48, 2705.

[5.55]   M.P. Harold and C. Lee, *Chem. Engng. Sci.* **1997**, 52, 1923.

[5.56]   H.W.J.P. Neomagus, G. Saracco, H.F.W. Wessel, and G.F. Versteeg, *Chem. Eng. J.* **2000**, 77,165.

[5.57]   G. Saracco, and V. Specchia, *Chem. Engng. Sci.* **2000**, 55, 3979.

[5.58]   J.F. Akyurtlu, A. Akyurtlu, and C.E. Hamrin, Jr., *Chem. Eng. Comm.* **1988**, 66, 169.

[5.59]   M.P. Harold and K.M. Ng, *AIChE J.* **1987**, 33, 9.

[5.60]   P. Cini and M.P. Harold, *AIChE J.* **1991**, 37, 7, 997.

[5.61]   M. Torres, J. Sanchez, J.A. Dalmon, B. Bernauer, and J. Lieto, *Ind. Eng. Chem. Res.* **1994**, 33, 2421.

[5.62]   K.L. Yeung, A. Aravind, R.J.X. Zawada, J. Szegner, G Gao, and A.Varma, *Chem. Engng. Sci.* **1994**, 49, 4823.

[5.63]   J. Szegner, K.L. Yeung, and A. Varma, *AIChE J.* **1997**, 43, 2059.

[5.64]   M.N. Tekic, R.N. Ratomir, and G. Ciric, *J. Membr. Sci.* **1994**, 96, 213.

[5.65]   R.P. Omorjan, R.N. Paunovic, and M.N. Tekic, *J. Membr. Sci.* **1998**, 138, 57.

[5.66]   R.P. Omorjan, R.N. Paunovic, and M.N. Tekic, *Chem. Engng. Proc.* **1999**, 38, 355.

[5.67]   Y. Lu, A. G. Dixon, W. R. Moser, and Y. H. Ma, *Catal. Today* **1997**, 35, 443.

[5.68]   J.S. Kim and R. Datta, , *AIChE J.* **1991**, 37, 11, 1657.

[5.69]   A.R. Garayhi, U. Flugge-Hamann, and F.J. Keil, *Chem. Engng. Technol.* **1998**,1, 21.

[5.70]   B. Park, V.S. Ravi-Kumar, and T.T. Tsotsis, *Ind. Eng.Chem. Res.* **1998**, 37, 1276.

[5.71]   V.S. Ravi-Kumar, T.T. Tsotsis, and M.Sahimi, *Ind. Eng. Chem. Res.* **1997**, 36, 3154.

[5.72]   L. Xu, M.G. Sedigh, M. Sahimi, and T.T. Tsotsis, *Phys. Rev. Lett.* **1998**, 80, 3511.

[5.73]   M.G. Sedigh, W.J. Onstot, L. Xu, W.L. Peng, T.T. Tsotsis, and M. Sahimi, *J. Phys. Chem. A* **1998**, 102, 8580.

[5.74]   G. Langhendries, G.V. Baron, and P.A. Jacobs, *Chem. Engng. Sci.* **1999**, 54, 1467.

[5.75]   A.A. Yawalkar, V.G. Pangarkar, and G.V. Baron, *J. Membr. Sci.* **2001**, 182, 129.

[5.76]   J.M. Sousa, P. Cruz, and A. Mendes, *Catal. Today* **2001**, 67, 281.

[5.77]   J.M. Sousa, P. Cruz, and A. Mendes, *J. Membr. Sci.* **2001**, 181, 241.

[5.78]   X. Tan and K. Li, *Chem. Engng. Sci.* **2000**, 55, 1213.

[5.79]   K. Li and X.Y. Tan, *AIChE J.* **2001**, 47, 427.

[5.80]   P.V. Shanbhag, A.K. Guha, and K.K. Sirkar, *J. Hazardous Mat.* **1995**, 41, 95.

[5.81]   P.V. Shanbhag, A.K. Guha, and K.K. Sirkar, *Ind. Eng. Chem. Res.* **1998**, 37, 4388.

[5.82]   V. Sinha, "Novel Membr. Reactor for Dissolved Oxygen Removal in Ultrapure Water Production", M. Eng. Thesis, National Univ. of Singapore, Singapore 2000.

[5.83]   L. Li, R.W. Borry, and E. Iglesia, *Chem. Engng. Sci* **2001**, 56, 1869.

[5.84]   A.M. Dean, *J. Phys. Chem.* **1990**, 94, 1432.

[5.85]   A. Heintz and W. Stephan, *J. Membr. Sci.* **1994,** 89,143.

[5.86]   A. Heintz and W. Stephan, *J. Membr. Sci.* **1994,** 89, 153.

[5.87]   G. Froment and K.B. Bischoff, *Chemical Reactor Analysis and Design*, Wiley, New York, 1990.

[5.88]   C. Venkateswarlu, M. Satyanarayana, and M. Narasinga, *Ind. Eng. Chem. Res.* **1958**, 50, 973.

[5.89]   J.T.F. Keurentjes, G.H.R. Janssett, and J.J. Gorissen, *Chem. Engng. Sci.* **1994**, 49, 4681.

[5.90]   Y. Zhu, R.G. Minct, and T.T. Tsotsis, *Chem. Engng. Sci.* **1996**, 51, 4103.

[5.91]   K. Okamoto, M. Yamamoto, Y. Otoshi, T. Semoto, M. Yano, X. Tanaka, and H. Kita, *J. Chem. Eng. Japan* **1993**, 26, 475.

[5.92]   R. Krupiczka and Z. Koszorz, *Sepn. Purif. Technol.* **1998**, 16, 55.

[5.93]   B. Park, Models and Experiments with Pervaporation Membrane Reactors Integrated with a Water Adsorbent System, Ph.D Thesis, University of Southern California, Los Angeles; USA 2001.

[5.94]   S.Y. Lim, B. Park, F. Hung, M. Sahimi, and T. T. Tsotsis, in press Chem. Engng. Sci., 2002.

[5.95]   J.M. Prausnitz, T.F. Anderson, E.A. Grens, C.A. Eckert, R. Hsieh, and J. O'Connell. *Computer Calculations for Multicomponent Vapor-Liquid and Liquid-Liquid Equilibria*, Prentice Hall, Englewood Cliffs, NJ, 1980.

[5.96]   G.J.S. van den Gulik, R.E.G. Janssen, J.G. Wijers, and J.T.F. Keurentjes, *Chem. Engng. Sci.* **2001**, 56, 371.

[5.97]   K. Kargupta, S. Datta, and S.K. Sanyal, *Biochem. Engng. J.* **1998**, 1, 31.

[5.98]   A.V. Skorokhodov, N.V. Men'shutina, and L.S. Gordeev, *Theor. Found. Chem. Engng.* **1998**,32, 48.

[5.99]   J.M. Engasser, „Reacteurs a Enzymes et Cellules Inmobilisées", in: *Biotechnologie*, Techniquc et Documentation Lavoisier, Paris, 1988.

[5.100] Z. Ujang and A. Hazri, *J. Membr. Sci.* **2000**, 175, 139.

[5.101] J.P. Tharakan and P.C.K. Chau, *Bioengng.* **1987**, 29, 657.

[5.102] L.G. Cima, H.W. Blanch, and C.R. Wilke, *Bioprocess Engng.* **1990**, 5, 19.

[5.103] H. Moueddeb, J. Sanchez, C. Bardot, and M. Fick, *J. Membr. Sci.* **1996**, 114, 59.

[5.104] P. Pavasant, L.M. Freitas dos Santos, E.N. Pistikopoulos, and A.G. Livingston, *Biotechnol. Bioengng.* **1996**, 52, 373.

[5.105] P. Pavasant, E.N. Pistikopoulos, and A.G. Livingston, *J. Membr. Sci.* **1997**, 130, 85.

[5.106] L.F. Strachan and A.G. Livignston, *J. Membr Sci.* **1997**, 128, 231.

[5.107] A.G. Livignston, J.P. Arcangeli, A. Boam, S. Zhang, M. Marangon, and L.M. Freitas dos Santos, *J. Membr. Sci.* **1998**, 151, 29.

[5.108] P.R. Salmon and C.R. Robertson, *Chem. Eng. J.* **1987**, 35, Bl.

[5.109] O. Miyawaki, K. Nakamura, and T. Yano, *J. Chem. Engng. Jpn.* **1982**, 15, 142.

[5.110] H. Ishizawa, T. Tanaka, S. Takase, and H. Hikita, *Biotechnol. Bioengng.* **1989**, 34, 357.

[5.111] T. Fuji, O. Miyawaki, and T. Yano, *Biotechnol. Bioengng.* **1991**, 38, 1166.

[5.112] P. Czermak, and W. Bauer, „Mass Transfer and Mathematical Modeling of Enzymatically Catalyzed Conversion in a Dialysis Membrane Reactor", in: *Engineering and Food: Ad-*

vanced Processes, Vol 3, W.E.L. Spiess and H. Shubert (Eds.), Elsevier Applied Science, New York, 1990.

[5.113]   G. Salzman, R. Tadmor, S. Guzy, S. Sideman, and N. Lotan, *Chem. Engng. Proc.* **1999**, 38, 289.

6 Economic and Technical Feasibility Issues of Membrane Reactor Processes

At various places throughout the first five chapters in the book we have, when it appeared relevant to the discussion, referenced studies which addressed issues pertaining to the economic/technical feasibility of membrane reactor processes. In this chapter we specifically focus our attention on these issues. In the discussion in this chapter we have, by necessity, drawn our information from published studies and reports. Several proprietary studies reportedly exist, carried out by a number of industrial companies, particularly during the last decade, which have evaluated the potential of membrane reactors for application in large-scale catalytic processes. By all accounts the conclusions reached in these proprietary reports mirror those found in the published literature. In the discussion which follows, we will first discuss catalytic and electrochemical reactors. We will then conclude with a discussion on membrane bioreactors.

6.1 Catalytic Membrane Reactors

In Chapter 2 we noted a number of small-scale membrane reactor-based processes for specialty chemicals production, which were reported to have found commercial application in the former Soviet Union. Small-scale Pd membrane reactors for methanol and methane reforming are reportedly currently undergoing pilot plant testing in Japan, for fuel cell applications, and as discussed in Chapter 2, two multi-year large-scale industrial efforts are currently ongoing aiming to commercialize synthesis gas production using solid oxide membrane reactors. These notable successes, and the great promise synthesis gas production efforts hold (Wilhelm *et al.* [6.1]), notwithstanding, the application of membrane reactors to large-scale catalytic reaction processes is still lacking. Insight on some of the challenges and barriers that still remain is offered in a number of recent published reports.

The earliest published such report, we are aware of, is by van Veen *et al.* [6.2], who have investigated the feasibility of membrane reactor (MR) technology using porous membranes. They evaluated three reactions, namely propane dehydrogenation, ethylbenzene dehydrogenation to styrene, and the water gas shift reaction (WGS). For propane dehydrogenation to propylene the authors compared the MR technology to two commercial catalytic dehydrogenation processes, namely the Catofin process (Lummus/Air Products) and the Oleflex process (UOP). (To put the whole discussion into a proper context, it should to be noted that, for the most part, propylene is, today, produced by steam cracker plants and refinery operations, and that catalytic technologies, in general, are not particularly attractive). The first process diagram they investigated is a modification of the adiabatic Oleflex process, which consists of four reactors, as shown in Figure 6.1. They considered four different process concepts: (1) A Knudsen membrane separator after the first, second, and third reactors; (2) a Knudsen membrane after the third reactor only; (3) three ideal membrane separators (i.e., permeable only to hydrogen) which remove all the hydrogen formed in the reaction after the first, second, and third reactors; (4) same as case (3) but with increased inlet temperature in the reactor.

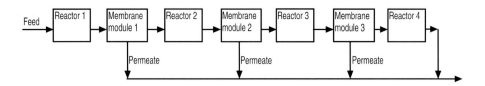

Figure 6.1. Process flow diagram including a membrane module after each reactor. From van Veen *et al.* [6.2], with permission from Elsevier Science.

Only cases (3) and (4) show any improvement in propylene yield over the base Oleflex process. The authors concluded, further, that moderate improvements in membrane perm-selectivity (say 10 for H_2 vs. C_3H_8) give marginal improvement in yield. Increasing the feed-side pressure, the amount of hydrogen in the feed, and using a sweep gas also had questionable value. Only lowering the pressure on the permeate side had some noticeable effect on the yield.

They also evaluated isothermal MR concepts and compared them in performance with the adiabatic Catofin and Oleflex processes. They studied two different type processes using Knudsen diffusion membranes: a process called CMRL, patterned after the commercial Oleflex process, with low propane conversion, and a process called CMRH, patterned after the commercial Catofin process with high propane conversion. They have calculated the return on investment (ROI) for all four processes. Though marginally better than the commercial processes, the ROI for all four processes evaluated is not very attractive. A sensitivity analysis indicates that for the ROI of the MR processes to be attractive a price difference between propane and propylene of more that \$300/ton is required. Though published calculations have only been performed for the propane/propylene pair, it is not unreasonable to assume that similar conclusions apply to other alkane/alkene pairs. Similar conclusions about catalytic alkane dehydrogenation have also been reached in a technical/economic evaluation study by Amoco workers and their academic collaborators (Ward *et al.* [6.3]).

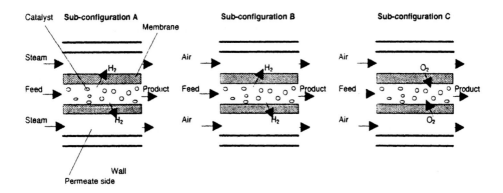

Figure 6.2. Membrane reactor sub-configurations. From van Veen *et al.* [6.2], with permission from Elsevier Science.

Van Veen *et al.* [6.2] also analyzed the application of membrane reactors to the reaction of ethylbenzene dehydrogenation to styrene. They studied (see Figure 6.2) various PBMR configurations utilizing a ceramic membrane. Three configurations were examined. In the first configuration A, hydrogen permeates through the membrane and is carried away with steam. In configuration B, hydrogen is swept away and burned with air to provide heat for the endothermic reaction (a concept proposed earlier by Itoh [6.4, 6.5]). This results in higher outlet temperatures, which result in higher conversions. The third configuration they examined uses oxygen instead of hydrogen permeable membranes. Air is used as the source of oxygen and permeates through the membrane to carry out the oxidative dehydrogenation of ethylbenzene. From the preliminary evaluations, configurations B and C did not prove promising, because of the loss of hydrogen, and potentially of styrene and ethylbenzene. Only configuration A was subsequently studied in greater detail. The conclusion was that the membrane reactors do not provide significant increases in the yield (primarily due to slow reaction kinetics) to provide enough gain in the ROI to justify the significant investment required to put together even a retrofit plant. For the typical plant size considered by van Veen *et al.* [6.2], the cost of the membranes alone was estimated to be ~$250 million. The economic evaluation for the water gas shift reaction was reported by van Veen *et al.* [6.2] to be the most promising of all applications. A number of other groups have also evaluated the same reaction and their results are discussed next.

The economic/technical feasibility of combining the WGS reaction and the hydrogen separation step in a single unit has been investigated by a consortium of European academic and industrial entities [6.6]. This group investigated the concept of the WGS membrane reactor (WGS-MR) for CO_2 removal in the IGCC power generation system by using microporous silica membranes. In order to establish full insight into the possibilities of the application of such a reactor, a multidisciplinary feasibility study was carried out comprising system integration studies, catalyst research, membrane research, membrane reactor modelling, and bench-scale membrane reactor experiments [6.7]. The study concluded that the application of the WGS-MR concept in IGCC systems is an attractive future option for CO_2 removal as compared to conventional options. The net efficiency of the IGCC process with integrated WGS-MR is 42.8 % (LHV) with CO_2 recovery (80 % based on coal input). This figure has to be compared with 46.7 % (LHV) of an IGCC without CO_2 recovery and based on the same components, and with 40.5 % (LHV) of an IGCC with conventional CO_2 removal. Moreover, an economic analysis indicates favorable investment and operational costs. The study concluded that development of the process is considered to be, in principle, technically feasible. However, the authors report that the currently available high temperature selective gas separation membranes (noble metal and SiO_2) are not capable of withstanding the harsh environments found in the WGS reaction step and, therefore, further development in this area remains essential.

More recently Criscuoli *et al.* [6.8] published a detailed economic feasibility study of the application of Pd membrane reactors to the WGS reaction. Conventionally the WGS reaction is carried out with two reactors in series, a high temperature shift (HTS) reactor followed by a low temperature shift (LTS) unit. The outlet from the LTS unit is treated in a separator to recover the hydrogen product. The authors considered as a base of comparison a conventional system, in which the separator recovers 85% of hydrogen as a high purity hydrogen stream (99.9999). For the WGS MR the calculations were carried out for two H_2O/CO molar ratios. A conventional industrial value of 9.8 (configuration MR1), and a second lower H_2O/CO molar ratio of 2 (configuration MR2) in order to account for the fact that the MR allows one to operate with a lower H_2O/CO molar ratio because the equilibrium is shifted by the selective hydrogen removal. Two other configurations were also investigated, where a palladium membrane separator is used first to separate the hydrogen from the reformate mixture (exiting the reformer) before it enters the Pd WGS membrane reactor for further processing. The authors in their calculations utilized a simple mathematical isothermal Pd membrane reactor model previously developed by them [6.9]. In Table 6.1 the results of the economic analysis are reported for a base case, for which the membrane thickness is taken to be 75μm and the hydrogen permeability through the membrane is taken to be $1.63 \times 10^{-8} mol \cdot m/m^2 \cdot s \cdot Pa^{0.5}$ at 597 K. As can be observed in Table 6.1 the Sep+MR2 configuration (which includes the membrane separator followed by the WGS membrane reactor with the low H_2O/CO molar ratio) allows the desired hydrogen recovery with the lowest capital and operating costs. However, all the MR-based systems are still more expensive than the conventional production systems. The difference in costs is the result of the high cost of the palladium membranes. Criscuoli *et al.* [6.8] also studied the effect of decreasing the membrane thickness (from 75μm down to 5μm), and increasing the membrane permeability by a factor of ten. The palladium membrane costs and the capital and operating costs decreased, but they were still higher when compared to the costs of the conventional production unit.

Table 6.1. Economic analysis of the WGS-MR process (US$). Adapted from Criscuoli *et al.* [6.8].

Configuration and parameters	Industrial plant (conv.: 98%)	MR1 (conv.: 99.3%)	MR2 (conv.: 99.94%)	Sep + MR1 (conv.: 99.25%)	Sep + MR2 (conv.: 99.4%)
Reactors	69,696	5,760,0	2,205,000	2,385,000	608,859
Catalyst	44,100	14,186	5,418	5,869	1,496
Compressor	-	325,245	325,245	325,245	325,245
Separation units	927,000	-	-	1,395,000	1,395,000
Capital costs	1,044,000	6,102,000	2,538,000	4,113,000	2,331,000
Depreciation (10% per year)	104,400	610,200	253,800	411,300	233,100
Capital charges (5% per year)	52,200	305100	126,900	205,650	116,550
Taxes (3% per year)	31,320	183,060	76,140	123,390	69,930
Insurance (3% per year)	31,320	183,060	76,140	123,390	69,930
Maintenance (3% per year)	31,320	183,060	76,140	123,390	69,930
ROI (20% per year)	208,800	1,220,400	507,600	822,600	466,200
Raw materials (per year)	1,170,000	1,170,000	1,170,000	1,170,000	1,170,000
Steam (per year)	325,245	325,245	66,908	325,245	66,908
Labor (20.7 US$/h)	183,066	183,066	183,066	183,066	183,066
Supervision 25% of labor (per year)	45,767	45,767	45,767	45,767	45,767
Membrane replacement (per year)	-	1,800,000	706,243	1,206,000	63,655
Electricity 0.026US$/Kwh (per year)	-	68,301	68,301	68,301	68,301
Operating costs (MUS$/per year)	2,187	6,273	3,348	5,076	3,195

The reason for that was that the effect of lowering the membrane thickness and permeability quickly saturated out, because the operation of the MR systems became limited by the catalyst activity.

The application of membrane reactors to methane reforming has also been evaluated in two recent studies. A technical and economic evaluation of the use of dense Pd-membrane in methane steam reforming has been presented by Aasberg-Petersen *et al.* [6.10]. They assumed a thin (2μm thick) Pd membrane, which exhibited perfect separation and, as a result, the pure hydrogen product was taken from the permeate side of the membrane. No sweep gas was used on the permeate side of the reactor. This necessitated compression of the low pressure hydrogen product. The authors concluded that membrane-based reforming using a dense film membrane became attractive only in the cases where electrical costs were low.

Onstot *et al.* [6.11] presented an evaluation of various plant designs incorporating high temperature mesoporous and microporous membranes for the dry-reforming reaction of methane for hydrogen production. In their paper the emphasis is not on the production of pure hydrogen, but on a hydrogen product of a certain purity to be used for power generation in either mobile or stationary applications. The focus in the study is on small to medium scale applications involving the reforming of landfill gas or other types of biogas (assumed to consist of equimolar amounts of CH_4 and CO_2, a typical composition for such a gas) to produce fuel hydrogen. They evaluated various design options in terms of their energy operating costs. Seven different plant designs were presented: a conventional hydrogen plant for the production of hydrogen; a conventional hydrogen plant equipped with a microporous carbon membrane-based CO_2 removal unit; two designs utilizing a high temperature, mesoporous ceramic membrane reactor in place of a conventional plug flow reformer; and a unique hydrogen plant design, which utilizes a yet-to-be developed high temperature microporous membrane. In addition, two cases in which excess high pressure hydrogen is made and utilized in a fuel cell were also presented.

Energy usage and operating economics were presented for all designs. The basis for all plants is the production of ten million standard cubic feet of hydrogen product per day (10 MMSCFD) of 97 % purity to be supplied at 38 °C and 7 bar, a product suitable for use in a high temperature fuel cell such as a molten carbonate (MCFC) or solid oxide fuel cell (SOFC). The membrane separation characteristics are shown to be critical, not only in terms of improving the reactor yield but also of impacting product purity and downstream purification process requirements. The energy requirements for all different designs are summarized in Table 6.2. The cost requirements for the various designs are shown in Table 6.3. Comparing the various designs one concludes that the lowest cost product is produced by Case 1a, the conventional hydrogen plant with the CO_2 Removal Module, while the highest cost product is produced by the Case 2 design, utilizing a membrane reactor equipped with a mesoporous membrane as the primary reformer. The values in these tables and the conclusion which process design is most efficient are strongly dependent on the assumption for the prices of the various commodities involved (the values for the base case of Tables 6.2 and 6.3 are typical for the design of a grassroots plant), which may vary significantly among various locations, and for more integrated facilities. A $0.01 per

KWH increase in the price of electricity, for example, greatly favors membrane reformer plants, while a comparable increase in the cost of oxygen makes all of the fuel cell designs non-competitive. It should be noted that capital costs have not been included in this analysis, nor has operation of the plants at space velocities significantly different than typically found in the industry been considered. As a result, plant designs for small amounts of hydrogen, such as for homes equipped with fuel cells to produce electricity for electrical cars, may operate at much lower space times and hence favor the ceramic membrane equipped plants. The Case 4 (Autoreformer) design never produced the lowest cost hydrogen product in this analysis, but its energy operating cost was always less than a conventional hydrogen plant, and has the prospect of large capital savings and ease of operation. What may be said without hesitation is that the carbon membrane-based CO_2 removal unit plants provided significant cost savings over conventional amine-based units.

Table 6.2. First law energy analysis of proposed hydrogen plant designs (adapted from [6.11]).

	Case 1 Conv. H_2 plant	Case 1a Conv. H_2 plant w/carbon membrane	Case 2 Reformer w/Al$_2$O$_3$ membrane	Case 2 Reformer w/Al$_2$O$_3$ and carbon membrane	Case 3 Conv. H_2 plant w/Fuel cell	Case 3 Conv. H_2 plant w/Fuel cell and carbon membrane	Case 4 Auto-reformer design
Feed	106.00	110.80	225.50	237.90	208.00	217.10	128.8
Fuel	82.40	86.50	155.80	164.40	161.50	169.50	78.07
Steam							
40 barg	-20.74	-22.12	-22.34	-23.57	-41.31	-43.36	-
14 barg	-9.47	-10.49	-15.01	-15.86	-18.58	-20.55	-
3.5 barg	72.21	9.50	140.97	5.01	141.61	18.61	-
Electricity	-	-	-114.05	-119.81	-114.05	-114.05	-
Energy usage, GJ/h	230.40	174.19	370.87	248.07	377.17	227.25	206.87
MJ/kgmol H_2 product	464.68	349.75	744.65	499.37	677.01	456.45	415.44

Negative numbers denote energy credits

Table 6.3. Estimated operating costs for various hydrogen plants (adapted from [6.11]).

	Case 1 Conv. H_2 plant	Case 1a Conv. H_2 plant w/carbon membrane	Case 2 Reformer w/Al_2O_3 membrane	Case 2 Reformer w/Al_2O_3 and car- bon membrane	Case 3 Conv. H_2 plant w/Fuel cell	Case 3 Conv. H_2 plant w/Fuel cell and carbon membrane	Case 4 Auto-reformer design
Feed	212.00	221.60	451.00	475.80	416.00	434.2	257.60
Fuel	164.80	173.00	311.60	328.80	323.00	339.00	156.14
Steam							
40 barg	-68.44	-73.00	-73.72	-77.78	-136.32	-143.09	-
14 barg	-28.41	-31.47	-45.03	-47.58	-55.74	-61.65	-
3.5 barg	180.53	23.75	352.43	12.53	354.03	46.53	-
Electricity	0	0	-2,004.98	-2,106.24	-2,004.98	-2,004.98	-
Oxygen	0	0	1,690.60	1,783.60	1,690.6	1,690.60	-
Operating cost US$/h	460.47	313.88	681.89	369.12	586.58	300.61	413.74
US$/kgmol H_2 product	0.88	0.59	1.47	0.91	1.30	0.75	0.79

Struis and Stucki [6.12] have evaluated the application of membrane reactors for methanol synthesis using methanol permselective Nafion® membranes. In their design calculations they utilize kinetic and membrane permeation data measured in their laboratory. They estimate that with 10 μm thin membrane under methanol synthesis plant technically relevant conditions (T = 200 °C, P = 40 bar, GHSV = 5000 h^{-1}), the single pass reactor yield improves by 40 %, and that the additional costs for the membrane materials correspond only to two production months. The ability of the Nafion® membranes to withstand such conditions for prolonged periods still remains, however, questionable.

Stoukides [6.13] has discussed the technical and economic constrains hampering the industrial application of electrochemical membrane reactors. Citing prior studies he notes that of all EMR those involving chemicals production with simultaneous generation of electricity stand the best chance of finding eventual commercial application. He and his coworkers have analyzed the electrochemical conversion of methane to syngas or C_2 hydrocarbons. The main findings are summarized in Table 6.4. Case 2 represents a cogenerative EMR producing electricity and synthesis gas, while Case 3 represents a cogenerative EMR producing electricity and ethylene. For comparison purposes Case 1 is a regular solid oxide fuel cell producing electricity, for which the products are CO_2 and H_2O. Case 4 is a hypothetical catalytic methane coupling plant with the same capacity for producing ethylene as Case 3. For Cases 1, 2, and 3 100 MW of electricity was produced. For Case 2, 3×10^8 kg of syngas (CO/H_2 ratio equal to 2.4/1) was produced annually, while for Cases 3 and 4 there was a production of 0.96×10^8 kg of ethylene annually. Note that the numbers on return on investment before taxes (ROIBT) for Cases 3 and 4 are very comparable. This is due to the much higher capital investment required for the EMR based plant. The conventional fuel cell shows the best ROIBT (though even for this case the numbers are marginal). Stoukides [6.13] appears to be rather pessimistic about the long-term prospects for commercialization of EMR. Some of the technical barriers/concerns he cites include high capital investment costs, concerns about long-term performance of electrodes and membranes, the need for high temperatures for the solid oxide EMR, and the generally fairly inexpensive products these EMR produce. Similar concerns probably apply to the more conventional high temperature catalytic MR.

Table 6.4. Economic summary of methane co-generative fuel cells (adapted from [6.13])

	Cost (in $ millions)			
Items	Case 1	Case 2	Case 3	Case 4
Fixed investment	162.1	180.9	255.4	114.6
Working capital	24.3	27.1	38.3	17.2
Total utilized investment (TUI)	186.5	208.1	293.7	131.8
Total production costs	56.3	87.3	91.3	59.8
Total credit	112.4	124.2	153.2	88.1
Income before taxes (IBT)[a]	56.1	38.9	62.0	28.3
Return on investment				
Before taxes (ROIBT)[b]	30.1%	17.7%	21.1%	21.5%

[a]IBT: Total credit – (Total production cost).
[b]ROIBT: IBT/TUI
Note: Case 1: regular methane fuel cell (produces CO_2 and H_2O); Case 2: co-generation of electricity and synthesis gas; Case 3: co-generation of electricity and ethylene; Case 4: production of ethylene only (regular catalytic reactor).

6.2 Membrane Bioreactors

As discussed in Chapter 4 there are already many industrial processes using MBR technology. There is a number of reasons why this is the case. MBR utilized either for the production of biochemicals or for environmental applications operate at low temperatures, where inexpensive commercial, organic, or polymeric membranes can be utilized. The MF or UF units, frequently coupled with the bioreactors in such MBR systems, are already proven and profitable technologies in their own right, and well accepted by the industry. MBR processes utilized in the production of fine chemicals or biochemicals, produce, in general, high value-added products. As noted above in the discussion on EMR, this is a key consideration for the success of membrane reactor based systems, which generally turn out to require a higher capital investment due to the membrane expense. It comes as no surprise that for such membrane bioreactors the economics are generally reported to be more favorable. Nevertheless, even for these applications a strong incentive still remains for lowering membrane costs. Tejayadi and Cheryan [6.14] have reported, for example, that for MBR involved in the production of various chemicals and biochemicals the membrane costs in 1995 still represented, on the average, about 28 % of the total capital investment, and 5 % of the operating costs.

Often the coupling of the membrane unit with the bioreactor results in significant synergy as in the study of O'Brien *et al.* [6.15] on the application of PVMBR to ethanol production, which we discussed in Chapter 3. The required bioreactor volume for the PVMBR system was smaller than that of the conventional system by a factor of 12. Nevertheless, it turns out that the PVMBR-based process is still 25 % more expensive than the classical batch fermentation process in terms of capital costs despite the substantial reduction in the required reactor volume. This cost differential is not only due to the membrane costs, which are, themselves, substantial, but also due to the cost of the additional hardware associated with membrane operation. The application of MBR for the ethanol production by fermentation faces marginal economics, since ethanol is a relatively cheap commodity chemical.

When producing high value biochemicals, the economic analysis is, generally, more favorable for the MBR systems, when compared with the more conventional units. Often selectivity rather than conversion may be the key, since higher selectivities typically result in the elimination of some of the purification steps which have a strong influence on the total production costs. The application of the multiphase/extractive enzyme membrane reactor for the production of a diltiazem chiral intermediate was reported in Chapter 4 [6.16]. The development of the industrial production of (2*R*,3*S*)-*trans*-methyl methoxy-phenylglycidate underwent through many steps, which included bench scale process reliability studies, pilot plant and scale-up investigations. From the cost analysis studies it was concluded that process economics were more sensitive to the process enantioselectivity attained rather than to productivity, since the largest operating cost is that of the racemic feed ester, and product yield is strongly dependent on enantioselectivity. To attain the most favorable process economics, high enantioselective process conditions were chosen (>99 % of enantioselective excess). As noted in Chapter 4, the process is currently commercial. A full-scale commercial facility consisting of twelve production scale membrane modules with a total membrane area of 1440 m^2 was reported operating in Japan [6.16]. The commercial production of this molecule using the MBR process attained a capacity of 75 metric tons per year in 1997 [6.16], representing more than a half of the diltiazem being sold in the USA.

The economic analysis of MBR processes is often a complex task, because it depends on many variables like the extent of reaction, number and configuration of the separation steps, etc. An interesting study was reported by Weuster-Botz *et al.* [6.17] for the synthesis of L-isoleucine. Isoleucine is an amino acid, which is used in the preparation of infusion solutions in parenteral nutrition therapies; its annual production is about 200 tons per year. Weuster-Botz *et al.* [6.17] carried out an economic analysis for the production of this molecule from glucose and D,L-α-hydroxybutyric acid by a *Corynebacterium glutamicum*. The main technical challenge with isoleucine bio-production is the purification step, because of L-valine, which is obtained as by-product, and which is very difficult to separate. This situation results in a very high down-stream processing cost, estimated to be between 15 % to 70 % of the total processing costs. Weuster-Botz *et al.* [6.17] in their analysis compared two different MBR configurations and processes for the production of 70 tons of L-isoleucine per year with respect to their downstream isoleucine separation and purification steps. The first MBR system coupled three parallel stirred-tank bioreactors with UF and RO membrane units, together with a crystallization step, carbon adsorption, electrodialysis, and a second crystallization step. The second MBR configuration was a simpler system coupling the three bioreactors with a UF membrane module, together with a crystallization step, carbon adsorption, and electrodialysis steps. The authors concluded that the efficiency and costs depend strongly on the extent of the reaction advancement. When isoleucine broth concentrations are higher than 20% the second and simpler MBR configuration shows more favorable economics. A reduction in processing costs from $ 166 to $ 76 per kg of L-isoleucine were obtained with this MBR configuration.

As noted in Chapter 4, other interesting biological molecules like monoclonal antibodies, which have production costs between $ 2 to $ 35/mg are also currently being produced by MBR technology. They are now commercial hollow fiber MBR processes available for the production of monoclonal antibodies, both, in the UK and in the USA. Many of the MBR approaches that have been used have significantly decreased the production costs for this type of chemicals. Shi *et al.* [6.18] reported that the use of hollow-fiber MBR operating with a conditioned medium containing cell growth factors circulating inside the fibers for cell nutrition purposes, not only facilitates and increases the production of monoclonal antibodies from hybridomas, but also results in production costs savings varying from 31 to 52.2 %, depending on the cell line producing the antibodies.

The profitability of a MBR process depends greatly on the membrane module configuration (flat, spiral wound, hollow-fibers, etc.) and on permeate flux and stability. In a related study Leach *et al.* [6.19] analyzed the costs of different methods commonly applied for the clarification of fermentation broths obtained during the production of antibiotics. They compared a centrifugation process, two conventional membrane modules, and a dynamic membrane system, which creates shear stress rates on the membrane surface in order to increase permeate fluxes (PallSep, see Chapter 4). The cost analysis was carried out on the basis of a concentration factor of 4 for the fermentation broth, and a total membrane surface area of 400 m^2. The results shown in Table 6.5 indicate that though the membrane separation processes may require similar capital costs, their corresponding energy consumption rates, which relate to the transmembrane pressure gradients and crossflow velocities needed in order to maintain a good permeate flow rate, are very different.

Table 6.5. Comparative costs (US$) analysis for clarification methods of a fermentation broth. Adapted from Leach *et al.* [6.19].

Process and parameters	PallSep (organic, sheets)	Flat (organic)	Ceramic	Centrifuge + rotary vaccuum filtration
Capital	>500,000	50,000–500,000	50,000–500,000	<10,000
Membrane (per year)	50,000–500,000	50,000–500,000	50,000–500,000	<10,000
Pre-treatment	<10,000	<10,000	<10,000	50,000–500,000
Energy	<10,000	50,000–500,000	50,000–500,000	<10,000
Environmental (disposal)	<10,000	<10,000	<10,000	50,000–500,000

As noted in Chapter 4, MBR wastewater treatment processes have already proven their commercial worth with many units already installed around the world. Stephenson *et al.* [6.20] carried out a comprehensive economic analysis of several MBR wastewater treatment processes. Their studies confirmed the finding that, as with the conventional MBR (and also with the PVMBR) systems, the key component of cost of the treatment unit are the membrane costs. In fact, over the last few years, both, the initial capital investment for the purchase of the membrane, and the costs of membrane replacement as a fraction of the overall costs are decreasing. These trends have been observed, for example, with many of the aerobic MBR industrial processes, like the Kubota process, which has cut the membrane costs by a factor of 4 since 1992. In addition, their operating costs related with membrane replacement, have decreased from ca. 54 % to ca. 9 % of total operating costs. Among the MBR systems Stephenson *et al.* [6.20] investigated and provided an economic analysis for is a hollow fiber MBR for the treatment of chicken processing waste. The capital expenditure of the plant based on a design flow of 1800 m^3 d^{-1} was reported to be $ 1.69 million, with operating costs of $ 0.22 per m^3 of treated waste (accounting for a payback period of three years). A plant for the daily treatment of 1000 m^3 municipal wastewater by a MBR with a submerged plate membrane system was reported as having a cost of around $ 1.2–1.7 million.

As it is true with the majority of industrial processes, MBR operating process costs strongly relate to the economy of scale. Davies *et al.* [6.21] have reported in an interesting study a comparison between a conventional activated sludge system (AS) and a MBR-based system for two different wastewater treatment capacities, namely 2,350 and 37,500 population equivalents (pe). Their analysis took into account many cost items, including labor, energy and equipment costs. Table 6.6 shows some of the main costs for the two different capacity MBR units (MBR1 and MBR2), and the two different capacity conventional activated sludge systems (AS1 and AS2). From their analysis one can conclude that at low capacity the MBR presents an advantage with respect to the AS system; however, that is not true for the higher capacity systems.

Table 6.6. Comparative cost (US$) analysis for two MBR and two activated sludge wastewater plants. Adapted from Davies *et al.* [6.21].

Process and Parameters	MBR1	MBR 2	AS1	AS2
Capacity (pe)	2,350	37,500	2,350	37,500
Max throughput (Ml.d^{-1})	1.4	22.5	1.4	22.5
Membrane modules	578,313	7,469,880	-	-
Sedimentation tanks	-	-	625,301	2,011,520
Total capital costs	738,554	8,786,174	1,180,969	4,749,710
Membrane replacement	36,145	555,004	-	-
Energy	12,048	128,816	15,663	178,398
Annual running costs	90,811	725,423	67,711	318,952
Total costs (US$/m^3)	13.9	7.5	15.9	5.9

Total capital costs for the MBR1 were reported to $ 738,554, 78 % of which was attributable to the membrane system. The costs for construction and the additional cost of the sedimentation tank for the activated sludge system offset the costs of the membranes, resulting in the AS1 system having a capital cost which is 160 % of the capital costs of the MBR1 plant. Nevertheless, at the larger scale the difference in the economy of scale between the total capital costs for construction and the membrane units resulted in the conventional system having a capital cost of approximately $ 4.7 million, which is only 54 % of the corresponding capital for the MBR plant.

Gander *et al.* [6.22] published recently a review of aerobic MBR, which also included cost considerations. These authors compared the performance of the two main configurations (side stream and submerged) of aerobic MBR, taking into consideration a number of cost factors, among which energy consumption is the most influential parameter. The energy costs relate with feed water pumping, recycling of the retentate, permeate suction, and aeration. Some of the results reported of this study are presented in Table 6.7.

Table 6.7. Comparative energy consumption of different aerobic MBR configurations (kWh·m^{-3} of product). Adapted from Gander *et al.* [6.22].

Process and parameters	1 Sub	2 Sub	3 Sub	4 SS	5 SS	6 SS
Membrane type	P&F POL	P&F POL	HF POL	T C	HF C	HF POL
Permeate flux (l m^{-2} h^{-1})	7.9	20.8	8	175	77	8.3
Energy consumption permeate	-	0.0013	0.0055	9.9	32	0.045
Energy consumption aeration	4.0	0.0091	0.14	2.8	9.1	10
Total energy consumption	4.0	0.022	0.14	13	41	10

Sub: submerged, SS: side stream, P&F: plate and frame, HF: hollow fibers, T: tubular, POL: polymer, C: ceramic.

The energy consumption for the side stream processes is much higher than that of the submerged configurations. A key part in this difference is due to the costs for aeration in the submerged units. The difference in energy costs between the two systems is also because of the fact that the permeate flow rates of the submerged processes are generally lower. In the side stream systems, it is usually pumping of the retentate recycle stream that accounts for the greatest energy expenses, contributing up to 60 % to 80 % of the total costs.

A cost analysis and comparison between two anaerobic processes for the treatment of wastewater was also carried out by Pillay *et al.* [6.23]. The first configuration consisted of a conventional system coupling a digester with a sedimentation unit. In the second configuration the sedimentation step was replaced by a membrane separation unit treating a side stream. Pillay *et al.* [6.23] report that for a 60 Ml d^{-1} plant the MBR process results in capital savings of about 27 % when compared with the classical configuration.

A cost analysis of an extractive membrane bioreactor (EMB) for wastewater treatment has been reported by Freitas dos Santos and Lo Biundo [6.24]. The EMB studied was similar with those reported in Chapter 4. Calculations were carried out for a feed flowrate of 1 m^{3}h^{-1} of wastewater polluted with dichloromethane at a concentration of 1 g l^{-1}. A minimum pollutant removal rate of 99 % and 8000 h of operation per year were considered. As expected, the analysis indicated that the costs are strongly dependent on the pollutant flow entering the bioreactor to be transformed. Two key parameters, namely the total membrane area required and the external mass transfer coefficient, were studied. The results show that the costs and membrane area decrease significantly as the mass transfer coefficient increases from 0.5×10^{-5} to 2.0×10^{-5} m·s^{-1} (these values are typical for large units, while laboratory measured values harbor around 5×10^{-7} m·s^{-1} [6.24]). Using a mass transfer coefficient of 1.0×10^{-5} m·s^{-1} the authors calculated the costs and the membrane area required for different wastewater flowrates. These results are shown in Fig. 6.3.

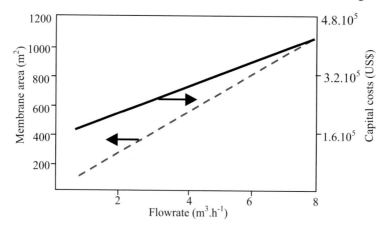

Figure 6.3. Capital costs for the EMB technology as function of the feed flowrate. Adapted from reference [6.24].

The EMB technology was also compared with other competing technologies. Data for steam stripping, activated carbon adsorption, and air stripping were obtained for year 1998 [6.24], whereas the pervaporation data used are from a 1990 study by Wijmans *et al.* [6.25]. The results are shown in Table 6.8.

Table 6.8. Comparative costs of different technologies for organics elimination in wastewater treatment (US$). Adapted from Freitas dos Santos and Lo Biundo [6.24].

Technology	Capital	Operating	Annualized costs (US$/m^3)
Carbon adsorption	480,000	64,000 (with regeneration) 128000 (without regeneration)	20-28
Steam stripping	240,000	40,000	10.8
Air-stripping + bioscrubbing	200,000-240,000	24,000	9
EMB	190,000	24,000	7.7
Pervaporation	180,000	50,000	2.5

As can be observed the costs for the EMB technology seem to be competitive with steam stripping, air-stripping, and pervaporation. Carbon adsorption costs are the highest, and this seems to be a less competitive technology. However, one must note that the organic pollutant (e.g., dichloromethane) studied does not adsorb very well onto activated carbon, and for a 99 % removal rate large quantities of the adsorbent are needed. The pervaporation costs seem to be the lowest. It is unclear, however, how robust pervaporation membranes are in treating wastewaters, which may contain other objectionable constituents in addition to the targeted pollutant. Both, the EMB and the pervaporation process possess the advantage of being modular processes. With the exception of EMB, all the other technologies produce a concentrated organic waste stream that has to be treated and disposed of. This step introduces additional problems of disposal and increases the costs for post-treatment. EMB technology, in contrast, produces a small aqueous waste by-product with a very low concentration of organics that can be easily disposed of.

Adham *et al.* [6.26] have recently investigated the feasibility of membrane bioreactor processes for water reclamation. Process evaluation was based on a survey of MBR cost estimates of the four major companies currently marketing MBR's, namely Zenon Municipal Systems (Canada), Mitsubishi Rayon Corporation (Japan), Suez-Lyonnaise-des-Eaux/Infilco-Degemont (USA), and Kubota Corporation (Japan). Based on the survey and a literature review of existing operations, Adham *et al.* [6.26] conclude that the MBR process offers several benefits over the conventional activated sludge process, including: smaller space and reactor requirements, better effluent water quality, disinfection, increased volumetric loading, and less sludge production. They present a cost evaluation of the Zenon MBR process (based on full-scale costs provided by the manufacturer) with two conventional wastewater treatment processes, namely oxidation ditch and conventional activated sludge, each serving as a pre-treatment to RO treatment. The design capacity selected was 1 million gallons/day, which is considered to be the most viable for use of MBR in the municipal services water reclamation market. The costs of the MBR process both in terms of capital costs but also total cost ($/1000 per gallon) are lower than those of the two conventional wastewater treatment processes. Adham *et al.* [6.26] conclude that MBR processes are a viable alternative, and that there is a significant market potential for the technology for many projects, pre-treatment to RO for water reclamation being one such example.

In conclusion the published studies, so far, indicate that catalytic and electrochemical membrane reactors still face an uphill battle towards commercialization, with key concerns being membrane cost and reliability. Membrane bioreactors are already commercial, but membrane costs and reliability are still an area where improvements are needed.

6.3   References

[6.1]   D.J. Wilhelm, D.R. Simbeck, A.D. Karp, and R.L. Dickenson, *Fuel Process. Technol.* **2001**, 71, 139.

[6.2]   H.M. Van Veen, M. Bracht, E. Harmoen, and P.T. Alderliesten, "Feasibility of the Application of Porous Inorganic Gas Separation Membranes in Some Large-Scale Chemical Processes", in: *Fundamentals of Inorganic Membrane Science and Technology*, A.J. Burggraaf and L. Cot (Eds), Elsevier Science, 1966.

[6.3]   T.L. Ward, G.P. Hagen, and C.A. Udovich, in: *Proc. 3rd ICIM*, Y.H. Ma, Ed.,Worcester, MA, USA, 335, 1994.

[6.4]   N. Itoh, *J. Chem. Eng. Japan* **1990**, 23, 81.

[6.5]   N. Itoh and R. Govind, *Ind. Eng. Chem. Res.* **1989**, 28, 1557.

[6.6]   M. Bracht, P.T. Alderliesten, R. Kloster, R. Pruschkek, G. Haupt, E. Xue, J.R. Ross, M.K. Koukou, and N. Papayannakos, *Energy Conversion and Management* **1997**, 38, S159.

[6.7]   E. Xue, M.O. Keefe, and J.R.H. Ross, *Catal. Today* **1996**, 30, 107.

[6.8]   A. Criscuoli, A. Basile, E. Drioli, and O. Loiacono, *J. Membr. Sci.* **2001**, 181, 21.

[6.9]   A. Criscuoli, A. Basile, E. Drioli, and O. Loiacono, *Catal. Today* **2000**, 56, 53.

[6.10] K. Aasberg-Petersen, C. Nielsen, and S. Jorgensen, *Catal. Today* **1998**, *46*, 193.

[6.11] W.J. Onstot, R.G.Minet, and T.T. Tsotsis, *Ind. Eng. Chem. Res.* **2001**, 40, 242.

[6.12] R.P.W.J. Struis and S. Stucki, *Appl. Catal. A-Gen.* **2001**, 216, 117.

[6.13] M. Stoukides, *Catal. Rev.- Sci.Eng.* **2000**, 42, 1.

[6.14] S. Tejayadi and M. Cheyran, *Appl. Microbiol. Biotechnol.* **1995**, 43 242.

[6.15] D.J. O'Brien, L.H. Roth, and A.J. McAloon, *J. Membrane Sci.* **2000**, 166, 105.

[6.16] J. Lopez and S.L. Matson, *J. Membr. Sci.* **1997**, 189, 211.

[6.17] D. Weuster-Botz, M. Karutz, B. Joksch, D. Schartges, and C. Wandrey, *Appl. Microbiol. Biotechnol.* **1996**, 46, 209.

[6.18] Y. Shi, J. Ploof, and A.Correia, *IVD Technol. Mag.* **1999**, May.

[6.19] G. Leach, I. Sellick, M. Collins, D. Caire, and C. Felisaz, *Proc. 2nd European Congress on Chemical Engineering*, Montpellier, France, 1999.

[6.20] T. Stephenson, S. Judd, B. Jefferson, and K. Brindle, *Membrane bioreactors for wastewater treatment*, IWA Publishing, London, 2000.

[6.21] W.J. Davies, M.S. Lee, and C.R. Heath, *Wat. Sci.Technol.* **1998**, 38, 421.

[6.22]   M. Gander, B. Jefferson, and S. Judd, *Sep. Purf. Technol.* **2000**, 18, 119.

[6.23]   V.L. Pillay, B. Townsend, and C.A. Buckley, *Wat. Sci.Technol.* **1994**, 30, 329.

[6.24]   L.M. Freitas dos Santos and G. Lo Biundo, *Environ. Progr.* **1999**, 18, 34.

[6.25]   J.G. Wijmans, J. Kaschemakat, J.E. Davison, and R.W. Baker, *Environ. Progr.* **1990**, 9, 262.

[6.26]   S. Adham, P. Gagliardo, L. Boulos, J. Oppenheimer, and R. Trussell, *Water Sci. Technol.* **2001**, 43(10), 203.

7   Conclusions

As the discussion, so far, has indicated membrane reactors have become a promising process option for a broad spectrum of applications, ranging from catalytic reactions like hydrocarbon dehydrogenation, to hormone synthesis, to the biological transformation of waste waters. Membrane reactor-based processes have already proven their economic and strategic worth to low temperature applications, like the bio-production of fine, high value-added chemicals, pervaporation, or wastewater treatment. This relates to the fact that these types of processes utilize polymeric and macroporous or mesoporous inorganic membranes, which are mature and readily available materials, and also have less severe requirements in terms of membrane housing and sealing. Good strides have also been made in recent years in a number of large-scale, high temperature applications, a good example being synthesis gas or hydrogen production, through partial oxidation or steam reforming of light hydrocarbons. Despite the progress realized until now in this area, further advances must await the development of more stable and affordable membranes. The expected benefits of such advances are, however, enormous given the size of the potential large-scale applications in the oil and petrochemical industry. In order for membrane reactor processes to make inroads into other application areas, beyond where they currently find use, careful process design and reactor analysis must be carried out in order to evaluate their potential advantages over the conventional processes currently in place. For the petrochemical industry the task is compounded by the fact that the existing classical processes are mature and highly optimized, and the generally low process ROI does not provide much leeway in terms of additional initial capital investment. Membrane costs and availability is a key factor here. As commented earlier, even for high value-added biological products membrane capital investment and operating costs are not negligible. For large-scale petrochemical applications such considerations remain probably the main factor hindering further significant progress.

Index